THE ORIGINS OF
Agriculture

Smithsonian Series in Archaeological Inquiry

Robert McC. Adams and Bruce D. Smith, Series Editors

The Smithsonian Series in Archaeological Inquiry presents original case studies that address important general research problems and demonstrate the values of particular theoretical and/or methodological approaches. Titles include well-focused, edited collections as well as works by individual authors. The series is open to all subject areas, geographical regions, and theoretical modes.

Advisory Board

THE ORIGINS OF
Agriculture

An International Perspective

EDITED BY C. WESLEY COWAN AND PATTY JO WATSON

with the assistance of Nancy L. Benco

Smithsonian Institution Press   Washington and London

© 1992 Smithsonian Institution

Library of Congress Cataloging-in-Publication Data
The Origins of agriculture : an international perspective / edited by
C. Wesley Cowan and Patty Jo Watson, with the assistance of
Nancy L. Benco.
p.   cm. — (Smithsonian series in archaeological inquiry)
"Most . . . chapters were originally prepared for a symposium held
in 1985 at the American Association for the Advancement of Science
meeting in Los Angeles"—P.
Includes bibliographical references and index.
ISBN 0-87474-990-5   0-87474-991-3 (pbk.)
1. Agriculture—Origin—Congresses.
2. Agriculture, Prehistoric—Congresses.
3. Man, Prehistoric—Food—Congresses.
4. Plant remains (Archaeology)—Congresses.
I. Cowan, C. Wesley, 1951–   .
II. Watson, Patty Jo, 1932–   .
III. Benco, Nancy L.
IV. Series.
GN799.A4075   1992
630'.9—dc20   91-32767   CIP

⊗ The paper in this book meets the requirements of the
American National Standard of Permanence of Paper for Printed
Materials Z39.48-1984.

For permission to reproduce illustrations appearing in this book,
please correspond directly with the authors. The Smithsonian
Institution Press does not retain reproduction rights for these
illustrations or maintain a file of addresses for photo sources.

Cover illustration: 1926 map depicting N. I. Vavilov's "centers
of origin of cultivated plants" from *Bulletin of Applied Botany and
Plant Breeding* (Leningrad) 16(2):134.

Publisher's Note: Chapter 6 is also included in Bruce D. Smith,
Rivers of Change: Essays on Early Agriculture (Washington, D.C.:
Smithsonian Institution Press, 1992).

To all those scholars, past and present, who have advanced our knowledge of the manifold and complex relationships between plants and people.

Contents

List of Illustrations

List of Tables

Acknowledgments

If the origins of agriculture remain somewhat shrouded by the millennia, the origins of this book have a more concrete beginning. Most of the chapters were originally prepared for a symposium held in 1985 at the American Association for the Advancement of Science (AAAS) meeting in Los Angeles. With the exception of the chapters by Bruce D. Smith and Emily McClung de Tapia, the rest were originally prepared for the AAAS symposium. All have been substantially updated since that time.

Considering the length of time that has elapsed since the Los Angeles symposium, we would like first and foremost to extend our deepest thanks to the individual authors who have endured what seemed like endless delays in the publication of papers written so long ago. Attempting to coordinate the efforts of scholars on two different continents and in three different countries has not been easy. We hope the results justify the wait.

When we submitted the manuscript that eventually resulted in the present volume, it was passed along to an anonymous reviewer by the Smithsonian Institution Press. The reviewer's comments were positive and constructive, and each paper has benefitted from those suggestions. Daniel Goodwin and the Press have been very patient and helpful, for which we thank them. Our editor at the Press, Nancy Benco, did a superb job of line-editing the individual chapter manuscripts, and working with the authors to produce the final revisions published in this book. The authors and the editors are deeply grateful to her.

Our individual institutions—the Cincinnati Museum of Natural History and Washington University—have been generous with resources for drafting, postage for bulky manuscripts, thousands of pages of xeroxes, overseas telephone calls, travel, and, above all, time. At Washington University, we are thankful to Jim Railey for undertaking the drafting of several maps and diagrams and to Dean Martin Israel, Faculty of Arts and Sciences, for providing funding for that and other pre-publication expenses. The final work on the book was done while the junior editor was a Fellow at the Center for Advanced Study in the Behavioral Sciences, Stanford, California; we are grateful for financial support to the Center from the National Science Foundation (BNS-8700864).

List of Contributors

C. Wesley Cowan is Curator of Archaeology and Chair of the Department of Anthropology at the Cincinnati Museum of Natural History. His research interests include the archaeology of eastern North America and the origins of agriculturally based economies. Most recently his research has centered on late prehistoric and protohistoric Fort Ancient societies in the central Ohio Valley.

Gary W. Crawford is Associate Professor and Chair of the Department of Anthropology at the University of Toronto. His research encompasses the processes of plant domestication and the origins of agriculture, computer applications to archaeology, and the development of public policy relating to archaeology. His current research focuses on the origins of Japanese agriculture.

Robin W. Dennell is Professor of Archaeology in the Department of Archaeology and Prehistory at the University of Sheffield. Dennell's research interests include paleoethnobotany and the origins of plant domestication in Europe and the Near East. For the past several years he has been involved in archaeological research in Pakistan.

Jack R. Harlan is Professor Emeritus in the Department of Crop Sciences at the University of Illinois, Champaign-Urbana. His research interests include the origins of domesticated cereal crops in the Near East and Africa.

Emily McClung de Tapia is Director of the Laboratoria de Paleoetnobotanica and Research Professor at the Instituto de Investigaciones Antropológicas, Universidad Nacional Autónoma de México, in Mexico City. Her research interests include paleoethnobotany and the origins of agriculture in Mexico and Central America.

Paul E. Minnis is Associate Professor of Anthropology at the University of Oklahoma. His general research interests include the prehistory of the American Southwest and ethnobotany. He is currently co-directing an archaeological project near Casas Grandes, Chihuahua, Mexico, and is also pursuing a comparative study of famine food use.

Naomi F. Miller is Research Specialist at the Museum Applied Science Center (MASCA) at the University Museum, University of Pennsylvania. Her research focuses on environment, agriculture, and plant use in the ancient Near East. She is currently working on archaeobotanical remains from the museum's excavations at Gordion, Turkey.

Deborah M. Pearsall is Associate Professor of Anthropology at the University of Missouri-Columbia.

Her research interests include paleoethnobotany and the origins of plant domestication in South America. She is currently working in Ecuador, investigating the emergence of agriculturally based economies.

Bruce D. Smith is Curator of North American Archaeology and Director of the Center for Archaeobiological Research at the National Museum of Natural History, Smithsonian Institution. Smith's research interests include the origins of plant domestication and agricultural economies, and the archaeology of Eastern North America, with an emphasis on Woodland and Mississippian period farming societies.

Patty Jo Watson is Professor of Anthropology at Washington University, St. Louis, Missouri. Her research interests include the origins of agriculture in Eastern North America. She is currently involved in the long-term study of human adaptation to the Green River valley of western Kentucky.

—— *1* ——

Introduction

Introduction

The transformation of human foragers into agriculturally based societies remains an enduring problem for scientists studying the human condition. Questions revolving around these transformations have been enough to launch entire expeditions to isolated valleys in the Near East and northern Mexico and to less remote places, such as the Green River valley of western Kentucky. Archaeologists, botanists, and crop scientists have collaborated to develop shadowy pictures of developmental processes that took place on every continent of the earth. This volume brings into focus these disparate pictures by presenting recent syntheses of agricultural beginnings at several places around the world.

Although worldwide in scope, this volume should not be considered an encyclopedic treatment describing all such transformations known (see Harris and Hillman 1989 for a more comprehensive and more technical set of papers). Rather, our intent is to make available a set of case studies from a variety of world regions, both tropical and temperate, to illustrate the diversity of specific details, as well as the processual similarities and differences, among several cases of shifts to horticultural and agricultural economies. We emphasize the New World evidence because it is less well-known internationally than such Old World cases as Europe and the Near East. It is also clear that the evidence for those transformations we associate with agricultural beginnings is most detailed and abundant in the New World.

This volume originated in a symposium we organized for the American Association for the Advancement of Science (AAAS) meetings in Los Angeles in 1985. The original versions of papers by Crawford, Dennell, Harlan, Miller, Minnis, and Pearsall were read at the symposium; those of McClung de Tapia and Smith were solicited later. We considered the AAAS meetings an appropriate forum for our symposium because few scholars outside the anthropological and botanical communities are familiar with the empirical data for agricultural origins, and because recent advances have made new information available that might prove of interest to other scholars. Indeed, our audience in Los Angeles consisted almost exclusively of scholars outside the disciplines that usually study agricultural origins.

With the exception of Harlan (a crop scientist), all of the authors in this volume are anthropologists.

C. WESLEY COWAN AND
PATTY JO WATSON

I

Without exception, all of the authors work with primary data bearing on the issue of early plant cultivation and are experts in the areas they represent here. Each chapter is as authoritative and up-to-date as possible, but also as clear and non-technical as possible. Each author summarizes the current archaeobotanical evidence for his or her area and offers interpretations of it.

Most chapters follow a relatively standardized outline. The history of investigations of plant domestication, the current archaeobotanical record, important sites, and current models treating the domestication process are discussed for each region. Although similar in basic format, the chapters vary considerably. This is, in part, a reflection of the writing style of the individual authors, but, more importantly, it is a reflection of the kinds and quality of available data. Harlan's chapter on African domestication, for example, does not contain the quantity of detail that Dennell is able to provide for the introduction and spread of domesticates in Europe. It was necessary for Dennell to sift a veritable mountain of information, gleaned from dozens of archaeological field projects, whereas Africa's archaeobotanical record is virtually unknown. In spite of these differences, the chapters provide accurate reflections of the current state of knowledge about early plant domestication in a series of different world areas. Although it was not a conscious goal of the authors or editors, we hope the inadequacies of the data from several areas will spur some much-needed research there.

Because the student and casual reader may not be familiar with all of the concepts and terms found in this volume, we devote the remainder of this first chapter to a summary of background information relevant to the data-oriented chapters that follow. It is not our primary purpose to review the history of intellectual thought surrounding agricultural origins or to provide a model of agricultural beginnings, but we believe the reader may find some historical perspective useful. We also include a brief discussion of the techniques used for obtaining data relevant to agricultural origins, as well as some consideration of the major problems in interpreting those data.

A Historical Perspective on the Origins of Agriculture

The Swiss botanist Alphonse de Candolle (1885) was the first to recognize that information from botany, philology, geography, and archaeology must be integrated and brought to bear if scholars were to understand agricultural origins. Using such data as were then available, de Candolle attempted to define the regions where most of the world's important crops were first domesticated.

The early- to mid-twentieth-century Russian botanist Nicolai Vavilov followed de Candolle's tradition of combining information from a variety of sources, including the rapidly developing field of genetics. Vavilov's work, supported by the Russian state, led him to many different parts of the world and enabled him to create formulations that are still influential.

Vavilov proposed that areas where both genetic and species diversity were highest were probably "centers" of domestication, and he identified several such centers or "hearths." Although many of his ideas are not in favor today, his search for regions where agriculture began provided the impetus for further investigations. The initial searches for the origins of both Near Eastern and Mexican crop plants, for example, were heavily influenced by Vavilov's early ideas.

As archaeobotanical data began to be accumulated from various continental regions, it quickly became apparent that the concept of a "center" of domestication was difficult to apply. When sites producing evidence of early domestication are spread out over an area that encompasses hundreds or thousands of square kilometers, how does one define a "center" (see also Harlan 1971)?

As the science of archaeology grew in the twentieth century, prehistorians discovered that they could contribute directly to the study of agricultural beginnings. Not only tools and architectural features could be brought to bear on the topic, but also the charred or, in some cases, desiccated remains of early cultivated plants could be recovered and studied to provide *direct* evidence of the domestication process.

In the years after World War II, two projects explicitly designed to gather information about early agriculture were fielded: Braidwood's Iraq-Jarmo Project in the Near East from 1948 to 1955 (Braidwood and Howe 1960; Braidwood et al. 1983) and MacNeish's search for the origins of corn in Mexico (MacNeish 1964). Both scholars were influenced by Vavilov's concept of domestication "centers" and de Candolle's multidisciplinary approach. In addition, Braidwood was seeking specific archaeological evidence of the "Neolithic revolution" hypothesized by V. Gordon Childe (1936).

Interest in agricultural origins—and in "environmental archaeology" in general—received tremendous impetus in the 1960s from growing global awareness of the human impact on the physical environment. This was particularly true in the United States where the 1960s saw the emergence of many subdisciplines of archaeology, most of which had an environmental and ecological emphasis. Paleoethnobotany was one of these and, thanks to the techniques routinely applied by today's generation of archaeologists to recover plant remains, more information related to the origins of agriculture has become available in the past decade than in the previous ten.

Paleoethnobotany

Paleoethnobotany is a word given currency in the 1960s by Hans Helbaek (1960), Richard Yarnell (1963), and others who perceived themselves as scholars applying ethnobotanical perspectives to the archaeological record of ancient plant use. That is, just as ethnobotany is the study of botanical knowledge and plant use cross-culturally (e.g., see Harshberger 1896; King 1984; Smith 1923, 1928, 1932, 1933; Stevenson 1915; Teit 1930), paleoethnobotany is the cross-cultural study of the interrelationships between prehistoric plant and human populations.

Until relatively recently, most plant remains recovered from archaeological contexts were the results of happenstance. If a large quantity of charred seeds, corn cobs, etc., was encountered during excavation, this material would be gathered up and saved; if something caught the eye of an excavator, it might be retained. These sorts of *macrobotanical* remains are obviously biased toward those plant tissues that are large and durable; in the field, small charred seeds and plant parts are not easily detected by the naked eye, and, as a consequence, specialized means are necessary to obtain such remains. The most common method by which these smaller macrobotanical plant fragments are concentrated and recovered is through the technique known as *flotation*.

Many different flotation systems have been devised in the past twenty years, but all of them depend upon the following basic principle: if archaeological sediment is released into a container filled with water, then the sediment sinks and the charred plant remains float. There is a considerable literature on flotation/water-separation (e.g., Dye and Moore 1978; Hastorf and Popper 1988; Minnis and LeBlanc 1976;

Pearsall 1989; Schock 1971; Struever 1968; Wagner 1982; Watson 1976; Wiant 1983). Some techniques are multiple-operator, machine-assisted assembly lines, whereas others are simple, one-person bucket or barrel operations. It is sufficient to note here, however, that all of these systems are aimed at retrieving charred macrobotanical plant remains, and that the basic purpose is the same for all: to concentrate and collect charcoal that is dispersed throughout archaeological deposits.

The development of flotation as a technique to recover microbotanical remains has revolutionized our ability to collect data that are directly relevant to the study of agricultural origins. Although the use of flotation has found worldwide acceptance, not all archaeologists employ it as a technique to retrieve plant remains. In some areas archaeological sediments may not be amenable to flotation. Even in cases where flotation is not physically possible, however, attempts are often made to recover plant remains. Dry or water sieving of sediments, for example, is routinely used in some European countries to obtain plant remains. Another reason for this circumstance is doubtless that the recovery and study of paleoethnobotanical data is time-consuming and expensive; many archaeological projects simply do not have the necessary time and money. On the other hand, it is commonly noted by paleoethnobotanists that many archaeologists have been slow to recognize the value of studying plant remains, and for this reason do not collect "float" samples or include archaeobotany in their research designs. Fortunately, this attitude seems to be disappearing.

When charred plant remains arrive at the paleoethnobotanist's laboratory, they are sorted, identified, weighed, counted, and described. Paleoethnobotanists use a variety of instruments to aid in these tasks, but most important is a high-powered binocular dissecting microscope; access to a scanning electron microscope, or SEM, is also frequently necessary. In addition, an accurate balance or scale and an adequate comparative collection, together with a set of key reference books specific to the region under investigation, are needed.

So far we have been referring to charred archaeobotanical remains, but in many parts of the world very old plant remains may be preserved in an uncharred, dry, or saturated state. In very arid and desert climates, like ancient Egypt or coastal Peru, dried plant parts may last indefinitely. Even in temperate climates, plants may be beautifully preserved in dry caves and rockshelters. Saturated sites (bogs, swamps, lakes, etc.) may also preserve fragile plant materials indefinitely. Obviously, identification and quantification of uncharred remains differs somewhat from work with charred, often distorted, flotation-derived fragments from an open archaeological site. Identification and dating problems are similar for both types of remains, however. There has been considerable success recently in the application of direct, accelerator mass spectrometer (AMS) dating techniques to small plant fragments, both charred and uncharred (e.g., Conard et al. 1984; Fritz and Smith 1988; Harris 1986; Smith and Cowan 1987).

Although considerable progress has been made in interpreting ancient plant remains, many serious problems have yet to be solved. There is still insufficient standardization in recovery techniques and in quantification procedures. This means that the results of paleoethnobotanical analyses are often difficult to compare directly.

There are also many problems in basic identification of charred remains because of the often seemingly unpredictable distortion and size alteration caused by the original charring process, and then the further fragmentation and wear and tear before the charcoal is retrieved from its archaeological context. However, paleoethnobotanists in one geographic area or those focusing on a specific archaeobotanical question often form very effective working groups to trade information, provide access to comparative collections, share new techniques, etc., as they seek to reach the same interpretive goals. These working groups have made substantial progress toward solving a number of common problems.

One such archaeobotanical question is that addressed in this volume: the evolution of domestic plants. Obviously the paleoethnobotanist must be able to differentiate between wild and domestic forms of the same species. This is sometimes possible on the basis of size, shape, or other morphological attributes (e.g., see chapters by Miller and Smith in this volume), or sometimes the plant is found outside its area of natural distribution, or it may be found in such abundance and in such contexts (storage pits or containers, for instance) as to indicate that it was domesticated, or at least cultivated. Most paleoethnobotanists distinguish between a species that is simply tolerated or encouraged and one whose reproduction is carefully controlled or managed. Ultimately, the distinction between the two—cultigens versus domesticates—is based upon a single important factor.

Whereas cultivated plants might continue to exist as a breeding population without the interference of humans, domesticated plants are so different genetically from their wild ancestors that their very existence depends upon human intervention in their life cycles. Ford (1985:6) has justifiably referred to domesticated plants as "cultural artifacts" resulting from human selection.

The careful reader will note in the following papers that the appearance of a domesticate in the archaeological record is the end result of generations of cumulative genetic transformations. Hence, the discovery of early domesticates defines only the end point of a process that might have taken hundreds or even thousands of years.

Thus, it is impossible to delineate a single point in the past for any continent or geographic region that marks the "beginnings" of agricultural economies there. Agriculturally based societies did not spring up overnight as the result of an idea proposed by some prehistoric genius. As will be seen in the chapters that follow, early agriculture is a process preceded by thousands of years of cultural and biological evolution.

Each of the following chapters summarizes the evidential base for a different part of the globe. We begin in the Old World with East Asia, then turn to the Near East, Africa, and Europe before moving to the New World where the Eastern Woodlands and the Southwest of North America, Mesoamerica, and South America are addressed. Although there is considerable variation in the quality and quantity of data for these eight regions, the authors have followed the same presentation format to enhance ease of comparison by readers not familiar with all eight. They also make frequent use of maps and tables to display relevant information in an efficient and readily accessible manner.

References Cited

Braidwood, R. J., and B. Howe (editors)
  1963   Prehistoric Investigations in Iraqi Kurdistan. Studies in Ancient Oriental Civilizations 31. University of Chicago Press, Chicago.

Braidwood, L. S., and R. J. Braidwood, B. Howe, C. A. Reed, P. J. Watson (editors)
  1983   Prehistoric Archeology along the Zagros Flanks. Oriental Institute Publications, vol. 105. University of Chicago, Chicago.

de Candolle, A.
  1885   Origin of Cultivated Plants. D. Appleton, New York.

Childe, V. G.
  1936   Man Makes Himself. Watts, London.

Conard, N., D. Asch, N. Asch, D. Elmore, H. Gove, M. Rubin, J. Brown, M. Wiant, K. B. Farnsworth, T. Cook
  1984   Accelerator Dating of Evidence for Prehistoric Horticulture in Illinois. Nature 308:443–46.

Dye, D. H., and K. H. Moore
  1978   Recovery Systems for Subsistence Data: Water Screening and Water Flotation. Tennessee Anthropologist 3(1):59–69.

Ford, R. I.
  1985   The Processes of Plant Food Production in Prehistoric North America. In Prehistoric Food Production in North America, edited by R. I. Ford, pp. 1–18. Anthropological Papers No. 75. Museum of Anthropology, University of Michigan, Ann Arbor.

Fritz, G. J., and B. D. Smith
  1988   Old Collections and New Technology: Documenting the Domestication of Chenopodium in Eastern North America. Midcontinental Journal of Archaeology 13(1):3–28.

Harlan, J. R.
  1971   Agricultural Origins: Centers and Non-centers. Science 14:468–74.

Helbaek, H.
  1960   The Paleoethnobotany of the Near East and Europe. In Prehistoric Investigations in Iraqi Kurdistan, edited by R. J. Braidwood and B. Howe, pp. 99–118. University of Chicago Press, Chicago.

Harshberger, J.
  1896   Purposes of Ethnobotany. Botanical Gazette 21(3):146–54.

Harris, D. R.
  1986   Plant and Animal Domestication and the Origins of Agriculture: The Contribution of Radiocarbon Accelerator Dating. In The Results and Prospects of Accelerator Radiocarbon Dating, edited by J. Gowlette and R. Hedges. Oxford University Press, Oxford.

Harris, D. R., and G. C. Hillman (editors)
  1989   Foraging and Farming; the Evolution of Plant Exploitation. Unwin Hyman, London.

Hastorf, C., and V. Popper
  1988   Current Paleoethnobotany. University of Chicago Press, Chicago.

King, F. B.
  1984   Plants, People and Paleoecology. Illinois State Museum Scientific Papers, vol. 20. Springfield.

MacNeish, R. S.
  1964   Ancient Mesoamerican Civilization. Science 143:531–37.

Minnis, P. E., and S. LeBlanc
  1976   An Efficient, Inexpensive Arid Lands Flotation System. American Antiquity 41:491–93.

Pearsall, D.
  1989   Paleoethnobotany. A Handbook of Procedures. Academic Press, San Diego.

Schock, J. M.
  1971   Indoor Water Flotation—a Technique for the Recovery of Archaeological Materials. Plains Anthropologist 16:228–31.

Smith, B. D., and C. W. Cowan
  1987   Domesticated Chenopodium in Prehistoric Eastern North America: New Accelerator Dates from Eastern Kentucky. American Antiquity 52(2):355–57.

Smith, H. H.
  1923   Ethnobotany of the Menomini Indians. Bulletin of the Public Museum of the City of Milwaukee 4(1):1–174.

  1928   Ethnobotany of the Meskwaki Indians. Bulletin of the Public Museum of the City of Milwaukee 4(2):175–326.

  1932   Ethnobotany of the Ojibwe Indians. Bulletin of the Public Museum of the City of Milwaukee 4(3):327–525.

  1933   Ethnobotany of the Forest Potwatomi Indians. Bulletin of the Public Museum of the City of Milwaukee 7(1):1–230.

Stevenson, M. C.
  1915   Ethnobotany of the Zuni Indians. Bureau of American Ethnology 30th Annual Report (1908–1909):35–102.

Streuver, S.
  1968   Flotation Techniques for the Recovery of Small-Scale Archaeological Remains. American Antiquity 33:353–62.

Teit, J. A.

1930   Ethnobotany of the Thompson Indians of British Columbia. Bureau of American Ethnology 45th Annual Report (1927–28):441–522.

Wagner, P. J.

1982   Testing Flotation Recovery Rates. American Antiquity 47(1):127–32.

Watson, P. J.

1976   In Pursuit of Prehistoric Subsistence. A Comparative Account of Some Contemporary Flotation Techniques. Midcontinental Journal of Archaeology 1:77–100.

Wiant, M.

1983   Deflocculants and Flotation: Considerations Leading to a Low-Cost Technique to Process High Clay Content Samples. American Archaeology 3:206–209.

Yarnell, R. A.

1963   Palaeo-ethnobotany in America. *In* Science in Archaeology, edited by D. Brothwell and E. Higgs, pp. 215–28. Thames and Hudson, London.

2

Prehistoric Plant Domestication in East Asia

GARY W. CRAWFORD

Introduction

In earlier syntheses on the origins of agriculture, China was a necessary subject of some deliberation, but Korean and Japanese agricultural origins were largely ignored (e.g., Harris 1977; Ho 1969, 1977). Discussions on Korea and Japan were of interest only to small circles of local archaeologists; few others seemed to think it worthwhile to pay attention to these two subareas of East Asia. After all, in neither Korea nor Japan were there apparent indigenous agricultural origins. At present, archaeologists are still debating the existence of food production in pre-Bronze and Iron Age Korea and Japan. The prevailing view is that agriculture began in Korea and Japan in the third millennium B.P. with the initiation of rice-based agrarian societies by both the diffusion of crops and ideas and the migration of people. Evidence is mounting, however, for earlier agricultural origins in these two areas.

By 1977 North China had been established as an independent center of agricultural origins (Ho 1977). Yet, at that time, there were only six radiocarbon dated sites with domesticated remains (Kabaker 1977:968–69). Chinese agricultural beginnings, then dated to about 7000 B.P., seemed to be later than their initiation in Southwest Asia. Both Southwest Asian and Chinese agriculture appeared to have begun in an *Artemisia* steppe, yet the immediate predecessors of Chinese agricultural complexes were unknown (Reed 1977:903–4). No early Holocene cultures comparable to the Natufian (Miller, this volume), for example, had been reported in China, although a long, rich archaeological record of early Holocene cultures was known for Japan. Data on the origins of rice agriculture in China were not available in 1977, although the minimal available data led Ho (1977) to suggest that a South China rice domestication center was likely.

Some questions posed more than a decade ago remain unanswered. Evidence of an early Holocene population from which agrarian society arose in North China is still inadequately documented (Chang 1986). In the last decade a number of phases immediately predating the Yangshao have been uncovered but they are substantially involved with food production (An 1989; Chang 1986; Yan 1989). The onset of food production in Southwest and East Asia is likely to have occurred in similar periods, sometime before 9000 B.P. The origins of rice husbandry are better known today, and Ho's (1977) suggestion of a South China center of rice domestication appears more likely. Many more sites—close to 30—with domesticated plant remains are known in Neolithic China. In Japan, the introduction, development, and spread of agriculture is an active area of research, as is the search for a mid-Holocene record of cultigens. In Korea, the archaeology of agricultural origins is probably the least developed of North Asia, although Korea's role in understanding crop dispersals to the east from China cannot be underestimated.

Northern China, Korea, and Japan are centers of diversity and potential sources of many cultigens.[1] Zeven and Zhukovsky (1975) compiled a list of 284 taxa of cultigens and managed plants in this part of Asia (Table 2.1). Not all are native to northeastern Asia, of course, and the relative economic importance of these plants has varied over time. Rice (*Oryza sativa*), an important food plant throughout much of the present world, became economically significant in northern China, Korea, and Japan only recently. Rice is not native to any of these regions either. Indigenous cultigens include broomcorn millet (*Panicum miliaceum*), barnyard millet (*Echinochloa utilis*), foxtail millet (*Setaria italica* ssp. *italica*), Chinese cabbage (*Brassica campestris*), *adzuki* bean (*Vigna angularis*), soybean (*Glycine max*), hemp (*Cannabis sativum*), great burdock (*Arctium lapa*), and buckwheat (*Fagopyrum esculentum*). Tree crops, such as peach (*Prunus persica*) and persimmon (*Diospyros kaki*), also attained cultigen status in this area. Barley (*Hordeum vulgare*) and bread wheat (*Triticum aestivum*) probably arrived from southwestern Asia by the third millennium B.P. Introduced cultigens were isolated from their parent gene pools and local forms and varieties eventually evolved. Much of the basic archaeological data on these plants is lacking, but the region is not completely without an archaeological record for domestication. Ho (1977), for example, relied almost entirely on ethnohistory for his discussions on early Chinese agriculture. He had little choice, but today more archaeological data are available. In this paper I review what *is* currently known about plant domestication in northern China, Korea, and Japan, with an emphasis on Japan.

The transition to food production in East Asia involves not only indigenous domestication and evolution of agricultural systems, but also the diffusion of cultigens within this geographical area and the introduction of cultigens from the west. Recent summaries of these processes can be found in Chang (1983), Ho (1977), and Li (1983) for China; Kim (1978), Nelson (1982), Pearson (1974), and Rowley-Conwy (1984) for Korea; and Crawford (1983, 1992), Crawford and Takamiya (1990), Esaka (1977), Kasahara (1984), Kotani (1981), Pearson and Pearson (1978), Rowley-Conwy (1984), and Tozawa (1982) for Japan.

Research is active in all three areas, but Japan, with a tradition of extensive salvage archaeology and rapid reporting of data, continues to be a window on Asia for western archaeologists seeking to understand Asian prehistory in general. In addition, recently collected data on cultigen prehistory in Japan are now providing a provocative view of the Asian Neolithic. The Jōmon of Japan is also proving to be of comparative importance in the study of the European Mesolithic (Dennell, this volume; Rowley-Conwy 1984) and the eastern North American Archaic and Woodland (Aikens 1981; Crawford 1983; Smith, this volume). Finally, populations of carbonized cultigens from Hokkaidō are being studied (Crawford 1986; Crawford and Yoshizaki 1987) and seem to provide the best comparative data on the taxonomy and evolution of East Asian cultigens available to western scientists at this time. Because few plant remains have been reported from northern China and Korea, this chapter focuses on recent developments in Japan and the relevance of the current data to problems elsewhere in eastern Asia.

Until the early 1970s few empirical data on Jōmon agriculture were available. Since then, considerable quantities of plant remains have been recovered from Jōmon and later occupations. After a brief discussion of East Asian Neolithic chronologies and a review of the early hypotheses, I explore revised concepts of Jōmon subsistence in light of the new data. In addition, I examine the first millennium A.D. complement of northern temperate Asian cultigens in order to shed light on the history of plant domestication in East Asia. The principal archaeologically visible cultigens are reviewed in terms of their botany, context, and the Asian Neolithic in general. Current perspectives are based mainly on my own research in Hokkaidō (Crawford 1983, 1992; Crawford and Takamiya 1990; Crawford et al. 1978), on the work of the Kōbunkazai Henshū Iinkai (Antiquities Editorial Committee), and on a project involving a ninth-century A.D. agricultural complex in Hokkaidō (Crawford and Takamiya 1990; Crawford and Yoshizaki 1987; Yoshizaki 1984). The latter project serves as a focus for the cultigen prehistory and botanical discussions to follow. Wherever possible, I have included references in English, and when an article is available in both Jap-

Table 2.1. Domesticated Plants and their Relatives in East Asia

English Common Name	Japanese Common Name	Latin Name
apricot	anzu	*Prunus armeniaca*
barley	ō-mugi	*Hordeum vulgare*
bean: adzuki (red)	azuki	*Vigna angularis* var. *angularis*
black gram or urd	ketsuru-azuki	*V. mungo*
mung	ryokuto	*V. radiatus* var. *radiatus*
soy bean	daizu	*Glycine max*
beefsteak plant	shiso	*Perilla frutescens* var. *crispa*
	egoma	*P. frutescens* var. *japonica*
bottle gourd	hyōtan	*Lagenaria siceraria*
buckwheat	soba	*Fagopyrum esculentum*
wild buckwheat		*F. cymosum*
giant radish	daikon	*Raphanus sativus*
great burdock	gobō	*Arctium lappa*
hemp	asa	*Cannabis sativa*
hops	karahanasō	*Humulus lupulus*
Job's tears	jyuzu dama	*Coix lachryma-jobi*
lacquer tree	urushi	*Rhus vernicifera*
melon	uri, makuwa-uri	*Cucumis melo*
millet:		
barnyard grass	ta-inubie, inubie	*Echinochloa crusgalli*
barnyard millet	hie	*E. utilis*
broomcorn proso	kibi	*Panicum miliaceum*
foxtail	awa	*Setaria italica* ssp. *italica*
mustard family	aburana-ka	Brassicaceae
Chinese cabbage	chūgoku kyabichi	*Brassica campestris* ssp. *chinensis*
Indian mustard	kyona/mibuna	ssp. *nipposinica*
Chinese cabbage	hakusai	ssp. *pekinensis*
rape, canola	natane	*B. napus*
paper mulberry	kaji-no-ki	*Broussonetia papyrifera*
pea	endō	*Pisum sativum*
peach	momo	*Prunus persica*
pear	seiyō-nashi	*Pyrus communis*
persimmon	kaki	*Diospyros kaki*
plum	sumomo	*Prunus salicina*
rice	kome	*Oryza sativa* ssp. *japonicum (sinica)*
		O. sativa ssp. *indica*
safflower	benibana	*Carthamus tinctorius*
sorghum	morokoshi	*Sorghum bicolor*
soybean	daizu	*Glycine max*
wheat, bread	ko-mugi	*Triticum aestivum*

anese and English, both are cited. This should make the available complex literature somewhat more useful to western students and researchers than otherwise might be the case. English common, scientific, and Japanese common names are given for plants discussed in the text. This should allow for clearer understanding for most readers, including Japanese readers who are more accustomed to using common rather than scientific names.

Environment

China, Japan, and Korea span a range of latitudes from tropical or subtropical to temperate, have a range of floral and faunal associations, and have climates controlled in part by the East Asian monsoon (Hsieh 1973; Liu 1988; Ren et al. 1985). China has the greatest extent of riverine and plain development of the three regions. East Asia tends to be isolated from

Central and South Asia by desert, steppe, and mountains. Korea and Manchuria are somewhat separated by the Changhai Mountains while Japan has the highest degree of isolation, separated from the mainland by the Japan Sea. North China and South China are distinctive in that the former is temperate while the latter is subtropical to tropical. In the interior, the Qin Mountains at about 30° north latitude create a topographical division between North and South China, forming a barrier to an easy north-south movement of people. In the lower Huai River, no such barrier to movement exists.

The potential vegetation of China (Fig. 2.1) is open to interpretation based on climatic and historic records. Human impact on habitats has been extensive so little natural vegetation remains. The discrepancies between two commonly cited vegetation maps of China illustrate problems with the reconstruction of potential vegetation. One map used by Hsieh (1973) and Kolb (1971) and another cited by Liu (1988) disagree on the location of the mixed-forest zone that lies between the deciduous forest and the broadleaf-evergreen forest zones, as well as on details of the boundaries of other zones. Figure 2.1 is based on Liu's (1988) depiction of Wu (1980) and on Ren et al. (1985), the latter of which includes a temperate-forest grassland between the steppe and deciduous-forest zone (Fig. 2.1, Zone 4).

China is unique among the three areas in having a well-developed steppe (Fig. 2.1) whose existence is due to a steep northeast-southwest moisture gradient, which results in a dry, continental climate in the interior (Liu 1988:1). The steppe-forest ecotone (Zone 4) crosses through the Wei River basin today, where some of the earliest evidence for plant and animal domestication is found in China. Ho (1977:426) argues that the semiarid steppe was the zone in which the earliest agriculture in North China arose. His evidence is in two forms: palynological and ethnohistoric. Unfortunately, at the time of Ho's synthesis, a dated pollen sequence was apparently not available. In fact, among 80 pollen records published for North China, few are radiocarbon dated (Liu 1988:5). The only data that Ho could use for his interpretation was the upper 20 meters of a loess profile in Wucheng County near the modern steppe-forest boundary (Ho 1977:421–23). The ethnohistoric data are late first millennium B.C. odes and songs whose contents include the names of plants and their habitats. Ho concluded that because trees and shrubs were confined to mountains, hills, slopes, and places near watercourses, North Chinese food production evolved in a steppe environment, making this area the only exception to the generalization that agriculture did not, and could not, first appear in a grassland area (Ho 1977:424, 426). This interpretation oversimplifies our understanding of the Holocene Loess Plateau vegetation and the context of the first plant husbandry in North China. There is simply not enough evidence to determine under what environ-

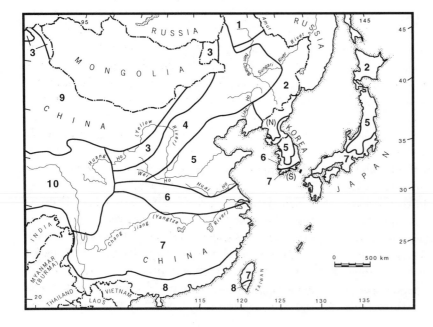

Fig. 2.1. Vegetation zones of East Asia. 1, boreal; 2, mixed conifer-deciduous; 3, steppe; 4, forest-steppe; 5, deciduous broadleaf; 6, mixed deciduous and broadleaf evergreen; 7, subtropical broadleaf evergreen; 8, tropical monsoonal rainforest; 9, desert; 10, highland.

mental conditions domestication began in North China. The steppe-forest ecotone location varied throughout the Holocene, as recent dated pollen sequences and archaeological data indicate. It would be safe to say that the location of the steppe-forest ecotone in China during the early and mid-Holocene is not known with any accuracy. Furthermore, an archaeological record for cultures immediately preceding the earliest known settlements on the Loess Plateau is poorly known and the earliest food producing communities on the plateau so far show little or no use of steppe resources.

An early Holocene mixed forest, which apparently dominated at Beizhuangcun in the Wei River basin until about 9000 B.P., was followed by a steppe dominated by *Artemisia* and Compositae (Liu 1988:13–14). The date for the beginning of the steppe at Beizhuangcun is only an estimate and the upper part of the pollen sequence is missing. At Xinlitun, near Beijing, increasing *Artemisia* and grass pollen seem to confirm increasing aridity in the latter mid-Holocene of North China (Liu 1988:15). Hardwood pollen dominated by oak is most common in the mid-Holocene at Xinlitun as well, indicating an expansion of the deciduous forest. In general, the steppe seems to have expanded to the southeast into the present forest zones of North China in the last glacial period. There is some question as to whether this expansion occurred during, or near the end of, the last glacial period (Liu 1988:17). A northwestern retreat of the steppe appears to have occurred in the mid-Holocene, followed by a return of the steppe sometime after 5000 B.P. The Xinlitun pollen diagram indicates that, although the deciduous forest expanded in the mid-Holocene, the steppe was probably an important component of local habitats. At the Fuhokoumen site in Hopei, now in the temperate forest-steppe zone (Zone 4), only deciduous forest zone animal remains were found (Chang 1986:191). Details of the chronology of the southeastern steppe boundary are still not well known.

Artemisia steppe likely existed in southwestern Japan from about 10,000–8,500 B.P. (Yasuda 1978:254). This period in southwestern Japan also characteristically has relatively high proportions of Compositae, grass, *Persicaria,* and *Lythrum* pollen (Zone RI). Oak pollen indicates the presence of woodlands of some form. Absolute pollen frequencies are relatively low as well (Yasuda 1978:172). Yasuda (1978:172) suggests that this vegetation is due to a sudden climatic warming that resulted in the formation of a grassland

during the transition from a late Pleistocene mixed forest to an early Holocene deciduous hardwood forest.

A mixed conifer-hardwood forest dominates eastern Manchuria, northern North Korea, and Hokkaidō, Japan (Fig. 2.1). Pine, fir, and spruce are prevalent conifers in each region while beech and oak are the common deciduous trees in Japan and Korea. Oak (*Quercus mongolica*) is the dominant deciduous tree in Manchuria. Deciduous broadleaf forests dominate the North China Plain, the central Korean Peninsula, and northern Honshū, Japan (Fig. 2.1). In China and Korea, oak predominates while in Japan the forest is mainly beech and oak. Some pine is present, along with a range of hardwoods.

A transitional zone (Zone 6) of deciduous and broadleaf evergreen trees lies between the Huai and Yangtze river basins, as well as in South Korea (Fig. 2.1). This zone is not usually portrayed on vegetation maps of Japan; it is usually subsumed within the broadleaf evergreen forest zone (Yasuda 1978). Yasuda (1978:138) believes that it is a separate element of the Japanese vegetation and that it was more extensive between 8500 and 6500 B.P. than at present. The broadleaf evergreen forest covers southwestern Japan, the southern tip of the Korean Peninsula, and most of South China. In Japan, several species of oak are common to coastal areas, while *Machilus* and *Castanopsis* species are the major tree taxa in interior regions. A tropical monsoonal forest is found in China, but not in Korea and Japan.

The modern potential vegetation of northeast Asia is obscured by agriculture and urbanization. The role of anthropogenic influences on Holocene vegetation is not documented well for China and Korea (Liu 1988; Pearson 1974), but Yasuda (1978) and Tsukada (1986) have examined both climatic and anthropogenic influences on vegetation in Japan. Evidence for swidden agriculture may occur as early as 7700 B.P. in Japan (Tsukada 1986; Tsukada et al. 1986). Human-caused changes linked to agriculture in Japan, however, are continuous only after 3200 B.P. By 1600 B.P., climate cannot be inferred from the pollen record due to extensive landscape modification by people (Yasuda:242, 250). By 3200 B.P. in southwestern Japan, rice pollen appears, signaling the development of intensified food production that spread rapidly to central Honshū by 2100 B.P. The area of the initial spread of rice agriculture in Japan coincides with the broadleaf evergreen forest zone (Yasuda 1978:257). With the spread of agriculture came the destruction of floodplain and hillside forests. At the same time, pine pollen increases (Yasuda 1978:242). At only one site, Torihama, is there evidence for extensive forest destruction any earlier (Yasuda 1978). Anthropogenic vegetation was, however, a localized but significant component of the environment surrounding and within mid-Holocene communities in Japan (Crawford 1983).

Throughout East Asia there is evidence for a warmer mid-Holocene. Seven of 80 pollen studies in North China span the Holocene, or much of it (Liu 1988). In these sequences Zone III, which is interpreted to be a warmer and probably more humid period, is characterized by maximum deciduous tree pollen frequencies. This zone has an upper boundary of about 5000–4300 B.P. in northeastern China and a lower boundary dating to about 8000 B.P. (Liu 1988:7, 15). Archaeological data further support the contention of a warmer mid-Holocene in China. Fauna, which are limited today to South China, are found in mid-Holocene sites in the north. Bamboo rat, water deer, and elaphure are identified at Yangshao sites (Chang 1986:79; Archaeological Institute, Chinese Academy of Science 1963:319). Chang (1986) also notes that water buffalo, elephant, rhinoceros, and tapir at some sites were north of their potential ranges today. At San-li-ho on the Shandong Peninsula the local environment was considerably wetter, with the remains of at least 20 alligators recovered (Chang 1986:162).

Zone II in North Chinese pollen sequences is presumably early Holocene and has high proportions of birch pollen, while Zone IV has high proportions of pine and low proportions of birch and other deciduous trees (Liu 1988). This is remarkably like the patterns seen in Japan with the spread of agriculture. In Japan pollen diagrams for Zones RIIa and RIIb represent a warm period that started about 8500 B.P. and lasted until about 3500 B.P. (Yasuda 1988:250), about 500 to 1,500 years later than suggested for China. Zone RIIa, the first part of a mid-Holocene warm period in Japan, consists of mainly deciduous hardwoods. The broadleaf evergreen forest was established at the beginning of Zone RIIb in southwestern Japan (Yasuda 1988:255). The RIIb period was warmer and wetter than present. By 3500 B.P. the mean annual temperature dropped while the humidity continued to rise.

Except for Japan, northeast Asian Holocene environmental history is not nearly as well documented as it is in North America, Europe, and Southwest Asia. Of particular interest is the history of the steppe-forest ecotone in Gansu, Shaanxi, Shanxi, Hopei, and Liaoning provinces. There appears to be little doubt that by the time substantial agrarian communities were well established on the North China Plain (by 8000 B.P.), a relatively warm period had begun. In Japan this warmer trend is associated with increasing humidity, which bracketed the beginning and end of the warm period by about one millennium. The role of anthropogenic influences on the desertification of North China and on the steppe-forest boundary is also in need of examination.

A variety of agricultural regions provides a rich potential for food production in East Asia. Three regions are recognized in Japan, two in Korea, and eleven in China, five of which are in the North Chinese region of concern to this paper (Kolb 1971; Tregear 1980; Trewartha 1963). In southern China temperate crops, including wheat, rapeseed, potatoes, and beans, are grown at elevations above 1600 m (Ren et al. 1985:343). The limit of rice production in the south is between 2400 and 2700 m above which oats and beans are grown on a limited scale (Ren et al. 1985:344). The agricultural regions do not correspond precisely to the East Asian vegetation zones illustrated in Figure 2.1. Factors such as technology and crop genetics, as well as growing season, rainfall, elevation, and snow cover, affect crop distributions.

In Japan north of 37° north latitude, in Manchuria, and in North Korea, dry crops predominate (Kolb 1971; Trewartha 1963). Oats, wheat, barley, millet, potatoes, soybean, and maize are the most common crops in northern Japan. Sorghum and millet are the most important crops in Manchuria. Maize and soybean production is also common. In North Korea sorghum, wheat, and soybeans are the primary crops. Rice is grown in these northern regions, but production is low.

South of 37° north latitude in Japan wheat and barley are produced, but rice, tea, and citrus fruits gain importance (Trewartha 1963). In the eastern portion of Zone 5 on the North China Plain, sorghum and wheat are the main crops (Tregear 1980). Barley, foxtail millet, corn, soybean, and cotton are produced as well. On the Loess Plateau in the western portion of Zone 4 the principal crops are winter wheat and broomcorn millet (Tregear 1980). Sorghum, sesame (*Sesamum indicum*), buckwheat, maize, and a variety of legumes are also grown. Rainfall is sparse on the Loess Plateau, coming mainly in the summer (Hsieh 1973). Rice is grown in irrigated areas that comprise less than 10 percent of the arable land on the plateau (Tregear 1980).

Immediately to the northwest of the deciduous forest zone (Zone 5) in China are two production systems: pastoralism and crop production. Crop production is limited to the area bounded by the eastern limit of Zone 9 (desert) and the northwest edge of Zone 5. Spring wheat and millet are the principal crops. Less than 10 percent of the grain crop is wheat. Other cultigens include sesame, rapeseed (*Brassica napus*), and sunflower (*Helianthus annuus*). North of this zone pastoralists raise horses, cattle, sheep, and goats.

Areas of greatest rice production are confined to Zones 6, 7, and 8 (Fig. 2.1). Zone 7 is the major rice growing region in Japan and was the region into which rice rapidly spread during the Yayoi Period (Akazawa 1982). Double rice cropping is limited to Zone 8 in China, where every other year three crops are often planted (Ren et al. 1985), and to southern Shikoku in Japan (Kolb 1971; Hsieh 1973). In many areas in East Asia the rice harvest is often followed by plantings of dry crops in drained rice paddies. Double cropping in China is most common in Zone 6 and in the Yangtze River basin (Tregear 1980). In Japan this practice takes place as far north as 37° north latitude, but difficulties arise in areas of heavy and continued snow cover (Trewartha 1963). The occurrence of double cropping increases to the south in Japan. The late fall plantings in the rice paddies include wheat, barley, rape, soybean and other legumes, sesame, and giant radish or *daikon* (*Raphanus sativus*).

Shifting or slash-and-burn agriculture is still practiced in the highlands of southern China and Japan. In the early 1960s about 152,000 households in Japan were engaged in this form of production (Trewartha 1963:209), mostly in the southwestern part of the country.

Chronology and Prehistory

The time span covered in this chapter ranges from about 8500 B.P. to 100 B.P. (Fig. 2.2). Chronological and processual problems, such as linearity of sequences and radiocarbon dating complexities, are simplified in Fig. 2.2; however, the chart serves as a useful guide for the discussions to follow. The time range is more a result of the availability of relevant data than it is of archaeological data pertaining to periods when actual domestication was taking place. Few data are available from times when domestica-

tion was occurring in northeastern Asia. As a result, I include all periods when any direct archaeological evidence of domesticated plants is available to elucidate problems and processes of domestication. The latest substantial archaeological data are from the 1100 B.P. Sakushu-Kotoni River site in Hokkaidō (Figs. 2.2–2.3). Some data from later sites in Japan are brought to bear on problems related to the Hokkaidō data. The Chinese sequence ends with the Han Dynasty because historic records from this period are useful and one of the last events of importance to this discussion occurs during this period—the expansion of wheat husbandry. The only data discussed in this paper that postdate the Bronze Age in Korea involve comparative wheat data from the 1100 B.P. Puyo site; cultigen remains from Korea are mainly Bronze Age.

China

In general, Chinese archaeologists assume that any site with ceramics and no evidence of metallurgy is Neolithic. Pre-8500 B.P. pottery-bearing sites are termed Early Neolithic. To date, no Early Neolithic sites have been identified on the Loess Plateau or on the North China Plain. Conversely, no Middle Neolithic sites, at least of the magnitude of the ones in North China, have been identified south of the Huai River. With the exception of the Zengpiyan Cave site (10,300–7000 B.P.) in Guilin (Chang 1986; Yan 1989), no evidence for domestication in South China exists before about 7000 B.P. At Zengpiyan dental and mandibular morphology of an estimated 67 pigs indicate that their domestication was taking place. Furthermore, 85 percent of the pigs did not live longer than two years. Chang (1986) is optimistic that plant husbandry was also established at Zengpiyan, but the evidence has not yet come to light.

So far, incontrovertible evidence for the earliest plant domestication in China is from the deciduous-forest zone of North China, in the context of well-established villages during the Middle Neolithic dating from 8500 to 7000 CAL.B.P. Chang (1986:89–90) reports four clusters of sites on the Loess Plateau that he assigns to the Peiligang culture: in southern Hopei (e.g., Cishan site); in Honan (e.g., Peiligang site); in the Wei River valley (e.g., the Laoguantai and Dadiwan sites); and in southern Shanxi (e.g., Lixiatsun site). In the northeast may be another Middle Neolithic cluster of sites that Yan (1989) places in the Xinlongwa culture. Four radiocarbon dates from Xinle suggest occupation there as early as 7500

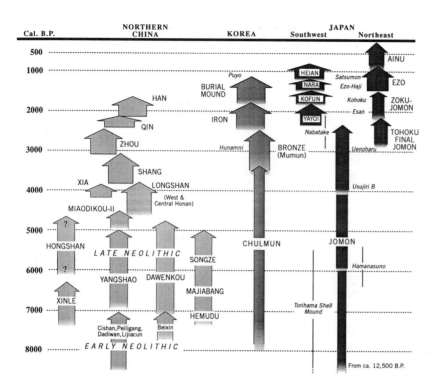

Fig. 2.2. Summary of Northeast Asian chronology. Arrow width is roughly proportional to the degree of dependence on food production. (Data for China and Korea from An 1989; Chang 1986; Nelson 1982; and Pearson 1982)

CAL.B.P., while the Xinlongwa site may date to the late eighth millennium B.P. (Chang 1986:175,181). Another Neolithic site in the same region dating to the eighth or ninth millennium B.P. is Cha-hai (Nelson 1990:239). Other possible Middle Neolithic occupations are in the Shandong, the northeast, east, and southern regions of the deciduous forest zone, and it seems reasonable to predict that the Shandong and Daxi Neolithic occupations extend back to at least the eighth millennium B.P. as well.

The Middle Neolithic occupations of the Loess Plateau consisted of villages supported by a mixed economy of hunting, gathering, fishing, and plant and animal husbandry (Chang 1986:90–95; Yan 1989). Sites are up to 140 m in diameter and in some areas are as dense as 1.5/km². Cemeteries and living areas of plaster-floored houses comprise these early communities (Chang 1986:90). Three cultigens are reported from four sites of this period: foxtail millet, broomcorn millet, and a Chinese cabbage (*Brassica campestris*) (Yan 1989). Foxtail millet is reported from the Cishan site while broomcorn millet has been identified at the Dadiwan, Peiligang, and Xinle sites (Chang 1986; Yan 1989). Chinese cabbage has been found at one Middle Neolithic site (Dadiwan). Except for the Cishan site, indications of seed quantities

are not reported. Some 80 pits at Cishan contained grain (Yan 1989). The foxtail millet is from the single sample that has been examined (Yan 1989). Husbanded animals during the Middle Neolithic are the pig (*Sus* sp.), dog (*Canis*), and, very likely, chicken (*Gallus* sp.) (Yan 1989).

The Late Neolithic, beginning about 7000 CAL.B.P., exhibits continuity with preceding phases. In the region of the Middle Neolithic cultures of the Huang and Wei rivers, the Yangshao culture persisted for some two millennia. The Dawenkou culture was established by this time on the lower elevations of the North China Plain east of the Yangshao culture. Developments continued in the north and east with the Xinle culture of Liaoning Province and the Hongshan culture of Liaoning Province and northern Hopei. In central coastal China, in the lower Yangtze and Huai rivers, the Majiabang and Hemudu cultures were established. The Yangshao is distributed over much of the Loess Plateau area of northern China. The period is best known to westerners through the excavations at the Banpo site. By the early 1960s over 400 Yangshao sites had been discovered (Archaeological Institute, Chinese Academy of Science 1963). Several temporal phases of the Yangshao have been defined; two are found at Banpo while four are rec-

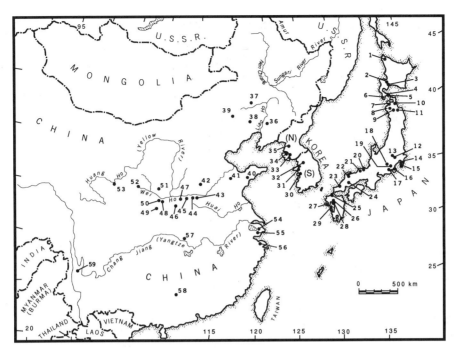

Fig. 2.3. Location of East Asian sites mentioned in the text. **Japan:** 1, Toyotomi; 2, Sakushu-Kotoni River, Idenshikōgaku, and Poplar Namiki; 3, Kashiwagigawa; 4, Usujiri; 5, Hamanasuno; 6, Kachiyama; 7, Kamegaoka; 8, Tareyanagi; 9, Aramachi; 10, Kazahari; 11, Ichinohe; 12, Daiichi Chuugakko; 13, Shimpukuji; 14, Shitakitabara; 15, Otsubo; 16, Yamagi; 17, Toro; 18, Sohara; 19, Idojiri; 20, Torihama; 21, Kuwagaishimo; 22, Megumi; 23, Ubuka Bog; 24, Nogoe; 25, Nabatake, Rokutanda, Itatsuke; 26, Wakudoishi; 27, Ikiriki; 28, Uenoharu; 29, Yamanotera and Kureishibaru. **Korea:** 30, Puyo; 31, Honamni; 32, Chit'am-ni; 33, Hunamni; 34, Soktali; 35, Chongok-ni. **China:** 36, Xinle; 37, Fuhokoumen; 38, Cha-hai; 39, Xinlingwa; 40, San-li-ho; 41, Dawenkou; 42, Cishan; 43, Peiligang; 44, Erlitou; 45, Miaodigou; 46, Xiawanggang; 47, Yangshao; 48, Jiangzhai; 49, Heijiawan; 50, Banpo; 51, Laoguantai; 52, Dadiwan; 53, Linxia; 54, Songze; 55, Majiabang; 56, Hemudu; 57, Kuanmiao-shan (Daxi Culture); 58, Zengpiyan Cave; and 59, Chien-ch'uan.

ognized at Jiangzhai (Cultural Relics Publishing House 1988). Yangshao sites are roughly five times the area of preceding Middle Neolithic communities. House structures, storage pits, and stone and bone technology are similar to those of preceding phases. New aspects include evidence of planned community layouts, pottery kilns, larger cemeteries often with multiple burials and at least one burial with rich grave goods, and the excavation of large ditches. Domesticated plant remains are reported from thirteen Yangshao period sites (Yan 1989). Foxtail millet is reported from seven sites, including Banpo. Broomcorn millet is reported only from Jiangzhai. Unidentified millet is reported from two sites in Honan Province (Yan 1989). Chinese cabbage is reported from Banpo and rice is reported from two Yangshao period sites, Heijiawan and Xiawanggang (Yan 1989). There is a single report of possible sorghum from Honan Province but, because no other Neolithic report of this crop has been made, it is in all probability a contaminant or a misidentification. Sorghum is not native to China and Ho (1977:475) contends that it did not become widely known in China until the thirteenth century A.D. Carbon isotope analysis indicates that the Yangshao diet may have been as much as 58 percent millet (Cai and Qiu 1984: cited in An 1989 and Yan 1989). This is similar to estimates of the maize (*Zea mays*) components of the Ontario Iroquois diet in North America (Schwarz et al. 1985).

By 6000 CAL.B.P., agricultural communities were established in south-central Ganzu Province. Chang (1986) prefers to group the several phases of the Gansu Province Late Neolithic with the Yangshao, while some prefer to identify the Gansu Province phases as separate cultures. Yan (1989), for example, lun.ps them within the Majiayao culture. Whatever the case, plant remains are similar to those reported from eastern Yangshao sites with the exception of hemp from the Linjia site and several reports of broomcorn millet. Five sites have foxtail millet while three have broomcorn millet. The broomcorn millet from Linjia consists of unthreshed inflorescences tied into bundles (Yan 1989).

A tradition contemporaneous with Yangshao, the Dawenkou culture, occupied the region east of the loess area to the coast, mainly on the Shandong Peninsula. Dawenkou is represented by about 100 sites, most of which are cemeteries (Chang 1986:159–60). House styles are not well known, although both pit houses and houses with floors at ground level are known; burials are the same as Yangshao interments (Chang 1986:162). Not surprisingly, because of the few domestic area excavations carried out, few plant food remains are reported from the Dawenkou sites. Foxtail millet is reported from two sites and broomcorn millet from one site (Yan 1989).

Local Neolithic phases are well known in southern Manchuria. Carbonized grains are reported from a posthole at the Xinle site dating to ca. 7500–6500 CAL.B.P. (Chang 1986:175–76). The grains are apparently broomcorn millet (Yan 1989). By 6500 CAL.B.P., food production was being carried out as far as 43° north latitude—the same latitude as south-central Hokkaidō, Japan.

Korea

In contrast to the food production of Neolithic China, Korean populations during the period from about 8000 B.P. to the fourth millennium B.P. (Fig. 2.2) are thought to have maintained a broad-spectrum subsistence base with millet husbandry appearing some time in the fourth millennium B.P. (Nelson 1982:115). This four- or five-millennia period in Korea is characterized by a high frequency of pottery known as Chulmun, or comb-pattern pottery (Nelson 1982; Pearson 1982). Little of it is actually comb-patterned but incising (trailing) is common. In contrast to Japanese mid-Holocene pottery, cord marking is rare, found only on ceramics in the south. Villages or hamlets of round, semisubterranean houses, as well as coastal shell middens, are known. In the Han River basin, site densities are as high as 1.0/2 km² (Nelson 1973). Houses are about 8.5 m² and 1 m deep, whereas in the Yalu River region houses are elongated with floor areas of about 20 to 30 m². The earliest potential cultigen report in Korea is from the undated Chulmun Chit'am-ni site (Di-Tap-Li), where two types of millet were found in the bottom of a dentate ware pot (Chard 1960; Kim 1978). The millets have not been specifically identified; cursory examination suggests they are either *Setaria italica* or *Panicum crusgalli* (Kim 1978). *P. crusgalli* is an outdated term for barnyard millet (*Echinochloa utilis*). Kim (1978) may have meant *Panicum miliaceum* (broomcorn millet). Similar cursory reports of buckwheat, sorghum, and barnyard millet in Hokkaidō have proven to be incorrect. Until proper identification of the grains is made, the Chit'am-ni report should be treated cautiously.

In southeastern Korea, bronze associated with plain Mumun pottery makes its appearance between 4000 and 3000 B.P. (Nelson 1982). Korean archaeologists place the beginning of the Bronze Age at 3000 B.P., when Mumun pottery and bronze are widely distributed throughout Korea (Nelson 1982). Unlike China, there are no bronze vessels, but weapons and symbols of status are the most common bronze items (Pearson 1982). The most common stone tool from this period is a crescent shaped knife (sickle) with two holes in it (Pearson 1982). The first rice in Korea is associated with radiocarbon dates of 3300 to 2850 B.P. at the Hunamni site (Choe 1982:524; Kim 1982:514; Nelson 1982:128; Pearson 1982). No rice is known from the Bronze Age in North Korea. Irrigation agriculture was probably established by the middle of the third millennium B.P. (Nelson 1982; Pearson 1982).

Japan

A tradition of mainly broad-spectrum foragers existed for much of the period from 12,000 B.P. until about 2400 B.P. in southwestern Japan, and until more recently in the northeast of that country (Fig. 2.1). The Jōmon period has been subdivided into six phases in the southwest and seven in the northeast. The justification for the dating of the individual phases is beyond the scope of this chapter; however, when a Jōmon site is discussed here, its phase assignment is provided together with pertinent dates. Considerable regional and temporal variation of the Jōmon is recognized, so no one settlement pattern or specific subsistence regime can be applied to the Jōmon as a whole (e.g., Akazawa 1982; Crawford 1983). Evidence for plant husbandry is present in some areas but lacking in others. As in Korea, coastal shell mounds are known for some areas and periods but not for others. At any rate, the Jōmon is superseded in the southwest by the Yayoi, a society that is characterized by substantial wet rice agriculture. The Hokkaidō Jōmon continues until about A.D. 500. Then it is followed by the Ezo period, a time of rapid change and increased interaction with the south. The Yayoi, too, is a period of culture change, in part due to processes involving increased contacts with the mainland. The southwestern sequence eventually led to a state-level sociopolitical organization, whereas Hokkaidō was the home of the historically documented Ainu. I have described the transition to agriculture in Japan as consisting of four phases (Crawford 1992). The first is mid-Holocene gardening and the second is the development of rice production in southwestern Japan. The third phase marks the development of agriculture in northwestern Honshū, while the final stage brings dry-land agriculture to Hokkaidō.

Dependence on food production generally increased through time in northern China, Korea, and Japan, excluding Hokkaidō. In Hokkaidō some populations were substantially agricultural during the early Ezo period, but the late Ezo or Satsumon populations and the Ainu seem to have depended less on plant husbandry (Crawford and Yoshizaki 1987), although this is still open to question. By the fourth millennium B.P. in China and the early first millennium A.D. in southwestern Japan and in Korea, dependence on domesticated resources was comparatively high. The Chulmun and Jōmon were primarily foraging stages, but some plant husbandry is evidenced in the Jōmon.

Initial Speculations (Japan)

The transition to wet-rice agriculture in Japan has been conservatively dated to between 2400 and 2300 B.P. This marks the beginning of the Yayoi, a wet-rice based tradition lasting nearly 700 years. As early as the 1950s and early 1960s at least ten cultigens had been identified from the Yayoi stage Toro site (Goto 1954). The plant remains include pear (Pyrus communis), soybean, and adzuki bean (Vigna angularis). Hops (Humulus sp.) is also reported but the plant is a native weed found throughout Japan. Buckwheat, in addition to five of these cultigens, was found at the Yamagi site (Goto 1962). At this point in the

investigation of Japanese agricultural history, the Yayoi clearly possessed something that Jōmon sites did not: cultigen remains in excellent context.

The Jōmon is enigmatic in many ways. Jōmon peoples were sedentary, lived in pit house villages of varying sizes, had elaborate material culture, and in some areas and periods had relatively dense populations. In spite of these characteristics, which are usually associated with food production, the Jōmon is generally viewed as a foraging tradition. Until the late 1970s in Japan, little research was being directed toward understanding the parameters of Jōmon subsistence. Identification of bone from prehistoric sites was not uncommon, however, and by 1975 remains of some 39 plant taxa from 208 sites had also been reported (Watanabe 1975). Beyond this, there was little empirical, ecofactual information pertaining to postulated Jōmon agriculture.

A few archaeologists in the early 1900s, such as Torii, Oyama, and Sumida, proposed that Jōmon technology was agriculturally related (see Kotani 1972; Sasaki 1971; Tozawa 1982; and Tsuboi 1964, for discussion). Tsuboi (1964) considered an apparently large population increase from Early to Middle Jōmon in the Chubu district to be evidence for a shift to food production. Fujimori (1965) added at least nine traits to the list of secondary evidence for agriculture: paucity of stone tools; functional differentiation of pottery; snake, human, and sun effigies on pottery; stone clubs; clay figurines; stone monuments; and the proliferation of digging tools and grinding stones. Several carbonized cakes were uncovered at these sites leading to speculation that the cakes were cultigen products. No plant remains were identified to support the hypothesis of Middle Jōmon agriculture, but Fujimori suspected that millet was grown, and others proposed that root crops were husbanded (Esaka 1977).

While Chūbu and northeastern Jōmon population densities were relatively high, the opposite is apparent in southwestern Japan. Koyama (1978, table 3) estimates that 82 percent of Jōmon sites are in northeastern Japan. The low site density in the southwest, according to Kondo (1964), suggests a scarcity of resources. Kondo (1964) proposes that rice, when it was introduced, solved a resource scarcity problem and was, therefore, rapidly accepted. Others, such as Nakao (1966) and Ueyama (1969) suggest that scarcity provided pressure to develop new resources. Nakao and Ueyama also noticed something else: the low density Jōmon occupations were situated in an environmental zone distributed not only over southwestern Japan, but also over southern China, northern Southeast Asia, northeastern India, and southern Korea—in the broadleaf evergreen forest zone (Shōyōjurin-tai) (Zones 7 and 8 in Fig. 2.1). Swidden systems were prevalent in these areas and it was supposed that the subsistence mode of what has come to be known as the "Shōyōjurin-tai Culture" was a solution to resource scarcity problems throughout this region (e.g., Ueyama 1969; Ueyama et al. 1976). Swidden was part of a postulated five-stage sequence leading to wet-rice agriculture. The second stage has been termed *hansaibai*, or semicultivation (Nakao 1966). Native plants, including nut trees, were semicultivated according to this model. Ueyama et al. (1976:88–89) suggest that nuts increased in size as a result of this process. Swidden agriculture lies developmentally between *hansaibai* and wet-rice husbandry in this model. Sasaki (1971) uses comparative ethnographic data to suggest the presence of slash-and-burn agriculture during the Jōmon (see also Pearson and Pearson 1978; Sasaki et al. 1982).

Secondary data have played a major role in establishing the pre-Yayoi agriculture hypotheses. The Middle Jōmon, particularly in Chūbu, characteristically had relatively elaborate material culture and apparently high population densities. Late and Final Jōmon swidden seemed possible in southwestern Japan in light of ethnographic data from similar environmental zones elsewhere in Asia. The few plant remains discovered by 1970, including the carbonized cakes from the Chūbu area, did little to confirm or deny the Middle Jōmon hypothesis. There is some evidence for bottle gourd and rice in later Jōmon sites, such as Shimpukuji, Wakudoishi, Rokutanda, Nogoe, Itatsuke, Yamanotera, and Kureishibaru (discussed in Kotani 1981), which supports the suspicion that plant husbandry was present by about 3000 B.P.

Archaeobotany and Recent Research

Since 1970 a few efforts to recover plant remains have nearly all added fuel to the pre-Yayoi plant husbandry debate. Flotation at the Final Jōmon Uenoharu site on Kyūshū recovered one barley grain and two rice kernels (Kotani 1972:231–32). One red bean is reported from the same site but is considered to be wild on the basis of its small size. No measurements are given, but the small size could also be a result of

carbonization. Kotani's work clearly demonstrated the value of recovering plant remains systematically to test subsistence related hypotheses in Japan—an observation made much earlier elsewhere in archaeological research (see, for example, Watson 1976).

Unfortunately, Kotani's early finds had little or no impact on the direction of research taken by Japanese archaeologists. Flotation is rarely conducted in Japan outside Hokkaidō today. It appears that influential local archaeologists have made their own conclusions about plant food subsistence and see no need to test what is self-evident. For example, Akazawa (1982) has constructed a model of differential acceptance of plant husbandry in two regions of prehistoric Japan. Akazawa claims that the prehistoric inhabitants of the Noto Peninsula had little interest in plant food and, therefore, resisted the introduction of rice. No systematic attempt to collect plant remains from the area was carried out, however; therefore, the model is untested. Yet this suggestion is cited in western literature as fact. Nevertheless, wherever purposive collections of prehistoric plant remains have been made in Japan since Kotani's work, results have tended to confirm the existence of some form of plant husbandry. A type of graded water sieving (*suisenbetsu*) used at the Nabatake site, for example, has resulted in the recovery of cultigens from as early as 2700 to 3000 B.P. (Kasahara 1982, 1984).

In the Early and Middle Jōmon, evidence points to the existence of gardening, although it was likely of minor economic value. The constituent of carbonized cakes from eight Middle Jōmon sites has been identified as beefsteak plant or *shiso* (*Perilla frutescens*), a cultigen in the mint family (Matsutani 1983, 1984). The Torihama Shell Mound, a stratified wet site in Fukui Prefecture, contains a wealth of plant remains, including cultigens. Bottle gourd (*Lagenaria siceraria*) (both seeds and parts of the rind), bean (either mung or black gram), *egoma* (a mint cultigen variety of *Perilla frutescens*), burdock, hemp, paper mulberry (*Broussonetia papyrifera*), and Chinese cabbage have been reported (Okamoto 1979, 1983; Umemoto and Moriwaki 1983). A single seed each of bottle gourd and beefsteak plant occur in Initial Jōmon levels. In addition, three broken melon-like cucurbit seeds were recovered. Nakanishi (1983) notes that the seeds do not resemble the local wild melon, *Tricosanthes*. From the published photographs the seeds appear to be those of melon (*Cucumis melo*). Five bottle gourd seeds and pieces of gourd rind are reported from one other Early Jōmon shell midden, Otsubo in Chiba

Prefecture. Finally, peach seeds are reported from the Ikiriki site near Nagasaki, Kyūshū (Minamiki et al. 1986). Peach is not native to Japan. The seeds are from water-logged, marine Early Jōmon deposits. The earliest specimens are from levels VIIIa and VIII'a. The top of level VIIIa is dated to 5660 ± 90 B.P., while two dates are available from level VIII'a: 5830 ± 30 and 5930 ± 30 B.P. (Minamiki et al. 1986). In the same deposits are large quantities of anchor stones indicating substantial seafaring activities that could have provided the mechanism for the introduction of peach from the mainland, where it was native. At any rate it would appear that gardening was a component of Jōmon subsistence in some areas of southwestern Japan by the Early Jōmon.

There is some evidence for early gardening in northeastern Japan as well. A systematic study of carbonized plant remains and related data from eight phases spanning 4,000 years points to barnyard grass and buckwheat husbandry, as well as increasing ecological disruption (Crawford 1983). The evidence for barnyard grass husbandry, an increase in caryopsis size by the end of the Middle Jōmon (Crawford 1983), is not conclusive so more research is required in the Middle and later Jōmon periods of southwestern Hokkaidō. A single buckwheat seed from near the floor of an early Jōmon pit house at the Hamanasuno site is still the only example of a seed of this cultigen in the Jōmon period. Confirmatory evidence for the presence of buckwheat in Early Jōmon Japan comes in the form of pollen from Ubuka Bog in southwestern Japan (Tsukada et al. 1986). Buckwheat pollen composes as much as 1 percent of the more than 1,000 grains counted in sediments radiocarbon dated to 6600 ± 75 B.P. (Tsukada et al. 1986:633).

Some data thought to be relevant to the Shōyōjurin-tai Culture hypothesis have been collected from Kuwagaishimo, a wet site near Kyōtō (Tsunoda and Watanabe 1976). Many of the remains of wild and weedy taxa are not carbonized so it is difficult to assess whether they represent utilization by the Jōmon. Acorn and a Japanese buckeye (Aesculus turbinata) are common among the uncarbonized remains. The remains are used to support the hypothesis that a leaching technology evolved in order to utilize these nuts, which contain tannic acid. Grinding stones are interpreted to be part of the leaching toolkit, not tools for grinding or bruising grain (Tsunoda and Watanabe 1976). The Kuwagaishimo material culture is thought to have been rooted in the Middle Jōmon of the Chūbu district and to have dif-

fused from there (Tsunoda and Watanabe 1976). Tsunoda and Watanabe (1976) believe that the Chūbu Middle Jōmon was characterized in part by intensive nut exploitation, not cultigen husbandry; furthermore, because diffusion, as they see it, moved from east to west, the west to east developmental sequence of agriculture that ultimately had been introduced from the mainland is contradicted. They are not willing to accept slash-and-burn agriculture in the Middle or Late Jōmon, but believe the intensive use of nuts and other plants to be a local variant of the semicultivation stage or hansaibai in the development of the final stage of the Shōyōjurin-tai swidden-based culture.

Using data from several wet sites, including Torihama, Nishida (1980, 1981, 1983) proposes that chestnut and walnut cultivation was the solution to apparent resource scarcity. His interpretation is based on plant remains that consist mainly of uncarbonized nut remains. Charcoal from these two nut trees is abundant at these sites as well. Nishida concludes that the charcoal quantities reflect the importance of these trees in an environmental context where they would not be abundant without human intervention. Nishida assumes that all the uncarbonized nut remains are food remains, an assumption not adequately explored. The Kuwagaishimo plant remains fit this model as well (Nishida 1981, 1983). Neither Nishida (1976) nor Tsunoda and Watanabe (1976) discuss the significance of carbonized rice grains, barley, and beans (probably adzuki) from the Late Jōmon deposits at Kuwagaishimo.

Finally, cultigens are reported from the stratified Nabatake site in Saga Prefecture, northern Kyūshū (Kasahara 1982, 1984). The site has Early, Middle, and Final Jōmon components, as well as early Middle Yayoi occupations. The three cultigens reported by Kasahara (1982) in the Jōmon levels (Final Jōmon) are rice, beefsteak plant, and foxtail millet. A single foxtail millet grain is from level 11 of Units CI-CIV (Kasahara 1982:378). Shiso (22 seeds) is found in levels 8, 10, and 11, and two rice grains are from levels 8 and 11. Final Jōmon radiocarbon dates at the site are 3000 ± 80 B.P. (level 13), 4030 ± 65 B.P. (level 12), and 2680 ± 80 B.P. (levels 10–11), 3230 ± 100, 2960 ± 90, and 2620 ± 60 B.P. (level 8) (Tosu-shi Kyōiku Iinkai 1982). Levels 8 to 11, those with cultigens, range in date from 3330 to 2560 B.P. Additional cultigens in the Yayoi levels include hemp, peach, and melon. Watanabe and Kokawa (1982) examined a sample of large seeds from Nabatake and report four seeds of mung bean or *ryokuto* (*Vigna radiatus*) from the Final Jōmon horizon as well.

Northeastern Spread of Plant Husbandry

Most researchers argue that, after rice husbandry was established in southwestern Japan in the Early Yayoi (Kanaseki and Sahara 1978), it diffused rapidly northeastward to about 35° north latitude during the Middle Yayoi. Not until the development of a rice variety that was capable of tolerating more northern climates did rice move to about 41° north latitude. Kondo (1964) and Sahara (1975) estimate that this third stage took place during the Middle to Late Yayoi. Akazawa (1982) believes that the flourishing Jōmon fishing societies in the east resisted rice agriculture, resulting in a cultural dichotomization (Akazawa 1982:203). Yayoi-influenced rice production moved northeastward, but initially only inland where "intensive plant collecting and/or incipient plant cultivation of native species" (Akazawa 1982:200) was taking place. The inland Jōmon populations, therefore, appear to have been more receptive to rice agriculture than their coastal counterparts. In recent years, however, prehistorians have viewed the rapid Early Yayoi spread as first a migration followed by the overwhelming reproductive success of the migrants (Brace et al. 1989; Hanihara 1987). Under these circumstances, receptivity of the Jōmon to rice may not have been as much of a factor as Akazawa suggests. Akazawa, however, provides valuable data about exploitation territory and technology regarding fish, but plant remains have not been collected from the same eastern coastal areas. Until they are collected, the reasons for the apparent differential acceptance of rice should remain open to question. In addition, the Tōhoku Yayoi was established in northern Honshū no later than three to four centuries after the Yayoi began in Kyūshū. I have questioned elsewhere the differential rate of Yayoi development in southwestern and northeastern Japan (Crawford 1992). Whatever the case, the Yayoi in northernmost Honshū maintained a local identity as a result of the acculturation of Final Jōmon populations (Crawford and Takamiya 1990).

Evidence contradicting the view of a wave of rice agriculture slowly progressing to the northeast by the Late Yayoi has been recovered in Tōhoku. Kuraku (1984) reports rice paddies and rice impressed pottery from the Tareyanagi site in Aomori Prefecture (first century A.D.) contemporaneous with Middle Yayoi in southwestern Japan. This rice may be descended from Jōmon rice in southwestern Honshū, but Hoshikawa (1984) hypothesizes that an early ripening rice variety was introduced from Kyūshū via the Japan Sea directly to northern Honshū, bypassing southwestern Honshū. Rice in Tōhoku could well have evolved from earlier rice in the area. A few rice grains and hull fragments are reported from a locality of the Final Jōmon Kamegaoka Culture in Aomori Prefecture in association with Ōbora A pottery (Sato 1984), which is contemporaneous with the Early Yayoi. Rice husks in pottery at the Aramachi site in Aomori Prefecture are a further indication that rice was present there about 2200 or 2300 B.P. (Ito 1984). The earliest rice in northeastern Japan has been accelerator radiocarbon dated to 2540 ± 240 B.P. (TO-2022). The rice is from the Late Jōmon component of the Kazahari site (D'Andrea 1992). Clearly, rice may have as long a history in the northeast as in the southwest. After the northern Yayoi was established, a dry farming system with barley, wheat, and millet gained success, probably at the expense of wet rice husbandry (Crawford 1992; Crawford and Takamiya 1990).

Elsewhere in northeastern Japan little is known about the subsistence of phases between the later Jōmon and the first historic period occupations. Prehistorians recognize non-Yayoi groups who are known as the Zoku-Jōmon, Continuing, or Epi-Jōmon (Fig. 2.2) in Tōhoku and Hokkaidō. Few villages are known from this period in Hokkaidō, the majority of sites being cemeteries. The Zoku-Jōmon and Tōhoku Yayoi were contemporary, with the two groups exchanging pottery, beads, and occupying mutually exclusive territories. Cultigens are reported from three Zoku-Jōmon sites: barley from K135 (Sapporo Station, North Entrance site) (Crawford 1987), hemp from Ebetsu Buto (Yamada 1986), and barley and rice from Mochiyazawa (D'Andrea 1992). The seeds are so few in number, however, that a case for Zoku-Jōmon plant husbandry is difficult to justify in view of their settlement pattern and technology. Village sites become common again about A.D. 600 with the development of the Ezo phase, generally acknowledged to be ancestral Ainu (Yoshizaki 1983). Until recently, the cultural sequence of Hokkaidō was described as a series of foraging societies culminating in the historically documented Ainu. In recent years, however, it has been discovered that food production played a significant role in early Ezo subsistence (Crawford 1992; Crawford and Takamiya 1990; Crawford and Yoshizaki 1987).

The Cultigens

The archaeobotanical assemblage from Ezo period Hokkaidō, besides confirming the existence of what appears to be an intensive phase of plant husbandry in some parts of Hokkaidō, provides an important body of data pertaining to prehistoric agricultural systems that are not based on rice in eastern Asia. Eleven or twelve cultigen taxa have been identified from the Sakushu-Kotoni River and other Ezo period sites in Hokkaidō (Crawford 1986; Crawford and Takamiya 1990; Crawford and Yoshizaki 1987). Sorghum and buckwheat are reported from one site each (Satsumae and K441 respectively) (Yoshizaki 1989). With a few exceptions, such as bottle gourd, Chinese cabbage, and paper mulberry, cultigens found at sites elsewhere in temperate East Asia are also found in Hokkaidō. The Hokkaidō data thus provide a focus for the discussion of cultigen history and descriptions. Crops in the grass family (Gramineae or Poaceae) are presented first, followed by a discussion of other cultigens listed in alphabetical order of common names.

Barley (*Hordeum vulgare*)

China is one center of diversity for barley (Harlan 1968; Zeven and Zhukovsky 1975:81). If barley was introduced to China from western Asia, then at least two millennia of isolation from western Asian gene pools would have provided ample time for the evolution of local genotypes and phenotypes. The possibility remains open, however, that barley was independently domesticated in China. *H. spontaneum*, the wild relative of cultigen barley, is reported in Tibet and Szechwan and may be the progenitor of Chinese naked barley (Chang 1983:78). Unfortunately, the archaeological evidence for the spread and diversification of eastern Asian barleys is nonexistent.

Takahashi (1955) has described the Asian forms of barley and at least one study identifies a separate Chinese subspecies, *H. vulgare* ssp. *humile* (Zeven and Zhukovsky 1975:81). This subspecies is a short, six-row barley. Takahashi's (1955) taxonomy does not include this subspecies, but he makes some observations on the distribution of certain characteristics of Asian barley that separates it into what he terms "Occidental" and "Oriental" groups. Three examples of traits with distinctive distributions are the alleles for tough rachis, frequencies of naked barley, and the occurrence of semibrachytic forms. Type E of three genotypes (Types E, W, and ew) of tough rachis barley dominates (95–100 percent) the "Oriental" region (China, South Korea, and southern Japan) while high frequencies of Type W (62–72 percent), similar to southwestern Asia, Europe, and the Soviet Union, occur in the "Occidental" region (Manchuria, North Korea, and northern Japan) (Takahashi 1955:240). Takahashi (1955) believes that this is a result of independent migrations into eastern Asia of two types of barley with separate evolutionary lines. Takahashi's "Oriental" line could be related to the barley that Chang (1977) suggests is derived from the plateau area of southwestern China.

Naked barley is found in two areas: the southern, warmer parts of the winter barley zones of central China, South Korea, and southern Japan, and the spring barley zones of northern Japan (Chang 1977:252). On average, half of the barley in these areas is the naked form (Chang 1977:252). Takahashi's map (1955:257) illustrates that only a small percentage of the Hokkaidō barley is hulled in contrast to the Tōhoku region where most of the barley is hulled. Both naked and hulled barley grains compose the Sakushu-Kotoni River site collection, but the proportions of naked versus hulled types have not been estimated yet. Based on my own observations, both types are present in significant quantities. The Ainu apparently grew both forms, but naked barley was more important to them (Hayashi 1975). A form of barley unique to eastern Asia is reported mainly in Japan and Korea where it was introduced from Japan at the turn of the twentieth century (ca. A.D. 1910) (Hayashi 1975:256). This is the semibrachytic form (*uzu*). A single gene is responsible for shortening plant parts without affecting their width, resulting in a variety of barley that tolerates heavy manuring in warm and rainy weather (Hayashi 1975:256). The semibrachytic form is restricted to southwestern Japan while only "normal" forms are found in Hokkaidō and Tōhoku (Takahashi 1955). Some normal forms are also found in southwestern Japan.

There is some disagreement on whether barley is

named on Shang oracle bones (Chang 1983:77). Ho (1977:448) believes that two characters for wheat are found on the oracle bones and that a character for barley is not. Whatever the case, there are no publications of confirmed archaeological finds of barley from China. Barley is reported from the late fourth millennium B.P. Hunamni site in Korea, so it is safe to assume that barley was grown in China during the early Shang Dynasty. Ho (1977:449) has already suggested that barley was introduced to eastern Asia by 4000 B.P. Two reports of barley suggest that it entered Japan by the beginning of the third millennium B.P. This is certainly possible because of its early appearance in Korea as well. The single carbonized grain from the Late Jōmon Uenoharu site in Kyūshū is a short grain measuring 3.8 mm in length by 2.0 mm in width. Two barley grains from the Late Jōmon Kuwagaishimo site near Kyōto are longer but still relatively short, measuring 5.0 by 3.3 mm and 5.4 by 2.5 mm (estimated from Plate 4 in Tsunoda and Watanabe 1976). Barley remains are not found in substantial quantities in Japan until the Kofun Period. By this time, sizeable populations of relatively short-grained barley are evident. A sample of 100 caryopses from the Daiichi Chūgakkō site in Kawagoe City, Saitama Prefecture, has average length, width, and thickness (range) of 5.0 (2.5–7.1), 2.3 (1.4–3.3), and 1.8 (1.2–2.7) mm, respectively (Naora 1956:271). A population of 1,000 kernels from the Shitakitabara site in Chiba Prefecture averages 5.0(3.6–7.1) by 2.7(1.8–3.5) by 2.2 (1.4–3.0) mm (Naora 1956:273, 282). One other group of barley grains (n=60) from the same site has mean dimensions of 6.3 (5.0–8.0), 2.6 (1.8–3.1), and 2.2 (1.5–2.8) mm.

Although our sample of archaeological barley is small, some conjecture is in order. The three samples from Saitama and Chiba prefectures segregate into two groups based on grain length. Shorter grains are relatively abundant in one sample from each prefecture. One sample from Chiba Prefecture contains no grains shorter than 5.0 mm, and grains longer than 7.0 mm compose more than 10 percent of the sample. The grain widths do not show a corresponding dichotomy, suggesting that grain length varies independently of width in these samples. For northern Japan, archaeological barley populations should be primarily long-grained if Takahashi (1955) is correct in his interpretation that only "normal" barley was grown there. Archaeological grain size (mainly breadth), however, can be affected by crop processing methods, such as sieving. Without knowing whether a grain sample was processed before carbonization, the representativeness of any one sample cannot be assessed. Furthermore, grain size variation in *uzu* barley has not yet been documented. We have identified some 30,000 barley grains from Sakushu-Kotoni River so far. Because the sample comes from a variety of contexts, it is likely representative of the whole population at the site. By contrast, we have no such assurances for Naora's (1956) data. The Ezo grains are generally "normal" size. A sample of 1,378 grains have mean (range) length, width, and thickness measurements of 5.3 (3.1–8.5), 2.5 (1.3–4.0), and 2.1 (0.9–3.2) mm, respectively. About 10 percent of the sample is shorter than 3.9 mm, suggesting that a proportion of the Sakushu-Kotoni River population is outside the range of variation of Takahashi's (1955) "Occidental" barley. Small examples of barley are evident in the first millennium A.D. in Japan. The one sample of predominantly "normal" barley from Hokkaidō suggests that the distributions of *uzu* and "normal" barley may well have been forming their historically documented pattern by the middle of the first millennium A.D. in Japan. The significance of the form of the Uenoharu and Kuwagaishimo grains is difficult to assess without comparative archaeological material from the same time period elsewhere in eastern Asia. Clearly, considerable work needs to be done on East Asian barley.

Barnyard Millet (*Echinochloa utilis*)

A common grass seed in northeastern Jōmon sites is barnyard grass or barnyard millet. Kidder (1959:54) mentions seeds of this grass from a Jōmon site and this report has occasionally been cited as evidence of millet husbandry by the Jōmon (e.g., Turner 1979). No confirmation of this identification has been made, however, nor is the context of the seeds reported. Research on several phases of Initial to Early Jōmon in southwestern Hokkaidō, however, supports the contention of its use by 5000 B.P. (late Early Jōmon) and the intensification of its contribution to the Jōmon by 4000 B.P. (the end of the Middle Jōmon) (Crawford 1983:31–34). An increase of nearly 15 percent in the size of the seeds over one millennium is cited as evidence for local domestication taking place. A carbonized mass adhering to a Middle Jōmon ceramic sherd contains at least one *Echinochloa* seed with portions of the palea and lemma (glumes enclosing the caryopsis or grain) still attached. The size and shape of the specimen is within the range of the cultigen barnyard millet. Barnyard grass caryopses are present in the Sakushu-Kotoni River samples but they are few in number and all are the size of wild specimens.

At the Ubuka Bog in southwestern Japan (Tsukada et al. 1986), grass pollen grains larger than 41 microns represent at most 5 percent of the pollen between 6600 and 3100 B.P. Such large grains suggest cereal plant husbandry during the Jōmon in southwestern Japan (Tsukada et al. 1986:633). More attention should be paid to the smaller carbonized grass seeds from Jōmon sites to test the existence of cereal plant husbandry from the Early Jōmon and subsequent periods. Two taxa closely related to barnyard grass are *Echinochloa crusgalli* var. *frumentacea* (or *E. frumentacea*) and *E. utilis* (Yabuno 1966). *E. utilis* is derived from *E. crusgalli*. The two taxa are in the same genome, one that does not include *E. frumentacea,* the Indian cultigen form (Yabuno 1966:320–21). The presence of barnyard millet in Ezo period Hokkaidō has not yet been confirmed but is reported from three historic sites in southwestern Japan (Matsutani 1984). The distinctions between *E. utilis* and broomcorn millet (*Panicum miliaceum*) are subtle. Several hundred grains in the size range of broomcorn millet at Sakushu-Kotoni River are relatively broad, have longer embryos, and are relatively smooth compared to broomcorn millet. Pending further study, I suggest that these are barnyard millet. A flotation sample collected from the Okawa site in Hokkaidō in 1989 contains the only confirmed example of barnyard millet in post-A.D. 600 Hokkaidō. Okawa, however, has a historic component and the grain could be recent.

Distinct from the food grain forms of *Echinochloa* are the weed forms associated with rice fields. Among the most noxious weeds of rice are members of the barnyard grass complex (Barrett 1983). One, *E. crusgalli* (L) Beauv. var. *crusgalli,* is a small-seeded weedy form while a large-seeded form is *E. crusgalli* (L) Beauv. var. *oryzicola* (Vasing) Ohwi, a crop mimic of rice (Barrett 1983:265). Archaeobotanical studies of subsistence systems based on wet rice in eastern Asia should pay close attention to this barnyard grass complex. Large seeds of *Echinochloa* should be carefully examined to distinguish *E. utilis* from *E. crusgalli* var. *oryzicola*. The two grains can be distinguished by their morphology.

There are several reports of Yayoi barnyard millet. Kasahara et al. (1986:121), for example, report 48 seeds from the Early Yayoi component and 3 seeds from the Final Jōmon level at the Megumi site in

Yonago city. It is important, however, to determine whether this is actually the crop form or rice paddy weed form of *Echinochloa*. If it is the paddy weed form, then the question of Yayoi barnyard millet production is still open.

Broomcorn or Common Millet (*Panicum miliaceum*)

Broomcorn millet is a tetraploid and the species is not known as a wild form, although there are several Asian species that may have contributed to the broomcorn millet genome (Smith 1976). Zeven and Zhukovsky (1975:32) note that China is the primary center of diversity for broomcorn millet. Botanical descriptions of broomcorn millet are not found in early Chinese documents (Chang 1983:66), but the plant is mentioned on oracle bones and in the *Book of Odes* (Ho 1977:438). Ho (1977) summarizes the philological evidence for *Panicum* and concludes that it was introduced to western and southern Asia from China. It has been identified at three Middle Neolithic and five Late Neolithic sites on the Loess Plateau (Yan 1989). Broomcorn millet has one of the lowest water requirements of any cereal (Smith 1976), insuring its success on the Loess Plateau in the early mid-Holocene. Its domestication probably began earlier than 8500 B.P., when it is first found as a domesticated form.

There is only one report of Jōmon broomcorn millet and it is from a Late Jōmon context at the Kazahari site, Aomori Prefecture (D'Andrea 1992). Similarly, the only Yayoi broomcorn millet recovered to date is from the same site (D'Andrea 1992). Broomcorn millet was an important crop at Sakushu-Kotoni River (over 64,000 specimens) during the ninth century A.D. It appears to have become important sometime between A.D. 300 and A.D. 800 in Japan.

The broomcorn millet from Hokkaidō is nearly all hulled, that is, only caryopses are present except for a few examples. A sample of 604 grains from the Kashiwagigawa site have mean length, width, and thickness measurements of 1.8, 1.8, and 1.4 mm, respectively.

Foxtail Millet (*Setaria italica* ssp. *italica*)

Foxtail millet also has a long history of use in eastern Asia. Three cultigen races are known: moharia from southwestern Asia and Europe, maxima from the Far East and Transcaucasian Russia, and indica from southern Asia (Rao et al. 1987). All available evidence points to its domestication in East Asia, although independent domestication elsewhere cannot be ruled out (Rao et al. 1987). Chang (1986) and Yan (1989) report foxtail millet from one Middle Neolithic and 15 Late Neolithic sites in North China. By 8000 B.P., foxtail millet seems to have become an important crop. Large quantities of carbonized remains of this grain were apparently found at Banpo (Ho 1977). Unfortunately, we lack any thorough botanical studies of these remains. According to the directors of the Banpo Museum, the plant remains have not been examined since they were excavated in 1956 (personal communication, 1986). As an archaeobotanist who has spent considerable time working on collections containing the three kinds of millets discussed in this chapter, I am concerned about casual reports such as these. The foxtail millet identification is likely correct, but other millets may be present in small proportions. So far, of 21 sites with millet in the Middle and Late Neolithic, only one site (Linjia) has both foxtail and broomcorn millet (Yan 1989). Until thorough, documented analyses are reported, these concerns will remain.

Historic documents show that foxtail millet was an important crop by the Zhou Dynasty (Chang 1983:66–67). The earliest confirmed report of foxtail millet from Korea is from the Bronze Age Hunamni and Honamni sites (Choe 1982:524; Kim 1982:514; Nelson 1982:128). Some foxtail millet is reported from the Final Jōmon levels at the Nabatake site in Kyūshū. A few reports of foxtail millet come from Yayoi sites (e.g., Kasahara et al. 1986). Cultigen *Setaria* may be present as early as the late Middle Jōmon in Hokkaidō. I recently reexamined carbonized grass specimens from Usujiri B and found at least nine specimens in one sample to be morphologically indistinguishable from foxtail millet seeds. In addition, foxtail millet is reported from the Late Jōmon component of the Kazahari site (D'Andrea 1992). Foxtail millet is the most numerous of the grains at Sakushu-Kotoni River; but, because the caryopses are so small (105 seeds from the site measure 1.2 by 1.0 by 0.7 mm; and 1,081 seeds from the Kashiwagigawa site measure 1.3 by 1.3 by 1.0 mm), the food value obtained from this grain was probably somewhat lower than that obtained from barley and broomcorn millet in the early Ezo. Two weedy species of *Setaria*, *S. glauca* and *S. italica* ssp. *viridis,* are the most common weedy grasses in the Sakushu-Kotoni River samples.

Rice (*Oryza sativa*)

Rice was domesticated in subtropical or tropical Asia rather than in the northeast, which is the subject of this chapter. Domestication seems to have occurred as the range of this grass expanded out of the highlands of Southeast Asia and southwestern China (Chang 1976a). The specific place or places of rice domestication, however, cannot be determined (Chang 1976a, 1976b; Oka 1988:110). *Oryza rufipogon* is accepted as the wild ancestor of cultigen rice (Oka 1988). This wild rice continuously grades from perennial to annual types. The latter is usually called "weedy rice" while the former is called "wild rice." Weed races (known as spontanea forms) hybridize freely with domesticated rice. Furthermore, the perennial and annual forms of *O. rufipogon* are distributed in a continuous belt from the foothills of the Himalayas to the Mekong region (Chang 1976b:101). Annual weedy rice, but not the perennial form, is also distributed in the Lower Yangtze River basin (Oka 1988; Yan 1989).

Three ecogeographic races are normally recognized: *Oryza sativa* var. *indica,* *O. sativa* var. *javanica,* and *O. sativa* var. *sinica* (*japonica*) (Chang 1976a). The *sinica* type, as Chang (1976a) refers to it, is traditionally known as *O. sativa* var. *japonica*. Following Oka (1988), I refer to it as "Japonica" and *O. sativa* var. *indica* as "Indica." Oka (1988:146) does not accept the *javanica* variety as a true type. It has never been formally defined and is essentially identical to Japonica. Oka groups Chang's *javanica* rices into a tropical subgroup of Japonica. Wild rice is not split into Indica or Japonica forms, although weedy rice does show some differentiation into similar types (Oka 1988:163). Gene flow between Indica and Japonica is restricted; F1 hybrids are sterile (Oka 1988). Rice produced in northern China, Japan, and Korea is all Japonica (Chang 1985; Oka 1988:154). Japonicas are also grown at higher elevations in northern Thailand and southern China where the lowland rice variety is Indica (Oka 1988:154). The evolution of the two types is as yet unexplained. Oka (1988:169) proposes that the founders of each type became established after the domestication of rice had begun. Japonica is usually considered to be short-grained while Indica is long-grained. This criterion is rather unreliable, however. The variation of grain size within the two groups gives a 39 percent chance of misclassification (Oka 1988:144).

The early evidence for rice domestication is rather

sparse. Chang's identification of the early rice from Non Nok Tha, Thailand (5500 B.P.), as a form intermediate between wild and domesticated could be construed to mean that rice agriculture was not part of the sixth millennium B.P. economy of northern Thailand (Yen 1982:52). Yen (1982:63) points out that this is an oversimplification; gene exchange between wild forms and forms grown in early rice fields was an ongoing process that could have resulted in the intermediate characters observed on archaeological specimens. Earlier rice comes from layer 4 at the Hemudu site (7000 CAL.B.P.), just south of Shanghai in the lower Yangtze River region. It is mainly long grain, described as subspecies *Xian Ding* (Yu 1976:21). Some short grain rice is also present in the same samples (Yan 1989:18). The mixtures are about 80 percent Indica-type and 20 percent Japonica-type (Oka 1988:136). Grass pollen as large as 49.5 microns has been recovered from the same level (Chekiang Provincial Museum 1978:107). The rice reported from the Luojiajiao site in Zhegiang Province from about 7000 B.P. is Japonica (Chang 1985). If rice was differentiated into Indica and Japonica varieties by 7000 B.P., the initial steps in the domestication process must have occurred much earlier than this. Before these reports, the earliest reports of short-grain rice, besides that from Hemudu, were from the Yangtze basin, dated to 3311 ± 136 CAL.B.C. and 2696 ± 190 CAL.B.C. (Chang 1983:73). The earliest rice in northern China is from the Yangshao site belonging to the Miaodikou II Phase (about 4200 to 5000 B.P.) reported by Andersson (1934). This rice is short-grained (Chang 1983:73). Accelerator radiocarbon dates on rice from the Andaryan site in the Philippines give the first solid date for rice in that area: 3400 ± 125 B.P. (Snow et al. 1986). This rice is also identified by T.-T. Chang as an intermediate form (Snow et al. 1986).

The weedy, spontanea form of rice has been recently found growing in the Yangtze basin (Yan 1989). Yan assumes that the weedy form is ancestral to domesticated rice and concludes, on the basis of this evidence, that rice could have been domesticated independently in the Yangtze basin. It is an oversimplification to assume that the weed form of *O. sativa* is ancestral to domesticated rice. The possibility that rice was independently domesticated in this region, however, should remain open.

Rice is known from the late fourth millennium B.P. Hunamni site in Korea (Choe 1982). In addition, a few examples of Late and Final Jōmon (about 3000 B.P.) rice have been reported from Nabatake (Kasahara 1982) and Uenoharu (Kotani 1972) in southwestern Japan. One measurable kernel from Uenoharu is relatively short-grained (4.4 by 2.2 by 1.8 mm). The Late Jōmon rice from the Kazahari site is also relatively short-grained. Data accumulated from over 100 Yayoi sites have yet to show the presence of any rice other than the short-grain Japonica variety whose length, width, and thickness measurements range from 3.7–5.3 by 1.3–2.0 by 3.0–3.7 mm, respectively (Sato 1971). Nineteen carbonized rice grains from Sakushu-Kotoni River are the northernmost prehistoric examples of this cultigen in Japan. The rice is relatively short-grained, averaging 4.1 by 2.6 mm. It must be kept in mind, however, that grain size is an unreliable characteristic for identifying rice varieties. The most probable route of entry of rice to Japan is from the lower Yangtze River in China, an area where both short- and long-grained rice were grown in prehistory, to South Korea and then to Kyūshū or directly to Kyūshū, bypassing Korea (Akazawa 1982).

Wheat (*Triticum aestivum*)

Wheat has a relatively long history in East Asia, although how long is currently open to question. Japan and Korea are considered to be secondary centers of diversity of this genus (Zeven and Zhukovsky 1975:33). East Asian wheat is fast ripening, has precocious forms, and crosses easily with rye, something southwestern Asian varieties do not do (Zeven and Zhukovky 1975:33). Oracle bone inscriptions indicate the presence of wheat in China by the late second millennium B.C., leading Ho (1977) to suggest that it must have been introduced sometime earlier. Chang (1983:77), however, does not agree that wheat is noted on oracle bone inscriptions. The earliest radiocarbon date associated with Chinese wheat is 3010 ± 90 B.P. at Chien-ch'uan (Chang 1977:455). Ho (1977:448), however, claims that archaeological wheat reports have not been verified. Not until spring wheat was introduced from Central Asia during the Han Dynasty did it surpass barley in importance (Ho 1969; Chang 1983:79). By 2300 B.P., wheat was being grown in Korea (Kim 1982). Today, spring wheat is grown mainly north of the Great Wall.

Japanese wheat has played an important role in modern wheat breeding. A dwarf variety of wheat known as Nōrin-10 (a catalogue number for the Department of Agriculture and Forestry or Nōrin) was used in crosses to produce a hardy dwarf wheat that has become an important crop in Mexico (Janick et al. 1981). This variety of wheat was reported as early as 1893 (Janick et al. 1981). It matures early, is short, and has stiff straw (Peterson 1965:223). Vavilov (1951:196) was also aware of small grained wheat in East Asia, and describes it as early ripening and low growing.

Interestingly, archaeological wheat from Japan and Korea is unusually small, indicating the importance of a variety of dwarf wheat in that part of Asia as early as the Kofun period. The evidence based on grain morphology also indicates that this wheat is neither club wheat (*Triticum aestivum* ssp. *compactum*) nor Indian dwarf wheat (*T. aestivum* ssp. *sphaerococcum*), the two dwarf varieties known to have been grown extensively in prehistory (Crawford and Yoshizaki 1987). The shape of the kernels is closer to the shape of Indian dwarf wheat grain from Pakistan and northern India, but this wheat never moved into East Asia (Zeven 1980). The Japanese wheat may be a form unique to Japan and Korea (Crawford and Yoshizaki 1987). It likely occurs in the Chinese archaeological record as well, because the source of the Korean and Japanese wheat is China.

Naora (1956) confusingly reports the presence of normal bread wheat, as well as compact wheat, from the Kofun period in Japan. His interpretation is based on twelve specimens of wheat from the Sohara site in Yamanashi Prefecture, six of which are identified by Naora (1956) as *T. vulgare* (*T. aestivum* in the modern terminology). The six examples of *T. vulgare* (*T. aestivum*) are smaller (averaging 4.0 by 2.5 by 2.0 mm) than the compact grains he reports (averaging 4.4 by 3.5 by 2.8 mm) (Naora 1956:283). All the specimens of wheat from the Sohara site seem to be the Japanese compact type. The wheat from Sakushu-Kotoni River currently numbers 5,666 specimens and is all free-threshing. The grains are unusually short compared to three other examples from Korea and Japan (Table 2.2). In general, however, except for some grains from Idenshikōgaku, the archaeological wheat from East Asia is all relatively compact. Only 38 of a total of 204 wheat grains from the Idenshikōgaku site, Hokkaido University, are compact; the rest are large. There is reason to believe the large grains are from a relatively late deposit at the site. Similar large wheat grains have been accelerator radiocarbon dated to 210 ± 50 B.P. at the Poplar Namiki site, Hokkaido University (Crawford and

Takamiya 1990). First millennium A.D. northern Japanese wheat may be an extreme form of the compact wheat that probably had a wide distribution in East Asia. A compact Japanese wheat is known from archaeological sites as late as the end of the Edo period (end of the nineteenth century A.D.) in southwestern Japan (Okada 1985).

Adzuki and Mung Beans (*Vigna angularis* and *V. radiatus*)

The *adzuki* and mung beans, both of which occur at Sakushu-Kotoni River, are members of the subgenus *Ceratotropis* of the genus *Vigna* (Marechal et al. 1978; Verdcourt 1970). *Ceratotropis* has an exclusively asiatic and oceanic distribution. The black gram or *urd* (*V. mungo*) is one other important cultigen in this subgenus. Black gram and mung are apparently derived from *V. radiata* var. *sublobata* (Verdcourt 1970), which is found today in eastern Africa and Madagascar (Marechal et al. 1978). *V. angularis* var. *nipponensis* is probably the wild ancestor of *adzuki*. Its distribution is limited to Japan (Marechal et al. 1978:214). *Adzuki* is the bean usually reported from prehistoric Japan and Korea. The only other archaeological mung beans (carbonized) reported in this area are from the Early Jōmon strata at the Torihama Shell Mound. Historic references to *adzuki* and mung beans in China are not found until the sixth century A.D. (Li 1983:47–48). Three probable mung beans and one other bean believed to be black gram were found in samples collected in 1975 (Umemoto and Moriwaki 1983). The hilum width and shape were used to distinguish the two taxa, but a scanning electron microscope examination of the seed coats of the archaeological specimens appears to be the primary basis for distinguishing the beans from each other and from other species and varieties of *Vigna*. In my opinion the patterns observed on the archaeological specimens are not clear in the published illustrations. The Torihama bean identifications should best be left open to question for the time being.

Adzuki and mung beans appear in historic records in eastern Zhou (Chang 1983:81), but there are no archaeological data pertaining to their putative domestication in China. In Japan the earliest appearance of the *adzuki* bean is in the Final Jōmon, where it appears in flotation samples from Uenoharu (Kotani 1972) and Nabatake (Kasahara 1984; Watanabe and Kokawa 1982). Most of the 61 beans from Sakushu-Kotoni River are *adzuki*. Some are mung, while a

Table 2.2 East Asian Wheat Measurements

Site and Location	Mean (Range) mm	Standard Deviation	N[a]
Sakushu-Kotoni River Site, Hokkaidō			
length	3.4 (2.1–5.5)	0.43	403
width	2.4 (1.4–4.2)	0.41	403
thickness	2.2 (1.1–3.5)	0.38	403
Idenshikōgaku, Hokkaidō			
length	5.2 (3.7–6.7)	0.70	131
width	3.4 (2.1–4.6)	0.51	131
thickness	2.8 (1.8–3.9)	0.47	131
Ichinohe, Iwate Prefecture			
length	4.1 (3.5–5.0)	0.29	48
width	2.8 (2.3–3.7)	0.33	48
thickness	2.2 (1.8–3.1)	0.28	48
Puyo, South Korea			
length	4.2 (3.2–5.3)	0.43	100
width	2.2 (1.4–3.5)	0.31	100
thickness	1.8 (1.2–2.3)	0.22	100

Sources: Wheat measurement information for the Sakushu-Kotoni River site and Idenshikōgaku site are derived from Crawford (1991); for Ichinohe from Sato (1986); and for Puyo from Naora (1956).
[a]This number represents the number of grains in the measured sample, not the total site sample. Wheat grains in the measured sample are all carbonized.

few are not identifiable to species. The Hokkaidō *Vigna* length, width, and thickness measurements are 6.0 (4.8–7.1), 3.9 (3.0–4.8), and 3.8 (2.3–5.0) mm, respectively.

Hemp (*Cannabis sativa*)

Hemp is a cultigen with a variety of uses, including oil and food from the fruit, fiber from the stems, and drugs from the flowers and fruiting tops (Bailey 1976:218). The genus, according to Bailey (1976), is monotypic, while Zeven and Zhukovsky (1975:130) and Beutler and Der Marderosian (1978) recognize a separate species, *C. ruderalis,* which may be derived from cultigen hemp. The Ezo phase sample consists of at least 160 achenes distributed in 14 samples at Sakushu-Kotoni River, which were derived mainly from one midden sample (124 specimens).

Hemp is indigenous to temperate central Asia (Bailey 1976:218; Simmonds 1976:203). Literary references to hemp date from 1122 B.C. in China (Li 1974:440). Li (1974:438) proposes that it was in use at least 3,000 years earlier in northern China on the basis of evidence for the existence of textiles and fiber production at Yangshao. The only direct evidence

for hemp husbandry in the North China Neolithic is from the western Yangshao (Majiayao) Culture at the Linxia site (5500–4500 B.P.) where carbonized hemp fruit was found in pottery on a house floor (Chang 1986:143). Fibres and achenes have been identified from the Early Jōmon levels at Torihama in southwestern Japan (Kasahara 1984).

Melon (*Cucumis melo*)

Three broken melon seeds from Sakushu-Kotoni River make this the northernmost archaeological occurrence of this taxon in Japan and the only prehistoric example from Hokkaidō. The generally accepted area of origin for melon is western Africa (Bailey 1976:342; Whitaker and Bemis 1976:67; Zeven and Zhukovsky 1975:30). Whitaker and Bemis (1976:67) consider India to be a secondary center of diversity, while Zeven and Zhukovsky (1975:30) describe China as a secondary center of diversity. Some 5,076 melon seeds have been found at 102 archaeological sites in Japan (Fujishita 1984:640). Thirty-six of the sites are Yayoi (1,557 seeds), while one, the Yamanotera site on Kyūshū, is Final Jōmon (1 seed). A melon seed, presumably *Cucumis* sp., is reported from the Early Jōmon levels at Torihama (Kasahara 1983). The northernmost reports before the Hokkaidō discovery were from Miyagi Prefecture, near Sendai, dating to the eighth and ninth centuries A.D.

Beefsteak Plant and *Egoma* (*Perilla frutescens*)

P. frutescens is grown today for its aromatic leaves and oil seeds, as well as its medicinal properties in China, Japan, and Korea (Li 1969). It is an annual herb with two forms: green-leaved (*egoma*) and red-leaved (beefsteak plant or *shiso*). Beefsteak plant is *P. frutescens* (L.) Britt var. *crispa* Benth [*P.crispa* (Thunb.) Nakai]. *Egoma* is classified as *P. frutescens* var. *japonica* Hara (see also Bailey 1976:846; Zeven and Zhukovsky 1975:34). *Egoma* is usually associated with slash-and-burn (swidden) agriculture, while beefsteak plant is usually a cultigen of house gardens (Matsutani 1983:183). *Egoma* and beefsteak plant seeds are distinguishable, but with some difficulty.

Seeds of *Perilla* are present in the Sakushu-Kotoni River collection, although not in great abundance (14 seeds). Its occurrence in other Ezo phase sites affirms this cultigen's presence by 800 A.D. in Hokkaidō. Its main area of diversity is China (Zeven and Zhukovsky 1975:34), but no prehistoric or early historic Chinese references to this plant exist. The earliest data are from Japan. A single beefsteak plant seed comes from the Initial Jōmon level 13 at Torihama and seven specimens are from Early Jōmon levels (Kasahara 1983). Seeds of beefsteak plant or the closely related *egoma* are a constituent of carbonized cakes from eight Middle Jōmon sites in Nagano, Giifu, Fukushima, and Nagano prefectures (Matsutani 1983:183). Beefsteak plant is a component of the Nabatake site Final Jōmon samples as well.

Other Cultigens

Buckwheat, bottle gourd, Chinese cabbage, great burdock, lacquer tree (*Rhus vernicifera*), and paper mulberry are reported from a few Jōmon sites. Yayoi sites also include peach, apricot (*Prunus armeniaca*), plum (*Prunus salicina*), pear, soybean, and pea (*Pisum sativum*) (Kasahara 1986; Kotani 1972:70–75; Naora 1956; Sato 1971). Buckwheat is the only taxon of this group thought to have been grown at Ezo phase sites. This is inferred from buckwheat pollen identified at six Ezo sites; it is recorded from six Zoku-Jōmon

sites as well (Okada and Yamada 1982:28). The same authors consider the appearance of such pollen significant because buckwheat is pollinated by insects (windborne pollen contaminates samples more easily than does insect-borne pollen) and the samples were taken from within house floors. Buckwheat pollen is present in sediments dating to 6600–4500 B.P. at Ubuka Bog (Tsukada et al. 1986). The earliest buckwheat pollen in northern Japan is from burned clay at the Late Jōmon (ca. 3000 B.P.) Kyūnenbashi site, Iwate Prefecture (Yamada 1980). Only one carbonized buckwheat seed is attributed to the Jōmon period, found in a late Early Jōmon pit house at the Hamanasuno site, Hokkaidō (Crawford et al. 1978; Crawford 1983). Extensive flotation sampling at the same and later sites in the area has produced no further examples. Uncarbonized buckwheat seeds are reported from Yayoi wet sites (Goto 1962). It is entirely possible that the rarity of carbonized buckwheat seeds at Jōmon and Yayoi sites is due to preservation factors. For example, among the earliest buckwheat in the Netherlands is a single carbonized achene from the twelfth-century Dommelin site (Van Vilsteren 1984:230). Van Vilsteren (1984) reports that uncarbonized examples are common in cesspits and that in these contexts it is never carbonized. In other words, the carbonized sample is not representative of buckwheat's actual importance. If the Hamanasuno buckwheat is not intrusive, it may not be showing up more often in Jōmon and later contexts as a result of preservation problems. However, hundreds of carbonized buckwheat achenes have been found at the Kachiyama historic (Japanese) site in Kaminokuni, Hokkaidō (Kaminokuni-cho Kyōiku Iinkai 1986). The wild parent of domesticated buckwheat, *Fagopyrum cymosum,* is a temperate eastern Asian plant (Campbell 1976:235) and is not native to Japan.

Chinese cabbage comprises four subspecies in China and Japan, two Chinese cabbages (*Brassica campestris* ssp. *chinensis* and *pekinensis*), Indian mustard (*B. campestris* ssp. *nipposinica* or *B. juncea*), and Chinese savoy (*B. campestris* ssp. *narinosa*) (Zeven and Zhukovsky 1975:29). Chinese cabbage was domesticated in East Asia, according to Zeven and Zhukovsky (1975) and Bailey (1976). McNaughton (1976:47) finds no evidence for its ancient cultivation in China, although the evolution of a subspecies in eastern Asia points to some historical depth for its use there. Li (1983:35), however, reports that Chinese cabbage seeds were recovered from the Banpo site. The National Museum in Beijing displays a ceramic sowing container from Banpo that contained Chinese cabbage seeds.

The paper mulberry, reported from Torihama (Kasahara 1983), is cultivated in warmer parts of Japan (Ohwi 1965). Zeven and Zhukovsky (1975) include this tree in their China-Japan center. Great burdock is cultivated in China and Japan today and Kasahara (1983:49) believes it was introduced from China. Ohwi (1965) does not list it in the flora of Japan. The lacquer tree is cultivated in southwestern Japan and China for its sap, which is used in lacquer production. Wooden artifacts and pottery from Jōmon sites occasionally are lacquered. Seeds of *Rhus* are common in flotation samples from the Kameda Peninsula (Crawford 1983), but the species is unknown.

An important aspect of northern Chinese plant husbandry was the domestication of fruit trees. These include apricot, pear, and peach, which are thought to have been domesticated long before the Zhou period (Li 1983:34). Most of these make their appearance in Japan during the Yayoi, along with soybean, pea, and plum (discussed in Kotani 1972:70–75). Peach, however, may have been introduced to Kyūshū as early as 6000 B.P. (Tarami-cho Kyōiku Iinkai 1986).

A carbonized sample of safflower achenes is reported from a Satsumon pit house at the Toyotomi site in northeastern Hokkaidō (Crawford 1985; Kohno 1959). An unknown quantity was originally collected, but just a few examples remain at the Asahikawa City Museum. These achenes were originally identified as buckwheat (Kohno 1959). One safflower achene has been found at Sakushu-Kotoni River. Safflower is native to the Middle East and the earliest record of its use is from Egypt about 1600 B.C. (Knowles 1976:31). The Far East is one of seven centers of diversity for this cultigen (Knowles 1976:132). The only other report of the early presence of safflower in Japan is Yamazaki's (1961) description of dye extraction in the eighth century A.D. in southwestern Japan.

Discussion and Directions for Future Research

The domestication and dispersal of crops in East Asia is a relatively untapped research area. In China, little is known of the period immediately preceding 8000 B.P., the period when domestication was well under

way. Nothing is known about these processes in Korea during the early Holocene. Although human settlement patterns and technological developments for the same period in Japan are well documented, plant remains have been collected from only a few Initial Jōmon sites. There is little evidence for early Holocene Japanese populations being other than foragers. For the time being, the earliest innovations in plant domestication in East Asia seem to have taken place in North China. Domestication appears to have occurred in a steppe-forest or forest zone. The closest analogue to North Chinese agricultural origins is found in the Near East. Before 10,000 B.P. in the Near East, habitation was localized in favorable zones and the steppe was little used (Moore 1989:624); agriculture evolved in unusually favorable circumstances, but shortly thereafter rainfall decreased, evaporation increased, and forests retreated (Moore 1985:11). Evidence for more steppe-like conditions in mid-Holocene North China suggest that the environment there was undergoing a similar transformation. The processes identified in the Near East may serve as valuable hypotheses to test in North China.

The few sites in Japan from which substantial plant remains have been recovered are producing provocative data regarding the evolution of plant husbandry in East Asia. Some reviewers (e.g., Rowley-Conwy 1984) reject the possibility of plant husbandry before 3000 B.P., whereas others (e.g., Kasahara 1984; Tozawa 1983) accept the presence of limited plant husbandry earlier than that. Reports of cultigens from other Early Jōmon sites, such as Hamanasuno, Otsubo, and Ikiriki, are suggesting that Torihama is not unique. Pollen evidence from Ubuka Bog in southwestern Honshū further supports the existence of Early Jōmon plant husbandry. The data indicate that buckwheat, hemp, great burdock, paper mulberry, bottle gourd, bean, Chinese cabbage, and perhaps millet were present in Japan or undergoing domestication during the Early Jōmon. The fact that nearly a decade of excavations at Torihama have resulted in so few examples of each plant indicates the lack of economic importance of plant husbandry at that time. In addition, the paucity of remains stresses the importance of extensive sampling of plant remains from Jōmon sites. Buckwheat is equivocally part of this Early Jōmon list, while barnyard grass is a more likely candidate for a husbanded plant in northern Japan. All pre-3000 B.P. records in Japan are of plants belonging to Zeven and Zhukovsky's China-Japan center, except for bottle gourd, black gram, and mung bean. Bottle gourd's wide distribution in both the Old World and New World by 3000 B.P. and its early presence in Japan indicate that it became part of the China-Japan center by at least 5000 B.P. The identification of mung bean and black gram in 6000 B.P. Japan is suspicious because no further archaeological examples of this bean are known from prehistoric eastern Asia. The problem may be a matter of distinguishing types of *Vigna,* assuming they are *adzuki,* or it may be that all early beans in eastern Asia are *adzuki,* derived from a local ancestor. I would urge a program of accelerator radiocarbon dating in Japan to test the age of many of these cultigens.

Barley, rice, foxtail millet, *adzuki* bean, and melon were added to the complement of cultigens by the Final Jōmon. Wheat seems to be a later addition, probably during the Yayoi period, but wheat did not become common until the Kofun period. Cultivation of apricot, soybean, pea, peach, persimmon, plum, pear, and watermelon is also apparent during the Yayoi. Safflower is known from historic records to have been present in southwestern Japan by the eighth century A.D., and archaeological data demonstrate its early (ninth century A.D.) appearance in Hokkaidō.

Data from Korea for foxtail millet, barley, wheat, rice, soybean, and *adzuki* are not out of line with the Japanese situation. Sorghum is reported from Korea as well, but it is rare in the Japanese archaeological record. Nevertheless, the range of archaeological cultigen remains from Korea is narrower than that from prehistoric Japan. The undated Chulmun millets are the earliest potential cultigens reported from Korea (Kim 1978). Rice, barley, millet, sorghum, and soybean are more securely identified and contextually placed at the Hunamni and Honamni Bronze Age sites (Choe 1982:524; Pearson 1982). The Bronze Age millets are foxtail and broomcorn (Kim 1982:514). Red (*adzuki*) bean is reported from the Bronze Age Soktali site and from an early first millennium A.D. shell mound in Kimhoe, where the earliest wheat in Korea is also reported (Kim 1982:514,517). By the end of the fourth millennium B.P. many of the cultigens reported in early Japan are present in Korea, but the research is unsystematic and does not tell us what was happening in Korea at the same time as the Early and Middle Jōmon in Japan. We should expect to see in Korea the early gardening of the cultigen mint *egoma* or beefsteak plant, bottle gourd, Chinese cabbage, and other cultigens reported from Jōmon sites.

Substantial Neolithic adaptations had evolved in North China by 8000 B.P., approximately contemporaneous with the Initial Jōmon of Japan and the earliest documented Chulmun of Korea. For the time being, foxtail and broomcorn millet, hemp, and Chinese cabbage are the only cultigens in early sites in Japan and Korea that are also reported in the Chinese Early Neolithic, and the Yangshao, Dawenkou, and Xinle cultures of the later Neolithic of China. Whether China will prove to be the source of all the other cultigens reported from Japan and Korea remains to be seen.

In a summary article on Jōmon subsistence, Richard and Kazue Pearson (1978) indicated the paucity of data and the points of view of Japanese prehistorians on Jōmon subsistence. They made five recommendations: flotation should be instituted on a large scale, bottom-land areas where agriculture is extensive today should be investigated for signs of prehistoric food production, pollen analysis in conjunction with archaeology should be furthered, ethnographic analogy should be judiciously applied, and archaeological cultures in Japan should be interpreted within an ecosystemic paradigm to reconstruct Jōmon subsistence and the organization of Jōmon production (1978:26–27). In the last decade, progress has been made on these recommendations, as well as on other dimensions of subsistence ecological research, but considering the provocative data now available, little actual headway is being made, and in northern China and Korea progress is particularly slow.

Archaeological evidence for plant husbandry in Japan is nearly as early as any from eastern Asia. The earliest collections comprise plants that are part of the China-Japan center of cultigen diversity defined by Zeven and Zhukovsky (1975). The only unequivocal non-eastern Asian cultigens in this early group are melon, bottle gourd, and mung bean. Bottle gourd is the only one in secure Early Jōmon contexts. The others, including hemp, paper mulberry, *adzuki* bean, beefsteak plant, great burdock, millet, and Chinese cabbage, are plants with a long history of use in Asia; besides *adzuki* bean and millet, little is known about the prehistory of these plants in Japan. The examples before the Late and Final Jōmon in Japan appear unusual because of their uniqueness, but this could well be due to a lack of research. The best available data to westerners from China at the moment are historic records. The earliest confirmed Korean data on plant husbandry are contemporaneous with the Late and Final Jōmon.

Further research on Jōmon sites in southwestern and northeastern Japan should be carried out, and the collection of plant remains from all sites, not just wet sites, should become routine. For the present, explaining plant husbandry origins in Japan, Korea, and China will be difficult until much more basic research is accomplished. Research today is also beginning to focus on the spread of agriculture to northern Japan. Some archaeologists are proposing diffusion of rice and other plant husbandry from Kyūshū directly to the north, bypassing the southwest entirely. Carbonized plant remains from several late first millennium A.D. occupations on Hokkaidō have produced evidence of a phase of extensive plant husbandry in a context where it had only been suspected before.

There is some urgency to increase the scale of research on the origins of plant husbandry in Asia. China, Korea, and Japan are all facing pressures on the archaeological database. In the 1984 field season on Hokkaidō, eighty sites comprising a total of 183,500 square meters were excavated at a cost (in U.S. dollars) of about $4 million, yet in the ten years preceding 1984, flotation samples had been taken from fewer than ten sites on the entire island and extensive collections of such remains from Jōmon sites in southern Japan number even fewer than that. It is imperative that this situation be remedied as soon as possible if we are ever to obtain an understanding of the complexities and achievements of ancient East Asian agriculture.

Acknowledgments

The research on the Sakushu-Kotoni River and other Japanese collections, as well as several fieldwork seasons, was supported by the Humanities and Social Sciences Committee of the Research Board of the University of Toronto, the Social Sciences and Humanities Research Council of Canada (SSHRC) (Grants No. 410-86-0769 and 410-89-0786), a bilateral exchange grant from SSHRC and the Japan Society for the Promotion of Science (Grant No. 473-86-0007), and two grants from Earthwatch. Masakazu Yoshizaki and the staff of the Salvage Archaeology Centre of Hokkaidō University made the research there possible and the success of this project is in no small way attributable to their cooperation. Hiroto Takamiya and Nicola Biggs provided library research assistance, while H. Takamiya, A. C. D'Andrea, K. Leonard, and Y. Tsubakisaka helped in the analysis of the plant remains. The manuscript was typed by Clara Stewart, with later drafts prepared by Joan Cheyne, Lillian Lee, and Madeleine Weiler. I also wish to thank Sarah Nelson, Patty Jo Watson, and Masakazu Yoshizaki for their comments on earlier drafts of this paper.

Notes

1. Many crops exhibit centers (regions) of genetic diversity. Some crops do not. Centers of diversity should not be confused with centers of origin. Crops still in their center of domestication often exhibit lower diversity than crops in secondary centers (areas to which the crops have diffused) (Harlan 1975:56). The reasons for diversity have relevance to prehistorians because a long history of continuous cultivation in an area and cultural diversity, which is associated with crop variety preferences, are two factors that encourage diversity (Harlan 1975:149).

References Cited

Aikens, C. Melvin
1981   The Last 10,000 Years in Japan and Eastern North America: Parallels in Environment, Economic Adaptation, Growth of Societal Complexity, and the Adoption of Agriculture. *In* Affluent Foragers, edited by S. Koyama and D. H. Thomas, pp. 261–73. Senri Ethnological Studies No. 9. National Museum of Ethnology, Senri, Osaka.

Akazawa, Takeru
1982   Cultural Change in Prehistoric Japan: Receptivity to Rice Agriculture in the Japanese Archipelago. *In* Advances in World Archaeology, vol. 1, edited by Fred Wendorf and Angela E. Close, pp. 151–211. Academic Press, New York.

An, Zhimin
1989   Prehistoric Agriculture in China. *In* Foraging and Farming, edited by D. R. Harris and G. C. Hillman, pp. 643–49. Unwin Hyman, London.

Andersson, J. G.
1934   Children of the Yellow Earth. Kegan Paul, Trench, Trubner and Co., London

Archaeological Institute, Chinese Academy of Science
1963   Xian Banpo. Archaeological Institute, Chinese Academy of Science.

Bailey, Liberty Hyde
1976   Hortus Third. MacMillan Publishing Company, New York.

Barrett, Spencer C. H.
1983   Crop Mimicry in Weeds. Economic Botany 37(3):255–82.

Beutler, John A., and Ara H. Der Marderosian
1978   Chemotaxonomy of *Cannabis* I. Crossbreeding between *Cannabis sativa* and *C. ruderalis,* with Analysis of Cannabinoid Content. Economic Botany 32(4):387–94.

Bishop, Carl W.
1933   The Neolithic Age in Northern China. Antiquity 7:389–404.

Brace, C. L., M. L. Brace, and W. R. Leonard
1989   Reflections on the Face of Japan: A Multivariate Craniofacial and Odontometric Perspective. American Journal of Physical Anthropology 78:93–113.

Cai, L., and S. Qiu
1984   Carbon 13 Evidence for Ancient Diets in China. Kaogu 10:949–55.

Campbell, C. G.
1976   Buckwheat. *In* Evolution of Crop Plants, edited by N. W. Simmonds, pp. 235–37. Longman, London.

Chang, Kwang-Chih
1977   The Archaeology of Ancient China. 3rd ed. Yale University Press, New Haven.

1981   In Search of China's Beginnings: New Light on an Old Civilization. American Scientist 60:148–60.

1986   The Archaeology of Ancient China. 4th ed. Yale University Press, New Haven.

Chang, T.-T.
1976a   The Origin, Evolution, Cultivation, Dissemination, and Diversification of Asian and African Rices. Euphytica 25:425–41.

1976b   Rice. *In* Evolution of Crop Plants, edited by N. W. Simmonds, pp. 98–104. Longman, London.

1983   The Origins and Early Cultures of the Cereal Grains and Food Legumes. *In* The Origins of Chinese Civilization, edited by David N. Keightley, pp. 65–94. University of California Press, Berkeley.

1985   Ethnobotany of Rice in Insular Southeast Asia. Paper presented at the 12th Congress of the Indo-Pacific Prehistory Association Penablanca, Cagayan, Philippines.

Chard, Chester S.
1960   Neolithic Archaeology in North Korea. Asian Perspectives 4:151–55.

Chekiang Provincial Museum, Natural History Section
1978   A Study of Animal and Plant Remains Unearthed at Ho-Mu-Tu. Kaogu 1:95–111.

Choe, C.-P.
1982   The Diffusion Route and Chronology of Korean Plant Domestication. Journal of Asian Studies 41(5):19–29.

Crawford, Gary W.
1983   Palaeoethnobotany of the Kameda Peninsula Jōmon. Anthropological Papers No. 73. Museum of Anthropology, University of Michigan, Ann Arbor.

1985   Hokkaidō no benibana no hakken (The discovery of safflower in Hokkaidō). Hoppōken 52:76–80.

1986   Sakushu-Kotoni River Site: The Ezo-Haji Component Plant Remains. Hokkaidō Daigaku Bungakubu, Sapporo.

1987   K135 Iseki Kara Shutsu Sareta Shokubutsu shushi ni tsuite (Plant seeds excavated from the K135 site). *In* K135 Site, edited by Sapporo-shi Kyōiku Iinkai, pp. 565–81. Sapporo-shi Kyōiku Iinkai, Sapporo.

1991   The North Asian Plant Husbandry Project. Report submitted to the Social Sciences and Humanities Research Council of Canada, Ottawa.

1992   The Transitions to Agriculture in Japan. *In* Transitions to Agriculture in Prehistory, edited by Anne Birgitte Gebauer and T. Douglas Price, pp. 117–32. Prehistory Press, Madison, Wisconsin.

Crawford, Gary W., William M. Hurley, and Masakazu Yoshizaki
1978   Implications of Plant Remains from the Early Jōmon, Hamanasuno Site. Asian Perspectives 19(1):145–48.

Crawford, Gary W., and H. Takamiya
1990   The Origins and Implications of Late Prehistoric Plant Husbandry in Northern Japan. Antiquity 64(245):889–911.

Crawford, Gary W., and Masakazu Yoshizaki
1987   Ainu Ancestors and Prehistoric Asian Agriculture. Journal of Archaeological Science 14:201–13.

Cultural Relics Publishing House
1988   Jiangzhai. Cultural Relics Publishing House, Beijing.

D'Andrea, A. C.
1992   Paleoethnobotany of Later Jōmon and Early Yayoi Cultures in Northeastern Japan: Northeastern Aomori and Southwestern Hokkaidō. Ph.D. dissertation, University of Toronto.

Esaka, Teruya
1977   Jōmon no Saibai-Shokubutsu to Riyō-Shokubutsu (Cultivated and utilized plants of the Jōmon). Dorumen 13:15–31.

Fujimori, Eiichi
1965   Idojiri. Chuo Koron Bijutsu Shuppan, Tōkyō.

Fujishita, Noriyuki
1984   Shutsudo itai yori mita uri-ka shokubutsu no shurui to henkan to sono riyōho (Archaeological Cucurbitaceae: types, temporal change and their use). In Kobunkazai no Shizen—kagakuteki Kenkyū (Natural scientific research of antiquities), edited by Kobunkazai Henshū Iinkai, pp. 638–54. Dōhōsha, Tōkyō.

Goto, S.
1954   Toro Iseki (The Toro site). Nihon Kōkogaku Kyōkaiken and Mainichi Shinbunsha, Tōkyō.

1962   Yamagi Iseki (The Yamagi site). Tsukiji Shokan, Tōkyō.

Hanihara, Kazuro
1982   Jinruigaku kara Mita Jōmon Jidai no Shokuseikatsu (Subsistence during the Jōmon period: an anthropological viewpoint). In Ki-kan Kōkogaku, pp. 64–66. Yuzankaku, Tōkyō.

1987   Estimation of the Number of Early Migrants to Japan: A Simulative Study. Journal of the Anthropological Society of Nippon 95:391–403.

Harlan, Jack R.
1968   On the Origin of Barley. U.S. Department of Agriculture Handbook 338:9–31.

1975   Crops and Man. American Society of Agronomy, Madison, Wisconsin.

Harris, David R.
1977   Alternative Pathways toward Agriculture. In Origins of Agriculture, edited by Charles Reed, pp. 179–243. Mouton Publishers, The Hague.

Hayashi, Y.
1975   Ainu no Nōkō Bunka (Ainu plant cultivation). Keiyōsho, Tōkyō.

Ho, Ping-Ti
1969   The Loess and the Origin of Chinese Agriculture. American Historical Review 75:1–36.

1977   The Indigenous Origins of Chinese Agriculture. In Origins of Agriculture, edited by C. A. Reed, pp. 413–84. Mouton Publishers, The Hague.

Hoshikawa, Kiyochika
1984   Wagakuni no Kodai inasaku ni tsuite no Sakumotsu Gakuteki na Kansatsu to Ni, San no Jikken (A few experiments and observations on our ancient rice agriculture). In Kobunkazai no Shizen-Kagakuteki Kenkyū (Natural scientific research of antiquities), edited by Kobunkazai Henshū Iinkai, pp. 611–16. Dōhōsha, Tōkyō.

Hsieh, Chiao-min
1973   Atlas of China. McGraw-Hill, New York.

Ito, Nobuo
1984   Aomori-Ken ni okeru Inasaku Nōkō Bunka no Keisei (The formation of rice agriculture in Aomori prefecture). In Hokuho Nihon Kai Bunka no Kenkyū (Research on culture of the northern Japan Sea), edited by Tōhoku Gakuin Daigaku, pp. 1–10. Tōhoku Bunka Kenkyūjo Report No. 16, Sendai.

Janick, J., F. W. Woods, R. W. Schery, and V. W. Ruttan
1981   Plant Science. W. H. Freeman and Company, San Francisco.

Kabaker, Adina
1977   A Radiocarbon Chronology Relevant to the Origins of Agriculture. In Origins of Agriculture, edited by Charles A. Reed, pp. 957–80. Mouton Publishers, The Hague.

Kaminokuni-cho Kyōiku Iinkai
1986   Kaminokuni Kachiyama Iseki. (The Kaminokuni Kachiyama site) Kaminokuni Kyōiku Iinkai, Kaminokuni-cho.

Kanaseki, T., and M. Sahara
1978   The Yayoi Period. Asian Perspectives 19(1):15–26.

Kasahara, Y.

1982   Nabatake Iseki no Maizo Shushi no Bunseki, Dōtei Kenkyū. (Analysis and identification of ancient seeds from the Nabatake site). *In* Nabatake, edited by Tosu-shi Kyōiku Iinkai, pp. 354–79. Tosu-shi Kyōiku Iinkai, Tosu-shi.

1983   Torihama Kaizuka (dai-6-ji hakkutsu) no Shokubutsu Shushi no Kenshutsu to Dōtei ni tsuite (The search for and identification of plant seeds from the Torihama Shell Mound, 6th excavation). *In* Torihama Kaizuka (The Torihama Shell Mound), edited by I. Okamoto, pp. 47–64. Fukui-Ken Kyōiku Iinkai, Fukui.

1984   Maizo Shushi Bunseki ni Yoru Kodai Nōkō no Kensho (2)—Nabatake Iseki no Sakumotsu to Zassō no Shurui oyobi Torai Keiro (Verification of ancient agriculture from the perspective of archaeological seed analysis—species and routes of introduction of crops and weeds at the Nabatake Site). *In* Kobunkazai no Shizen-Kagakuteki Kenkyū (Natural scientific research of antiquities), edited by Kobunkazai-Henshū Iinkai, pp. 617–29. Dōhōsha, Tōkyō.

Kasahara, Y., A. Fujizawa, and M. Buda

1986   Yonago-shi Megumi Iseki no Shushi Bunseki Dōtei (Analysis and identification of seeds from the Megumi site, Yonago city). *In* Megumi Iseki (The Megumi site), edited by Yonago-shi Kyōiku Iinkai, pp. 98–128. Yonago-shi Kyoiki Iinkai, Yonago-shi.

Kidder, J. E.
1959   Japan before Buddhism. Praeger, London.

Kim, Jong-bae
1978   The Prehistory of Korea. University of Hawaii Press, Honolulu.

Kim, W.-Y.
1982   Discoveries of Rice in Prehistoric Sites in Korea. Journal of Asian Studies 41:513–18.

Knowles, P. F.
1976   Safflower. *In* Evolution of Crop Plants, edited by N. W. Simmonds. Longman, New York.

Kohno, Hiromichi
1959   Hokkaidō shutsudo no Ōgata U Jikei Tekki ni tsuite (A large 'u'-shaped iron tool excavated in Hokkaidō). Hokkaidō Gakkai Daigaku Kōkogaku Kenkyūkai Renrakushi.

Kolb, Albert
1971   East Asia. Metheun & Company, London.

Kondo, Yoshio
1964   Yayoi Bunka-ron (An essay on the Yayoi Culture). Nihon Rekishi 1:139–88.

Kotani, Yoshinobu
1972   Economic Bases during the Later Jōmon Period in Kyūshū, Japan: A Reconsideration. Ph.D. dissertation, University of Wisconsin. University Microfilms, Ann Arbor.

1981   Evidence of Plant Cultivation in Jōmon Japan: Some Implications. *In* Affluent Foragers, edited by S. Koyama and D. H. Thomas, pp. 261–73, Senri Ethnological Studies No. 9. National Museum of Ethnology, Senri, Osaka.

Koyama, S.
1978   Jōmon Subsistence and Population. Senri Ethnological Studies 2, Miscellanea I. National Museum of Ethnology, Senri, Osaka.

Kuraku, Yoshiyuki
1984   Tōhoku Chihō ni Okeru Kodai Inasaku o Saguru (The search for ancient rice agriculture in the Tōhoku region). *In* Kōbunkazai no Shizen-Kagakuteki Kenkyū (Natural scientific research of antiquities), edited by Kōbunkazai Henshū Iinkai, pp. 603–10. Dōhōsha, Tōkyō.

Li, Hui-Lin
1969   The Vegetables of Ancient China. Economic Botany 23:253–60.

1974   An Archaeological and Historical Account of *Cannabis* in China. Economic Botany 28(4):437–48.

1983   The Domestication of Plants in China: Ecogeographical Considerations. *In* The Origins of Chinese Civilization, edited by D. N. Keightley, pp. 21–64. University of California Press, Berkeley.

Liu, Kam-biu
1988   Quaternary History of the Temperate Forests of China. Quaternary Science Review 7:1–20.

Marechal, R., Jean-Michel Mascherpa, and Francoise Stainer
1978   Étude taxonomique d'un groupe complexe d'éspeces des genre *Phaseolus* et *Vigna* (Papionaceae) sur la base de données morphologiques et polliniques, traitées par l'analyse informatique. Boissiera 28:1–273.

Matsutani, Akiko
1983   Egoma-Shiso. *In* Jōmon Bunka no Kenkyū, vol. 2. Seigyō (Research on the Jōmon Culture, Vol. 2. Occupations), edited by Shinpei Katoh, Tatsuo Kōbayashi, and Tsutomu Fujimoto, pp. 50–62. Yuzankaku, Tōkyō.

1984   Sōsadenkenzo ni yoru Tanka Shujitsu no Shiki-betsu (Identification of carbonized seeds with the scanning electron microscope). *In* Kōbunkazai no Shizen-Kagakuteki Kenkyū (Natural scientific research of antiquities), edited by Kōbunkazai Henshū Iinkai, pp. 630–37. Dōhōsha, Tōkyō.

McNaughton, I. H.
1976   Turnip and Relatives. *In* Evolution of Crop Plants, edited by N. W. Simmonds. Longman, New York.

Minamiki, M., S. Nohjo, S. Kokawa, S. Kosugi, and M. Suzuki
1986   Shokubutsu Itai to Kokankyo (Plant remains and ancient environment). *In* Ikiriki Iseki (The Ikiriki site), edited by Tarami-cho Kyōiku Iinkai, pp. 44–53. Tarami-cho Kyōiku Iinkai, Tarami-cho.

Moore, Andrew M. T.
1985   The Development of Neolithic Societies in the Near East. *In* Advances in World Archaeology, vol. 4, edited by Fred Wendorf and Angela E. Close, pp. 1–69. Academic Press, New York.

1989   The Transition from Foraging to Farming in Southwest Asia: Present Problems and Future Directions. *In* Foraging and Farming, edited by David R. Harris and Gordon C. Hillman, pp. 620–31. Unwin Hyman, London.

Nakamura, Jun
1984   Kodai Nōkō toku ni Inasaku no Kafun Bunseki-gakuteki Kenkyū (Ancient agriculture: rice pollen analysis). *In* Kōbunkazai no Shizen-Kagakuteki Kenkyū (Natural scientific research of antiquities), edited by Kōbunkazai Henshū Iinkai, pp. 581–602. Dōhōsha, Tōkyō.

Nakanishi, Naoyuki
1983   Shushi no Hyōchaku to Kōkogaku (Archaeology and stranded seeds). *In* Torihama Kaizuka (The Torihama Shell Mound), edited by I. Okamoto, pp. 28–36. Fukui-ken Kyōiku Iinkai, Fukui.

Nakao, Sasuke
1966   Saibai Shokubutsu to Nōkō no Kigen (Cultigens and the origin of agriculture). Iwanami Shinsho, Tōkyō.

Naoki, Kojiro
1964   Kokka no Hassei (The emergence of the state). Nihon no Rekishi 1:189–233.

Naora, Nobuo
1956   Nihon no Kodai Nōkō Hattatsushi (The history of the development of ancient Japanese agriculture). Saera Shobo, Tōkyō.

Nelson, Sarah M.
1973   Chulmun Period Villages on the Han River in Korea: Subsistence and Settlement. University Microfilms International, Ann Arbor.

1982   Recent Progress in Korean Archaeology. *In* Advances in World Archaeology, vol. 1, edited by Fred Wendorf and Angela E. Close, pp. 99–149. Academic Press, New York.

1990   The Neolithic of Northeastern China and Korea. Antiquity 64:234–48

Nishida, Masaki
1976   Shokubutsu Shushi no Dōtei (Seed identification). *In* Kuwagaishimo Iseki (The Kuwagaishimo site), edited by B. Tsunoda and M. Watanabe, pp. 244–49. Heian Hakubutsukan, Kyōto.

1980   Jōmon-jidai no Shokuryō Shigen to Seigyō-Katsudo—Torihama Kaizuka no Shizen Ibutsu o Chūshin to shite (Food resources and subsistence activity during the Jōmon period—based on the natural remains from the Torihama Shell Mound). Quaternary Anthropology 2(3):3—56.

1981   Jōmon-jidai no Ningen-shokubutsu Kankei—Shokuryō Seisan no Shutsugen Katei (Human-plant relationships in the Jōmon period and the emergence of food production). Bulletin of the National Museum of Ethnology 6(2):234—55.

1983   The Emergence of Food Production in Neolithic Japan. Journal of Anthropological Archaeology 2:305—22.

Ohwi, Jisaburo
1965   The Flora of Japan. Smithsonian Institution, Washington, D.C.

Oka, H. I.
1988   Origin of Cultivated Rice. Japan Scientific Societies Press, Tōkyō.

Okada, Atsuko
1984   Tōhoku Chihō to Ezo Chii (The Tōhoku district and the Ezo region). Sōzō no Sekai 49:97--105.

Okada, Atsuko, and Goro Yamada
1982   Hokkaidō ni Okeru Nōkō no Kigen ni Kansuru Yosatsu (A short note on the origin of agriculture in Hokkaidō). *In* Hokkaidō ni Okeru

Nōkō no Kigen (The origin of agriculture in Hokkaidō), edited by T. Umebara, pp. 26—34. Hokkaidō University, Sapporo.

Okada, F.
1985   Ido Umetsuchi no Shokubutsu Shushi Chōsa (Investigation of plant seeds from well fill). *In* Heian Kyōtō-shigai no Chōsa, edited by Heian Gakuen, pp. 55—79. Heian Gakuen, Kyōto.

Okamoto, I.
1979   Torihama Kaizuka (The Torihama Shell Mound). Fukui-Ken Kyōiku Iinkai, Fukui.

1983   Torihama Kaizuka (The Torihama Shell Mound). Fukui-Ken Kyōiku Iinkai, Fukui.

Pearson, R.
1974   Prehistoric Subsistence and Economy in Korea: An Initial Sketch. Asian Perspectives 17:93—101.

1982   The Archaeological Background to Korean Art. Korean Culture 3(4):18—29.

Pearson, Richard, and Kazue Pearson
1978   Some Problems in the Study of Jōmon Subsistence. Antiquity 52(204):21—27.

Peterson, R. F.
1965   Wheat. Leonard Hill Books, London.

Rao, K., E. Prasada, J. M. J. de Wet, D. E. Brink, and M. H. Mengesha
1987   Infraspecific Variation and Systematics of Cultivated *Setaria italica,* foxtail millet (Poaceae). Economic Botany 41(1):108—16.

Reed, C. A.
1977   Origins of Agriculture. Mouton Publishers, The Hague.

Ren, M., R. Yang, and H. Bao
1985   An Outline of China's Physical Geography. Foreign Languages Press, Beijing.

Rowley-Conwy, Peter
1984   Postglacial Foraging and Early Farming Economies in Japan and Korea: A West European Perspective. World Archaeology 16(1):28—42.

Sahara, Makoto
1975   Nōgyō no Kaishi to Kaikyū Shakai no Keisei (The beginning of agriculture and the establishment of stratified society). Nihon Rekishi 1:113—82.

Sasaki, Komei
1971   Inasaku Izen (Before rice agriculture). Nihon Hōsō Shuppan Kyōkai No. 147, Tōkyō.

Sasaki, Komei, Tatsuo Kobayashi, and Makoto Sahara
1982   Nōkō no Hajimari o Megutte (On the beginning of agriculture). *In* Rekishi Koron Books 10: Nōkō Bunka to Kodai Shakai, pp. 18—47. Yuzankaku, Tōkyō.

Sato, Toshiya
1971   Nihon no Kodai-Mae (Ancient rice in Japan). Yuzankaku, Tōkyō.

1984   Kamegaoka Iseki Sawane chiku B-ku Shutsudo no Ine Eika narabi ni Tanka Mairyu (Rice caryopses and carbonized rice from the Kamegaoka site, Sawane Locality B). *In* Kamegaoka Iseki (The Kamegaoka site), edited by Aomori Kenritsu Kyōdōkan, pp. 218—24. Aomori Kenritsu Kyōdōkan, Archaeology Report 16, Aomori.

1986   A-6 Tateana Jukyō Shutsudo Komugi-yō Ibutsu (Wheat-like remains excavated from pit-house A-6). *In* Ichinohe Baipasu Kankei Maizo Bunkazai Chōsa Hokokushō III (Ichinohe bypass salvage archaeology report III), edited by K. Takada, pp. 341—46. Ichinohe-chō Kyōiku, Iwate. Iinkai.

Schwarz, Henry P., Jerry Melbye, M., Anne Katzenberg and Martin Knyf
1985   Stable Isotopes in Human Skeletons of Southern Ontario: Reconstructing Palaeodiet. Journal of Archaeological Science 12:187—206.

Simmonds, N. W.
1976   Hemp. *In* Evolution of Crop Plants, edited by N. W. Simmonds, pp. 203—4. Longman, London.

Smith, P. M.
1976   Minor Crops. *In* Evolution of Crop Plants, edited by N. W. Simmonds, pp. 301—24. Longman, London.

Snow, Bryan E., D. E. Nelson, Richard Shutler Jr., J. R. Southon, and J. S. Vogel
1986   Evidence of Early Rice Cultivation in the Philippines. Philippine Quarterly of Culture and Society 14:3—11.

Sugihara, Sosuke
1977   Nihon Nōko Shakai no Keisei (The formation of Japanese agricultural society). Yoshikawa Hirofumikan, Tōkyō.

Takahashi, R.
1955   The Origin and Evolution of Cultivated Barley. Advances in Genetics 7:227—66.

Tarami-cho Kyōiku Iinkai
1986   Ikiriki Iseki (The Ikiriki site). Tarami-cho Ky-
        ōiki Iinkai, Tarami-cho.

Tosu-shi Kyōiku Iinkai
1982   Nabatake. Tōsu-shi Kyōiku Iinkai, Tōsu-shi.

Tozawa, Mitsunori
1982   Jōmon Nōkō-ron (An essay on Jōmon agricul-
        ture). In Nihon no Kōkogaku o Manabu (2):
        Genshi—Kodai no Seisan to Seikatsu (Learn-
        ing Japanese archaeology (2): Origins—ancient
        production and lifeways), edited by Hatsue
        Otsuka, Mitsunori Tozawa, and Makoto
        Sahara, pp. 173—91. Yuhikaku Sensho,
        Tōkyō.

1983   Jōmon Nōkō-ron (An essay on Jōmon agricul-
        ture). In Jōmon Bunka no Kenkyū Vol. 2:
        Seigyō (Research on the Jōmon culture, Vol.
        2: Occupations), edited by Shimpei Katoh,
        Tatsuo Kobayashi, and Tsutomu Fujimoto,
        pp. 254—66. Yuzankaku, Tōkyō.

Tregear, T. R.
1980   China: A Geographical Survey. Hodder and
        Stoughton, London.

Trewartha, G. T.
1963   Japan: a Physical, Cultural, and Regional Ge-
        ography. University of Wisconsin Press, Mad-
        ison.

Tsuboi, Kiyotari
1964   Jōmon Bunka-ron (An essay on the Jōmon cul-
        ture). Nihon Rekishi 1:104—38.

Tsukada, M.
1986   Vegetation in Prehistoric Japan. In Windows
        on the Japanese Past: Studies in Archaeology
        and Prehistory, edited by Richard Pearson, pp.
        11—56. Center for Japanese Studies, Univer-
        sity of Michigan, Ann Arbor.

Tsukada, M., Yorko Tsukada, and Shinya Sugita
1986   Oldest Primitive Agriculture and Vegetational
        Environments in Japan. Nature 322:632—34.

Tsunoda, B., and M. Watanabe
1976   Kuwagaishimo Iseki (The Kuwagaishimo
        site). Heian Hakubutsukan, Kyōto.

Turner, Christy G. II
1979   Dental Anthropological Indications of Agri-
        culture among the Jōmon People of Central
        Japan. American Journal of Physical Anthro-
        pology 51(4):619—36.

Ueyama, Shumpei
1969   Shōyōjurin bunka: Nihon Bunka no Shinso
        (The broadleaf evergreen zone culture: deep
        structure of the Japanese culture). Chuokoron-
        sha, Tōkyō.

Ueyama, Sumpei, Komei Sasaki, and Sasuke Nakao
1976   Zoku—Shōyōjurin Bunka: Higashi Ajia
        Bunka no Genrui (The broadleaf evergreen
        zone culture—a sequel: the origins of East
        Asian culture). Chuokoronsha, Tōkyō.

Umemoto, K.
1984   Jōmon-ki no Ryokuto-rui ni tsuite (Jōmon
        period mung beans). In Kōbunkazai no
        Shizen-Kagakuteki Kenkyū (Natural scientific
        research of antiquities), edited by Kōbunkazai-
        Henshū Iinkai, pp. 655—56. Dōhōsha,
        Tōkyō.

Umemoto, K., and B. Moriwaki
1983   Jōmon-ki Mame-ka Shushi no Dōtei. (Identi-
        fication of Jōmon period beans). In Torihama
        Kaizuka, edited by I. Okamoto, pp. 42—46.
        Fukui-Ken Kyōiku Iinkai, Fukui.

Van Vilsteren, V. T.
1984   The Medieval Village of Dommelin: A Case
        Study for the Interpretation of Charred Seeds
        from Post Holes. In Plants and Ancient Man,
        edited by W. Van Zeist and W. A. Casparie,
        pp. 227—35. A. A. Balkema, Rotterdam.

Vavilov, N. I.
1951   The Origin, Variation, Immunity and Breed-
        ing of Cultivated Plants. Chronica Botanica
        13(1/6):1—364.

Verdcourt, B.
1970   Studies in the Leguminosae—Papilionoideae
        for the 'Flora of Tropical East Africa: IV.' Kew
        Bulletin 24: 507—69.

Watanabe, Makoto
1975   Jōmon Jidai no Shokubutsu Shoku (Plant food
        of the Jōmon period). Yuzankaku, Tōkyō.

Watanabe, M., and S. Kokawa
1982   Nabatake Jōmon Banki (Yamanotera-so) kara
        Shutsudo no Tanka Gobō, Azuki, Egonoki, to
        Mitanka Meron Shushi no Dōtei (Identification
        of carbonized gobō, azuki, egonoki, and un-
        carbonized melon seeds from the Final Jōmon
        (Yamanotera level) at Nabatake). In Nabatake,
        edited by Tosu-shi Kyōiku Iinkai, pp. 447—
        54. Tosu-shi Kyōiku Iinkai, Tosu-shi.

Watson, Patty Jo

1976   In Pursuit of Prehistoric Subsistence: A Comparative Account of Some Contemporary Flotation Techniques. Midcontinental Journal of Archaeology 1(1):77—100.

Whitaker, Thomas W., and W. P. Bemis

1976   Cucurbits. In Evolution of Crop Plants, edited by N. W. Simmonds, pp. 64—69. Longman, London.

Wu, Zhengyi

1980   The Vegetation of China. Science Press, Beijing.

Yabuno, Tomosaburo

1966   Biosystematic Study of the Genus Echinochloa. Japanese Journal of Botany 19(2):277—323.

Yamada, Goro

1980   Iwate-ken Kitakami-shi Kyūnenbashi Iseki no Kafun Bunseki ni tsuite (Analysis of pollen from the Kyūnenbashi site, Kitakami, Iwate Prefecture). In Kyūnenbashi Iseki, Report Number 6, edited by the Kitakami Kyōiku Iinkai, pp. 63—75. Kitakami Bunkazai Chōsa Hōkoku, No. 29, Kitakami.

1986   Hokkaidō ni Okeru Senshi Jidai no Shokubutsusei Shokuryō ni tsuite (Prehistoric vegetable foods in Hokkaidō). Hokkaidō Kōkogaku 22:87—106.

Yamazaki, Akira

1961   Kusaki-zome (Dyeing with natural materials). Getsumei-kai, Kanagawa.

Yan, Wenming

1989   The Origins of Agriculture and Animal Husbandry in China. Paper presented at the Circum-Pacific Prehistory Conference, Seattle.

Yasuda, Yoshinori

1978   Prehistoric Environment in Japan. Institute of Geography, Faculty of Science, Tōhoku University, Sendai.

Yen, Douglas E.

1982   Ban Chiang Pottery and Rice. Expedition 24(4):51—64.

Yoshizaki, Masakazu

1983   Hokudai Kñai no Iseki (The Hokkaidō University campus sites). Hokkaidō Daigaku, Sapporo.

1984   Kōkogaku ni okeru Ezo to Ezo-chi (1): Kodai Ezo-chi no Bunka (The Ezo and Ezo region from an archaeological perspective (1): the ancient Ezo culture). Sōzō no Sekai 49:80—97.

1989   K441 Iseki Kita 34 Chome Chiten Shutsudo no Shokubutsu Shushi (Plant seeds recovered from the K441 site, North 34 Street locality). In K441 Iseki Kita 34 Chome Chiten, edited by Sapporo-shi Kyōiku Iinkai, pp. 70—79. Sapporo-shi Kyōiku Iinkai, Hokkaidō.

Yu, Xiuling

1976   A Few Observations on the Rice Grains and Bone "xi" Unearthed from the Fourth Layer of the Hemudu Site. Wenwu 8:20—23.

Zeven, A. C.

1980   The Spread of Bread Wheat over the World since the Neolithicum as Indicated by its Genotype for Hybrid Necrosis. Journal d'Agriculture Tropicale et de Botanique Appliquee 27:19—53.

Zeven, A. C., and P. M. Zhukovsky

1975   Dictionary of Cultivated Plants and their Centers of Diversity. Center for Agricultural Publishing and Documentation, Wageningen, The Netherlands.

3

The Origins of Plant Cultivation in the Near East

NAOMI F. MILLER

Between about 11,000 and 6000 B.C. a series of irrevocable changes took place in the societies of the ancient Near East. Before this period, groups of mobile foragers lived by hunting and gathering. By the end of it, much of the region was inhabited by settled villagers who relied primarily on farming and stock-raising for their livelihood. The cultivation and eventual domestication of plants played a major role in this process.

One can think of plant cultivation as a single point along a continuum of human/plant interaction. At the least intrusive end is the simple harvesting of crops, which can have an effect on a wild plant population; the use of fire and other simple techniques also encourage particular wild plants. At a greater degree of interference is cultivation, in which a crop is intentionally planted. At the most intrusive end is domestication, in which a plant has evolved into a new form under continued manipulation by humans so that it may have lost the ability to reproduce itself (Helbaek 1969:403).

In the Near East, two aspects of human plant exploitation were distinctive. First, the native grasses were highly productive. Wild wheat and barley grew in dense stands and served as a valuable food source even before cultivation had begun. It was this use of wild grasses for *food* that probably led to their early domestication. Second, early reliance on domesticated grasses was facilitated by the adoption of complementary sources of protein, namely, leguminous crops, such as pea and lentil, and domesticated animals. Domestication followed increased wild plant use by sedentary human populations by several thousand years. The beginning of plant cultivation prior to domestication, and the association between plant cultivation and changing patterns of human/plant interaction can, however, be approached only indirectly.

In this chapter I review the geographical, archaeological, and botanical background of plant cultivation, note some of the more recent archaeobotanical research on the origins of plant cultivation, and discuss certain aspects of the theories of agricultural origins that were developed during the 1960s, 1970s, and 1980s. Finally, I offer a critique of the use of archaeological plant remains as evidence for those theories.

Geography

This chapter focuses on those parts of Iran, Iraq, Turkey, Syria, Jordan, Lebanon, and Israel where the early development of food production took place

(Fig. 3.1). The region is characterized by several major land forms (M. Zohary 1973; Fisher 1978). The Afro-Arabian tableland borders the area to the south and is today largely desert. The Irano-Anatolian folded zone, which includes the Zagros, Taurus, and Syro-Palestine ranges, forms a mountainous arc where most of the arboreal vegetation now grows. Between these two zones is a region of rolling terrain and alluvial plains. A portion of this region in Turkey, Iraq, and Iranian Kurdistan is sometimes referred to as the "hilly flanks of the Fertile Crescent" (Braidwood and Howe 1960:3). The two major rivers of Mesopotamia, the Tigris and the Euphrates, have their origins in the mountainous regions and flow in a generally southeasterly direction through the Syrian steppe and the Mesopotamian lowland.

The tremendous variability in topography has a strong influence on climate and vegetation. The coastal regions in the west are characterized by a Mediterranean regime of hot, dry summers and cool, wet winters. A lowland area known as the "Syrian Saddle," which separates the Anti-Taurus from the Syro-Palestine ranges, allows the Mediterranean influence to extend inland. The inland regions, however, have a more continental climate.

Generally, there are altitude and latitude clines in temperature and available moisture. This variability is reflected in the distribution of the natural vegetation. Following M. Zohary (1973), one can define several different types of vegetation. Because of severe deforestation and degradation of the landscape, such reconstruction of the natural flora is based on discontinuous patches of relatively undisturbed vegetation that still exist today.

Oaks (*Quercus calliprinos* Webb, among others) and Aleppo pine (*Pinus halepensis* Mill.) predominate in

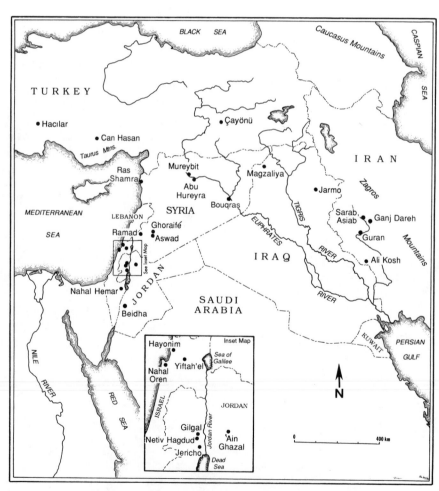

Fig. 3.1. The Near East and archeological sites mentioned in the text.

the Mediterranean coastal forest west of the great bend in the Euphrates. In addition to oak, pistachio (*Pistacia* spp.) and carob (*Ceratonia siliqua* L.) are important trees that have edible fruit and nuts. Lebanon cedar (*Cedrus libani* A. Rich.) is found only at higher elevations. The drier inland regions of the folded zone are dominated by oak (*Quercus brantii* Lindl.), but pistachio continues to be an important component. In the more arid regions of southern Iran where it is too dry for oak, pistachio-almond steppe forest predominates. The forest grades successively into steppe forest, steppe, and desert in northern Syria and Iraq. Riverine associations characterized by willow (*Salix* sp.), poplar (*Populus* sp.), and tamarisk (*Tamarix* sp.) cut across environmental zones. The wild cereal grasses and legumes are at home in the open oak forest and steppe forest regions (Harlan and Zohary 1966; Zohary 1989; Zohary and Hopf 1973).

Moisture is the limiting factor for agriculture over much of the region. The minimum rainfall needed for winter crops is approximately 250 mm. Fields may be planted in locally moist plots to improve harvest security.

Ancient Climate

Reconstructions of climate and vegetation for the period of the earliest cultivation and adoption of agriculture are critical elements for models of agricultural origins in the Near East. Several recent syntheses have described the climate of the Near East at the end of the Pleistocene and in the early Recent period (Bottema and van Zeist 1981; Wright 1977, 1980, 1983; van Zeist and Bottema 1982). Much of the Near East had little or no forest cover during the glacial period, probably because of higher aridity and/or lower temperatures than today. Although treeless steppe characterized the Zagros Mountains, the Levant remained forested, and the amount of forest cover in Syria and Turkey fluctuated (Bottema and van Zeist 1981). Pollen evidence suggests that the forest (primarily oak) gradually expanded from refugia in the west and eventually reached the Zagros Mountains. Modern analogues to this type of forest cover contain wild cereal grasses, so it is likely that the grass habitat expanded as well (Wright 1983). Of special interest is the situation in northwest Syria where the climate from about 10,000 to 7000 B.C. was apparently moister than at present and the vegetation was dominated by deciduous oak (van Zeist and Woldring 1980).

History of Investigations

The first systematic attempt to recover empirical evidence for the transition to food production was Robert Braidwood's Iraq-Jarmo Project (Braidwood and Howe 1960; Braidwood et al. 1983). The project included specialists in geology, botany, and zoology. Hans Helbaek, the project's botanist, wrote an important early survey of the botanical evidence for domestication (Helbaek 1960). Although some of the results of the project have been superseded by the discovery of earlier agricultural settlements, the Jarmo project was a model of interdisciplinary research; all subsequent work on agricultural origins in the Near East can trace an intellectual descent from it.

Important work on early agriculture continued during the 1960s. In Iran, Hans Helbaek (1969) analyzed painstakingly collected plant remains from the Deh Luran Plain (Hole et al. 1969). Braidwood and his colleagues transferred their operations to Çayönü in southeastern Turkey (van Zeist 1972). Several projects were carried out under the inspiration of Eric Higgs and the auspices of the British Academy Major Research Project in the Early History of Agriculture (French et al. 1973; Moore et al. 1975; Noy et al. 1973). The Tabqa Dam Project of the 1970s yielded a wealth of archaeobotanical material from sites in northwestern Syria (Cauvin 1977; Moore et al. 1975). Much of this work has only recently been completed. Although some archaeologists are concerned primarily with the origins and development of early village life (e.g., Cauvin 1978), the archaeobotanical study of early agriculture can easily be accommodated within such research projects since farming in the Near East seems to have been associated with sedentism from the beginning. A summary of work through the 1960s is available in Renfrew (1969) and Flannery (1973). Henry (1989) and Moore (1985) have recently written archaeological reviews of the Natufian and Neolithic periods. Botanical reports or notes are available for the early village sites listed in Table 3.1.

Archaeological Background

The earliest evidence of morphologically new plants occurs in the archaeological record at about 9000 B.C.

Table 3.1. Botanical Reports from Epipaleolithic and Aceramic Sites

Site	Period	Reference
Levant		
Nahal Oren	Kebaran, Natufian, PPNA, PPNB	Noy et al. 1973
Hayonim	Natufian	Hopf and Bar-Yosef 1987
Wadi Hammeh 27	Natufian	Edwards et al. 1988
Netiv Hagdud	PPNA	Kislev et al. 1986
Gilgal	PPNA	Noy 1988
Jericho	PPNA, PPNB, Neolithic	Hopf 1983, Western 1971
Nahal Hemar	PPNB	Kislev 1988
Yiftah'el	PPNB	Kislev 1985
'Ain Ghazal	PPNB	Rollefson et al. 1985
Beidha	PPNB	Helbaek 1966
Syria		
Abu Hureyra	Late Epipaleolithic, Aceramic Neolithic	Moore et al. 1975, Hillman et al. 1989
Mureybit	Late Epipaleolithic	van Zeist and Bakker-Heeres 1986b
Aswad	Aceramic Neolithic	van Zeist and Bakker-Heeres 1985
Ramad	Aceramic Neolithic	van Zeist and Bakker-Heeres 1985
Ghoraifé	Aceramic Neolithic	van Zeist and Bakker-Heeres 1985
Ras Shamra	Aceramic Neolithic, Neolithic	van Zeist and Bakker-Heeres 1986a
Bouqras	Neolithic	van Zeist and Waterbolk-van Rooijen 1985
Turkey		
Çayönü	Aceramic Neolithic	van Zeist 1972, Stewart 1976
Hacılar	Aceramic Neolithic, Neolithic	Helbaek 1970
Can Hasan III	Aceramic Neolithic	French et al. 1972, Hillman 1978
Iraq		
Jarmo	Aceramic Neolithic	Helbaek 1960, Watson 1983, L.S. Braidwood et al. 1983:541
Magzaliya	PPNB	Lisicyna 1983
Iran		
Ganj Dareh	Aceramic Neolithic	van Zeist et al. 1986
Ali Kosh	Aceramic Neolithic, Neolithic	Helbaek 1969

Domesticated barley has been reported from the Levant at two Pre-Pottery Neolithic A (PPNA) sites, Netiv Hagdud (Kislev et al. 1986) and Gilgal (Noy 1988), and from Iran in the earliest levels at Ganj Dareh (van Zeist et al. 1986), although these early specimens may be morphologically indistinguishable from the wild type (Kislev 1989). Domesticated einkorn and emmer wheat have been found in Syria and Turkey at Aswad, Çayönü, and Neolithic Abu Hureyra. Emmer wheat gave rise to durum wheat by the eighth millennium B.C., and bread wheat appeared in the seventh millennium B.C. (van Zeist 1986a). Since plant domestication represents the culmination of a long process of human/plant interaction, we know that important changes in plant use had already taken place during the preagricultural period, sometimes called the Epipaleolithic. During

that time, people in the Near East began to occupy permanent or semipermanent villages. Subsistence strategies apparently changed as well, with increased consumption of plants and mollusks. Kent Flannery (1969) calls this change in the prehistoric menu the "broad spectrum revolution," although recent studies have questioned this characterization of the preagricultural diet (Edwards 1989; Henry 1989).

The Epipaleolithic refers to the (presumed) foraging cultures of the Late Glacial and Early Holocene periods. Local sequence names in the Levant include the Kebaran (to ca. 11,000 B.C.) and Natufian (ca. 11,000–9400 B.C.) and in the Zagros the Zarzian and Karim Shahirian, which are roughly contemporary with the Levantine sequence. Early farming societies are referred to as "Neolithic." The earliest Neolithic societies that did not use ceramic vessels, although many had some knowledge of the properties of clay, are called Pre-Pottery Neolithic in the Levant (PPNA, ca. 9400–8500 B.C.; PPNB, ca. 8500–6700 B.C.) and simply the Aceramic Neolithic elsewhere. An extended discussion of local sequences and chronological problems is beyond the scope of this chapter, but they have been discussed by others (Aurenche et al. 1987; Bar-Yosef and Vogel 1987; Braidwood and Howe 1960; Henry 1989). The early farming societies adopted various combinations of crops and animals in different parts of the Near East. Village life based on the complete Near Eastern complex of wheat, barley, pulses, sheep, goat, pig, and cattle took several thousand years to develop (Table 3.2).

The Nature of the Evidence

Several lines of evidence contribute to the study of agricultural origins. Prehistoric tools and facilities provide indirect evidence of plant use, human skeletal remains are used in dietary reconstructions of the Epipaleolithic, and ecological and botanical studies of plant remains found on archaeological sites shed light on the transition to food production. Finally, the archaeological context in which plant remains are found must also be considered.

Plant processing equipment and facilities became important elements of Epipaleolithic material culture. Grinding stones, some of which were used for pigments (Moore et al. 1975:58), could also have been used for grain or acorn processing, and flint sickle blades were used for cutting grasses. Roasting pits, which are present on some sites, could have been used to process grain. Storage technology developed as well. Although pottery had not yet been invented, underground pits were used to solve the problem of preserving seasonally abundant, storable plant resources, particularly wild cereals.

Flannery used the term "preadaptation" for the technological changes that preceded and permitted reliance on agricultural production. Until recently, the association between increasing dependence on plant foods and the development of new food processing technologies has been somewhat conjectural. Now, however, it has been borne out by several studies showing that human skeletal remains bear traces of an individual's dietary history. For example, the consumption of stone-ground foods has been shown to lead to a rapid wearing down of teeth. While this pattern is typical of later agricultural villagers of the Near East, it first appears in the skeletons of the late Epipaleolithic (P. Smith 1972). Bone strontium analysis provides additional, although somewhat contro-

Table 3.2. Simplified Chronology for the Epipaleolithic and Neolithic in the Near East

Calibrated Date[a] B.C.	Levant	Syria/Anatolia	Zagros	Uncalibrated Date[b] (B.C.)
		Pottery Neolithic		
6700	6000
8000	PPNB		Aceramic Neolithic	7000
		Aceramic Neolithic		
8500			7600
9000	PPNA		(Proto-Neolithic)	8000
9400	. . . .	. . . .	8300
11,000				9000
	Natufian		Karim Shahirian	
12,000	. . . .		. . . .	10,000
	Geometric Kebaran			
. . . .	Epipaleolithic	Zarzian	11,000
				12,000
	Kebaran			13,000
				14,000
				15,000

Source: The information on local sequences was compiled from Aurenche, Évin, and Gascó (1987) and Bar-Yosef and Vogel (1987).
[a]Calibrated radiocarbon dates are interpreted from Stuiver et al. (1986, fig. 7).
[b]Uncalibrated dates are based on Libby half-life (5568 years).

versial, evidence for dietary change. Since plants and mollusks both have higher proportions of strontium relative to calcium, herbivores and shellfish eaters exhibit higher bone strontium than carnivores. An analysis of a small series of Levantine Epipaleolithic skeletal remains suggests that the individuals had consumed more plants compared with others from earlier periods (Schoeninger 1981, 1982; also compare with Sillen 1981). This finding is further substantiated by the presence of grinding and storage facilities at the sites. For reviews of the Near Eastern skeletal evidence, see P. Smith et al. (1984) and Rathbun (1984).

The archaeobotanical evidence for early plant domestication in the Near East (Table 3.3), although incomplete, is fairly clear (van Zeist 1986; Zohary and Hopf 1988). The wild ancestors of early domesticated cereals are inferred on genetic, morphological, and phytogeographical grounds (Harlan and Zohary 1966; Zohary 1989; Zohary and Hopf 1988) (see Figs. 3.2–3.4). Rapid evolution toward morphologically domesticated types may explain why intermediate forms are rarely encountered in the archaeobotanical record (Hillman and Davies 1990). Two-row barley (*Hordeum distichum* L. emend. Lam.) evolved from the two-row wild form (*H. spontaneum* C. Koch), which grows today in an arc stretching from the Levant to western Iran. The most primitive domesticated wheat, einkorn (*Triticum monococcum* L.), evolved from a wild form (*T. boeoticum* Boiss. emend. Schiem.) whose native habitat extends from southeastern Turkey to western Iran. Both wild and cultivated types of einkorn are genetically diploid. The origin of the tetraploid emmer wheats is still problematic. Wild emmer (*T. dicoccoides* Korn), the closest relative of the domesticated variety (*T. dicoccum* Schubl.), grows in the Levant. Some consider it to be a weedy form that is derived from domesticated emmer, but others believe it is the wild ancestor. Both wild and domestic emmer have two genomes, A and B. The A genome is derived from einkorn, but the source of the other is uncertain (Feldman 1976). In contrast to einkorn and emmer, which are both hulled wheats, durum, or hard wheat (*T. durum* Desf.), is a free-threshing tetraploid that is closely related to emmer. A later hybridization between cultivated emmer and a wild, goat-face grass, *Aegilops squarrosa* L. (= *T. tauschii* [Coss.] Schmalh), gave rise to the free-threshing hexaploid bread wheat (*T. aestivum* L.; see also Kislev 1984). A third cereal, rye (*Secale cereale* L.), may also have originated in the

Near East (Hillman 1978), but it was established as a crop relatively late.

Fortunately for the archaeobotanist, the forms of domesticated cereals are distinct from those of wild types. The cultigens have larger and plumper grains than their wild counterparts. In addition, the wild forms have brittle stalks, or rachises, which shatter into individual internodes when they are ripe, allowing the plants to propagate themselves in the wild. Human harvesting practices may have selected for a tougher rachis, so that the domesticates should be recognizable by the presence of intact rachis segments. Unfortunately, the rachises of the more primitive domestic wheats—einkorn and emmer—have a tendency to break up with threshing, and those of some wild barleys may not shatter at all (Kislev 1989). Therefore, although domestication involved morphological changes to both fruit and stalk, many of the reported examples of domesticated cereals from early sites can be identified only on the basis of grain size and shape (e.g., van Zeist 1972; van Zeist and Bakker-Heeres 1985). Several researchers are trying to refine the criteria for identifying archaeobotanical specimens (Kislev 1989; Körber-Grohne 1981).

Legumes also appear early in the archaeobotanical record, sometimes in greater quantity than the cereals. This is true of both small-seeded legumes, such as clover (*Trifolium*) and medick (*Medicago*), and pulses, such as lentil (*Lens culinaris* Med.), pea (*Pisum sativum* L.), chickpea (*Cicer arietinum* L.), and vetches *Vicia sativa* L., *V. ervilia* (L.) Willd., and *V. faba* L.). The cultivated legumes are generally not readily distinguishable from their wild counterparts, since many characteristics, such as size and surface texture, overlap. For example, while domesticated lentil tends to be larger than the wild type, this is not always the case. Further complicating matters, the rigors of ancient processing and archaeological recovery may destroy distinctive characteristics, such as the surface texture of the seed coat of pea. Regardless of present-day differences, therefore, the earliest cultivated legumes would not be morphologically distinct from the wild forms (Zohary and Hopf 1973). Lentil, pea, and chickpea occur at early agricultural sites. The vetches, most of which are grown today for fodder, are relatively common at Epipaleolithic sites. A fava-like vetch occurs in PPNA levels at Jericho (Hopf 1983), and a large quantity of domesticated fava bean (*V. faba*) dating to the sixth millennium B.C. has been reported from Yiftah'el in Israel (Kislev 1985).

Table 3.3. Seeds of Likely Crops from Epipaleolithic and Early Farming Sites

| | Cereal | | | | | Pulse | | | | | | Other |
| | Wheat | | | Barley | | | | | | | | |
	Einkorn	Emmer	Bread/Hard	2-row	6-row	Lentil	Pea	Vetch	Fava bean	Chick-pea	Lupine	Flax
Epipaleolithic												
Hayonim	·	·	·	·	·	·	·	·	·	·	•	·
Nahal Oren	w	·	·	w	·	·	·	•	·	·	·	·
Abu Hureyra	w	·	·	w	·	+	·	•	·	·	·	·
Mureybit	w	·	·	w	·	+	cf	+	·	·	·	•
PPNA												
Netiv Hagdud	·	·	·	w/c	·	·	·	·	·	cf	·	·
Jericho	+	+	·	+	+	+	·	·	·	·	·	·
PPNB												
Jericho	•	•	•	•	+	•	•	·	+	+	·	•
Nahal Hemar	·	•	·	•	·	•	•	·	+	·	·	·
'Ain Ghazal	·	•	·	•	·	•	·	·	·	•	·	•
Beidha	+	·	cf	w/c	·	+	·	·	·	·	·	w?
Magzaliya	+	+	·	+	+	+	·	•	·	·	·	·
Other Aceramic												
Aswad	+	•	•	w/c	·	+	•	•	·	•	·	w
Çayönü	•	•	·	·	·	+	•	•	·	+	·	·
Ganj Dareh	+	·	·	•	·	+	·	·	·	·	·	·
Ali Kosh	+	+	·	•	·	+	+	·	·	·	·	+
Jarmo	+	+	·	+	+	+	+	·	·	·	·	·
Hacılar	w?	•	·	w	w	+	·	·	·	·	·	·
Can Hasan III	•	•	•	•	·	+	·	•	·	·	·	·
Abu Hureyra	•	•	·	·	·	+	·	·	•	+	·	·
Ghoraifé	+	•	·	w/c	+	+	+	+	+	+	·	+
Ramad	·	•	+	+	+	•	+	+	·	·	·	+
Ras Shamra	·	+	·	+	+	+	·	·	·	·	·	+
Neolithic												
Jericho	•	•	·	•	+	•	+	·	·	·	·	·
Ras Shamra	·	•	·	+	+	•	+	·	·	·	·	+
Ramad	+	•	·	+	•	•	+	+	·	+	·	•
Bouqras	•	•	•	•	•	•	+	·	·	·	·	·
Hacılar	•	•	•	•	·	•	·	•	·	·	·	·
Ali Kosh	+	·	·	·	+	+	·	·	·	·	·	+

w = wild; • = numerous; w/c = wild or cultivated; + = present or quantity not reported; cf = tentative identification.

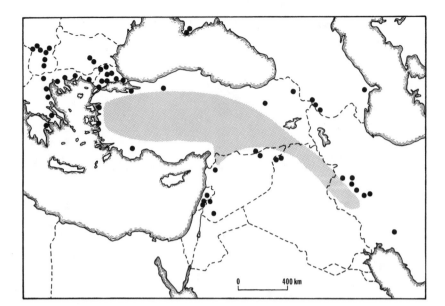

Fig. 3.2. Present-day distribution of wild einkorn wheat, *Triticum boeoticum*. Dots represent places where weedy forms have been seen. (After Harlan and Zohary 1966; Zohary 1989)

While researchers learn a great deal about crop evolution from a study of the plants themselves, they assess the role of plants in the ancient economy through a combination of botanical and archaeological analyses. The remains of plants, whether they are domesticated or not, have a cultural context if they are found on an archaeological site. Their cultural significance depends on how they were used and how they came to be preserved in the archaeological record. In the 1960s, researchers explicitly considered the possibility that at least some weed seeds were eaten, and generally assumed that the seeds of cultivated plants represented accidentally charred food remains. During the 1970s, studies demonstrated that residues from different stages of crop processing could leave an assortment of weed and cultigen seeds (Dennell 1974; Hillman 1981). More recently, I have suggested that some seeds from later agricultural sites came from burning dung as fuel (Miller 1984; Miller and Smart 1984), and this interpretation may be applicable to earlier remains as well. As I will discuss below, these considerations have important implications for theories of agricultural origins.

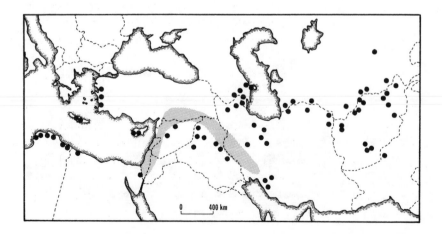

Fig. 3.3. Present-day distribution of wild barley, *Hordeum spontaneum*. (After Harlan and Zohary 1966; Zohary 1989)

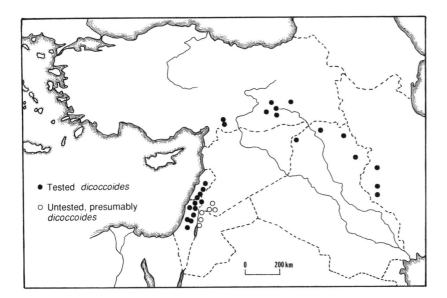

Fig. 3.4. Present-day distribution of wild emmer wheat, *Triticum dicoccoides*. (After Harlan and Zohary 1966; Zohary 1989)

● Tested *dicoccoides*

○ Untested, presumably *dicoccoides*

0   200 km

Previous Archaeobotanical Work

Epipaleolithic Near Easterners apparently began eating the wild ancestors of plants that their descendants ultimately domesticated. Any discussion of domestication, therefore, must start with the botanical evidence for plant use.

In addition to gaps in the archaeobotanical record resulting from the relatively small number of excavated archaeological sites, archaeobotany suffers from several other disadvantages. First, because systematic archaeobotanical studies did not begin until the 1950s, few incipient or early agricultural sites excavated have been tested for plant remains. Second, because so much work is recent, relatively few final archaeobotanical reports have been published. Third, even when great care has been taken to recover charred seeds, many ancient sites have yielded low densities of carbonized plant remains (e.g., Noy et al. 1973).

Prior to 1970, the most complete paleoethnobotanical study of an early agricultural community was done by Hans Helbaek (1969) for Ali Kosh in southwestern Iran. The study provided the botanical basis for much of the subsequent discussion about agricultural origins and development in the Near East, especially Flannery's work (1969). Domesticated plants and husbanded animals occurred in the earliest levels. Most of the plant remains were recovered from ash lenses and midden (Hole et al. 1969). Extraordinary quantities of small-seeded legumes (94 percent of the

seeds, about 29,000 seeds) occurred in samples from the Bus Mordeh phase (beginning of the eighth millennium B.C.). In the subsequent Ali Kosh phase (end of the eighth millennium B.C.), cultigens, primarily emmer, comprised 40 percent of the seed remains, and small-seeded legumes dropped to 19 percent. In the final Mohammad Jaffar phase, small-seeded legumes once again made up a substantial portion of the remains (59 percent, about 5,000 seeds). Hole, Flannery, and Neely's (1969:342–54) interpretation of the Ali Kosh remains emphasized aspects of the seed assemblage that favored progressive agricultural development. As these figures suggest, however, the assemblage does not necessarily exhibit straight-line development.

A summary and discussion of plant remains from other early sites up to the early 1970s appears in Renfrew (1969, 1973) and Flannery (1973). Therefore, I confine the remainder of my discussion to sites with botanical remains reported in the last twenty years.

Among the earliest macroscopic plant remains recovered to date from the Near East are those from Nahal Oren in Israel (Noy et al. 1973). The site is located in the Wadi Fellah at an elevation of about 100 m between the uplands and the Mediterranean coast. The earliest levels at Nahal Oren, which date to the Kebaran period, had very low seed densities. The single largest category was vetch, which comprised 27 out of a total of 64 seeds. Three wheat grains identified as *Triticum dicoccum*, the domestic type, were reported, but they are now known to be intrusive (Gowlett and Hedges 1987). The Natufian levels yielded 25 seeds, including 11 vetch remains, but no emmer. PPNA and PPNB levels produced similar low densities of seeds, including vetch and emmer. The PPNB levels are distinguished by a comparatively large number of plant types, including several nuts and fruits. Another Natufian site, Hayonim, which is 13 km east of the Mediterranean coast and about 250 m above sea level, revealed the predominance of a large-seeded legume, lupine (*Lupinus pilosus* L.).

Mureybit (van Zeist and Bakker-Heeres 1986b) and early Epipaleolithic Abu Hureyra (Hillman et al. 1989), which date to between the ninth and eleventh millennia B.C., are located about 30 km apart on the Euphrates river in northwestern Syria at an elevation of about 300 m. Both have evidence for the ancient use of wild einkorn about 100 km from its present habitat. The middle levels of Mureybit are characterized by the presence of "fire pits," which may well represent grain roasting facilities. Two of the samples from these levels show high concentrations of nearly pure wild einkorn. The archaeological context and the purity of these samples together support the conclusion that einkorn was intentionally harvested for food. A slightly later sample consists primarily of *Polygonum,* a plant that never became a Near Eastern cultigen but appears to have been collected for food at Mureybit. The remaining samples are mixed assemblages of weedy types. Abu Hureyra exhibits a slightly wider range of materials from this time period. Although wild-type einkorn is quite common, the assemblage includes mixed samples of seeds from many wild plant types, most of which were never domesticated (Hillman et al. 1989). Hillman et al. (1989) suggest that the weed seeds are most likely food remains, although they could easily be interpreted as the remains of burned trash, fuel, or contaminants of harvested cereals (see below). Both sites have some tree fruits (pistachio and hackberry), which is consistent with the recent interpretation that the climate of northwestern Syria was relatively moist during the Epipaleolithic (van Zeist and Woldring 1980).

Aswad (van Zeist and Bakker-Heeres 1985), another ninth millennium B.C. site, is located near a former lake in the steppe region of Syria. It has yielded the earliest domesticated emmer found to date (van Zeist and Bakker-Heeres 1985; see Kislev 1989 for an alternative view), as well as probably cultivated pea and lentil. Farming, however, seems to

have begun in the Levant, and only later spread beyond the Jordan Valley to such sites as Aswad (van Zeist 1986; Bar-Yosef and Kislev 1989). Of all the sites considered here, Aswad most closely resembles the Bus Mordeh phase at Ali Kosh in that small-seeded legumes represent a disproportionately large component of the total seed assemblage (55 percent for Phase I at Aswad versus 94 percent for the Bus Mordeh phase at Ali Kosh; at Çayönü they are well under 1 percent; and at Mureybit they represent less than 2 percent). Later "Neolithic" Abu Hureyra, which is roughly contemporary with Aswad, also has domesticated plants, including einkorn, emmer, hulled and naked six-row barley, chickpea, and lentil. Plant assemblages from sites in the steppe region (Aswad, Abu Hureyra, and Ali Kosh) are similar to one another in that they contain a variety of weedy types.

Jericho is located at the northern end of the Dead Sea about 300 m below sea level. The area is extremely arid today and probably was dry in the past as well. A large, freshwater spring at the base of the mound site has provided water for millennia. Although early reports (Hopf 1969) contended that farming in such an arid climate would have required at least simple irrigation techniques, recent geological work suggests that crops could have been planted to take advantage of water run-off in nearby wadi bottoms (Hopf 1983). Although a small Natufian occupation has been found at Jericho, the earliest plant remains date to PPNA and PPNB times. They were collected before flotation had become the standard procedure for recovering plant remains, so the assemblage is skewed in favor of concentrations of seed remains. Although the seeds of several crops are usually mixed, the number of weed impurities is quite low, suggesting that the seeds were cleaned crop or food remains. The presence of cultivated einkorn and emmer was documented with latex impressions of spikelets taken from mudbrick found at the site.

Several PPNB sites, including Ramad, Jericho, and Nahal Hemar, have yielded evidence of flax. Charred seeds from Ramad probably represent a domestic type of flax (*Linum usitatissimum*) (van Zeist and Bakker-Heeres 1985), as do two impressions of flax capsules (fruit cases) from Jericho (Hopf 1983). Although flax may have been domesticated for its oily seed, it may also have been the first plant in the Near East to be domesticated for non–dietary purposes. Linen textile fragments dating to PPNB were found preserved in dry deposits at Nahal Hemar cave (Schick 1986).

'Ain Ghazal is a PPNB site located at the outskirts of Amman, Jordan, on the edge of the Syrian steppe. The domesticated plants found there consist primarily of pea and lentil (Rollefson et al. 1985). A decline in the diameter of construction timbers suggests that large, old-growth trees were cut down in earlier times but that the forest did not regenerate, since, later on, only smaller, immature trees were available as construction material (Köhler-Rollefson 1988).

Ganj Dareh, an eighth- and possibly ninth-millennium B.C. site, is located in the folded mountain zone of Iran at an elevation of 1400 m. All levels contain domesticated barley (van Zeist et al. 1986). The earliest occupation, level E, is characterized by fire pits reminiscent of those found at Mureybit. Level E seems to be a temporary, perhaps seasonal, camp without permanent architecture, and is followed by four occupation levels with solid architecture.

Unlike most of the other sites, Çayönü in southeastern Turkey is located in the oak woodland zone at about 830 m. The earliest levels, which date to the early ninth millennium B.C., contain domesticated einkorn. Morphologically wild emmer and peas from the site may also represent cultivated crops (van Zeist 1972). Weed seeds are only sparsely represented, presumably because neither dung nor herbaceous plants were used for fuel. Archaeological charcoal includes pieces from trees that would have been growing nearby: oak, ash, pistachio, and almond in the forest, and tamarisk near watercourses. Although it was found in the heart of the natural habitat of wild einkorn, the one-seeded variety present at Çayönü probably originated in the west (van Zeist 1972).

These sites span the millennia during which agriculture became established. Although we have only a small sample of plant materials used by ancient peoples, it seems that the cultivation of different crops spread across a wide area in several environmental zones during the course of several thousand years. I have concentrated on the Near Eastern staple crops because of their early importance in the archaeobotanical record. By the time the Sumerian scribes began to record what they ate—after 6,000 years of agricultural endeavor—Near Easterners were growing a wide variety of cereals, pulses, fruits, and vegetables (see Sumerian Agriculture Group 1983; Zohary and Hopf 1988).

✗ Theories of Agricultural Origins in the Near East

Modern archaeobotanical research on agricultural origins, which began with the Iraq-Jarmo Project, was designed to test several previous theories. Braidwood (Braidwood and Howe 1960) chose to work in the hilly flanks region because he assumed that the earliest farmers would have first cultivated the cereals in their natural habitat. It could not be assumed, however, that the natural habitat areas of the past were the same as those of today, since, according to some theories, there had been major post-Pleistocene climatic changes (Childe 1969[1952]:25). Therefore, one of the project's major goals was to establish an environmental baseline for the period of incipient and early farming; this research suggested that the climate was not substantially different from that of today.

A second goal was to find firm evidence for early agriculture through an archaeological survey and the excavation of Jarmo, an early village site located in the natural habitat zone of wild einkorn and barley. Dating to the seventh millennium B.C., the earliest levels at Jarmo already contained domesticated plants and animals. Therefore, questions about the *origins* of plant cultivation could not be addressed directly at the site.

According to Harlan and Zohary (1966), wild harvests are so plentiful in the natural habitat zone that one could question why people would go to the trouble of cultivating plants there at all. This issue was also addressed by Binford (1968), who agreed that people would not have begun cultivating plants in the natural habitat zone. Rather, the relatively populous plant-collecting communities living in the natural habitat zone would have expanded into areas marginally suited for wild cereals, where they would have encountered food shortages and would have been forced to take up farming to increase their food supply. Binford's population pressure model was elaborated upon by Flannery (1969), who suggested that subsistence changes that took place prior to agriculture—during the "broad spectrum revolution"—could have been a response to population growth in the marginal zone. His key argument was that, as plentiful as wild cereals were in a natural habitat zone, they were second choice foods in a marginal zone because they required more energy to find and process. The chief advantage of a plant-based diet was that it could support a higher population. As the marginal zones became more densely populated, however, people began to plant crops intentionally to ensure an adequate food supply.

If, however, the natural habitat zone was significantly larger at the end of the Pleistocene, wild cereals could well have become a *preferred* food source because of their reliability and ease of collection. In fact, "broad spectrum revolution" is a bit of a misnomer for early Holocene dietary changes. As Henry (1989) points out, the Natufian diet strongly favored just a few species, mainly the wild cereals and gazelle. Paleoenvironmental research during the past ten years suggests that during this critical time, much of the Levantine and Syrian steppe was indeed moister than it is at present (van Zeist and Woldring 1980; Henry 1989), and the boundaries of the ancient natural habitat zone are not the same as they are today. Bohrer (1972) even suggests that human disturbance of the oak forest could have encouraged the growth of wild cereals, since their present habitat is coppiced oak scrub. At the present time, the evidence is insufficient to ascertain whether wild cereals grew in vast, dense stands in the present-day marginal zone, or in favorable, but restricted microenvironments.

The concern with identifying the natural habitat and marginal zones is something of a red herring. Since population pressure is a function of the resource base and not just of absolute population density, pressures to increase food supplies could be as significant in the natural habitat zone as in the marginal zone. Although the population pressure argument is still fairly popular (Binford 1983; Cohen and Armelagos 1984; Smith and Young 1972, 1983), the extreme version based on demographic disparities between the natural habitat and marginal zones is no longer tenable.

Nutritional Considerations

Many theoretical models of agricultural origins focus on nutrition and sedentism because Natufian culture, with its increasingly plant-based diet and sedentary communities, was the direct precursor of later agricultural societies (e.g., Bar-Yosef and Kislev 1989; Flannery 1969; Hassan 1976; Henry 1989). Plant cultivation (at least of einkorn and emmer) seems to have originated in the Levant and spread to the Zagros (Hole 1984; van Zeist 1986).[1] Skeletal evidence also suggests that Natufian populations began to rely heavily on plants, while villagers in the Zagros continued to emphasize animal products (Schoeninger

1981). Cohen and Armelagos (1984) considered a decline in the nutritional and health status of preagricultural populations to be a test implication of population pressure theories of agricultural origins, but different subsistence strategies in the Zagros and the Levant would have had different nutritional consequences.

Two critical nutrients of any diet are carbohydrates and protein. The cereals are a good source of carbohydrates and they are also fairly high in protein. This is particularly true of einkorn, whose starchy endosperm comprises a smaller proportion of the grain than that of the plumper tetraploid and hexaploid wheats, and thus contains more protein (Akroyd 1970). Unlike animal protein, however, cereal protein is incomplete. It is low in lysine, one of the essential amino acids required by the human body to manufacture protein.

It is here that legumes come to play a role in our understanding of early Holocene subsistence. Legumes are a significant part of the archaeobotanical assemblage at both preagricultural and early agricultural sites. Although cereals are low in lysine, pulses, such as lentil, chickpea, and pea, are rich in it; in addition, at least one clover has a relatively high lysine content (FAO 1970). Although it has generally been assumed that both pulses and small-seeded legumes contributed to the Epipaleolithic diet, the archaeological context of carbonized seeds does not allow us to assume that these items were consumed by people. The pulses were eventually domesticated, but the small-seeded legumes were not. One of the ironies of the archaeobotanical record is that domesticated pulses flourish at the moment that domesticated animals, a more complete source of protein, enter the PPNB subsistence system.

Although most models assume human populations try to maintain as balanced a diet as possible, they differ in their interpretation of subsistence practices. This stems, in part, from the difficulty of reconstructing ancient diets from archaeobotanical remains (Dennell 1979). If plant remains cannot be directly translated into a dietary reconstruction, how else can one evaluate ancient eating habits? Human skeletal remains provide a complementary line of evidence. In their review of Natufian remains, P. Smith et al. (1984) conclude that nutritional and health stresses did not change very much between Natufian and early agricultural times in the Levant. In particular, they find no evidence for nutritional stress due to periodic food shortages. Rather, there is a noticeable decline in nutritional health after the PPNB, following the establishment of agriculture.

Significance of Sedentism

The presence or absence of population pressure in the marginal zone defined the terms of the debate about the origin of agriculture for many years. As early as 1973, however, Flannery (1973) stressed that in the Near East sedentism seemed to be an important precursor to domestication, and probably to cultivation as well (see also Bar-Yosef and Kislev 1989). In this view, sedentism would have been permitted by the high productivity and storability of the wild cereals.

Sedentism did not develop uniformly across the Near East. Very early sedentism with increased dependence on plants is characteristic of the late Epipaleolithic in the Levant, where evidence for domesticated animals prior to PPNB times is scanty (for alternative views, see Legge 1972 and Moore 1982). Transhumance based on pastoral production seems to have been more common in the Zagros (e.g., at Ali Kosh). The integration of pastoral and agricultural economies probably contributed to the ultimate success of the early village farming way of life in the Near East.

There are several ways to view sedentism as an impetus to plant cultivation. Sedentary peoples may cultivate food in order to ensure a *reliable* food supply or to *increase* their food supply to satisfy growing social or dietary needs.

First, even without population growth, a sedentary population will eventually deplete the densest stands of wild cereals, making it necessary to engage in supplemental seed planting to ensure locally abundant and reliable harvests (Hayden 1981; Hassan 1981). In addition, there is no way to ensure good wild harvests on the same plot of land from one year to another. Presumably, people would initially have grown whatever cereal was at hand. Only later, as population densities increased, would they have selected the most productive crops available. This behavior could account for the rapid spread of emmer from its center of origin in the southern Levant at the expense of the apparently less productive einkorn (see Gill and Vear 1980:61).[2] Einkorn may have been reliable enough for early sedentary peoples but not productive enough for later, more populous villages.

Second, sedentarization may involve social and economic changes. A sedentary lifestyle requires forms of social control and mechanisms for dispute adjudication that differ from those appropriate to a mobile lifestyle (Bender 1978). The economic advantages of sedentism based on the harvesting and storage of a year's supply of grain, for example, would be obviated if villages were chronically rent by disputes. Unlike mobile foragers, sedentary people are usually not free to pick up and move if they get into an argument with their neighbor. In addition, they must maintain ties with other villagers for trade or marriage exchanges. Group burials seem to reflect social distinctions within Natufian communities in the Levant, and special treatment accorded to some individuals within burial groups seems to reflect status differences beyond age and sex (G. Wright 1978). Bender (1978) argues that dispute adjudication, social control, and increasingly complex social networks were maintained with feasting and gift-giving, which, in turn, were supported by surplus food production provided by plant cultivation.

Finally, sedentarization may have led to a reduction of certain factors, such as miscarriage and abortion, that limited population growth. As populations grew, people would have had to find, and ultimately produce, more food. Because it promoted population increase and exhaustion of the wild food supply, sedentism is a critical element of population pressure models of agricultural origins (e.g., Smith and Young 1972, 1983).

It is not yet possible to refute any of these theoretical positions on the basis of present archeological evidence. The importance of any one factor would tend to vary in different cultural settings. It may be most accurate to integrate aspects of all three factors, although in the Near East the effects of high population densities do not become apparent until after cultivation begins. An archaeobotanical approach can help us evaluate several aspects of these models of early cultivation and agriculture.

The Role of Archaeobotanical Evidence

Archaeobotanical evidence is used in three ways in investigating agricultural origins in the Near East. First, it has been used to generate models of ancient diet. Second, it provides a means to assess the environmental conditions and effects of early agricultural practices. Third, insofar as charred plant remains originate in fodder or dung fuel, archaeobotanical evidence may tell us more about animal husbandry practices than about either human food or the environment.

Older interpretations of botanical remains suggest that subsistence activities became more broadly based and more labor intensive as traditional sources of food, primarily large game animals, became scarce. Flannery extrapolated the idea of the "broad spectrum revolution" from plant and animal remains at the early agricultural site of Ali Kosh (Flannery 1969). In his view, the enormous quantities of hard-to-process, small-seeded legumes represented an important protein source at Ali Kosh during the Bus Mordeh phase, an archaeobotanical example of a broad resource base. Even unnutritious wheat spikelet forks were believed to have been eaten (Helbaek 1969). I suggest that high numbers and high proportions of small-seeded legumes relative to cultivated pulses and cereals need not be surprising. First, the assemblage could represent the burned debris of crop processing activities; small-seeded legumes could, after all, be quite prevalent in grain fields and be removed at a settlement through winnowing (Hillman 1981). Second, many of the small-seeded legume plants produce more seeds than the large-seeded legumes and cereals (Stevens 1932). Third, dung and fodder are possible sources of legumes and other weedy plants (Miller 1984; compare Moore 1982). In all three cases, one would expect an archaeobotanical assemblage to have a relatively high proportion of weed seeds and rachis fragments relative to cultigens. In fact, Helbaek (1969) noted that small-seeded legumes and spikelet forks were more highly correlated with one another within samples than were cereals and spikelet forks, a result expected for the remains of animal fodder or food processing debris, not human food. These possibilities are all quite plausible explanations for the plant assemblage at Ali Kosh, where the inhabitants had access to domesticated wheat for carbohydrate and domesticated sheep and goats, as well as wild animals, for protein.

One cannot explain the diverse charred seed assemblages from Epipaleolithic and Aceramic steppe sites, such as Abu Hureyra and Aswad, as the fuel residues of the dung of domesticated animals. Consequently, Flannery's model might be supported by the plant assemblage from Epipaleolithic Abu Hureyra, which Hillman et al. (1989) believe represent the remains of human food. As suggested above, however, plausible alternative explanations for such assemblages are that the seeds came from plants burned directly as fuel or were byproducts of harvested cereals (see also Hillman 1981).

A second use of archaeobotanical data has been to help characterize the environmental setting of early field systems. For example, Aswad, which is located in a region outside the present natural habitat of wild emmer, yielded evidence of very early domesticated emmer but none of its wild predecessor (van Zeist and Bakker-Heeres 1985). The presence of a wild plant outside its range would be prima facie evidence for its cultivation, if one could directly infer the ancient range of the plant. Interpretations do, however, change as new evidence refines paleoenvironmental reconstructions. Thus, the early occurrence of wild einkorn at Mureybit in northwestern Syria was used as evidence supporting the marginal zone hypothesis until a new environmental reconstruction suggested that the ancient climate was wetter at that time than it is today (van Zeist and Woldring 1980).

Another ecological approach views farming in the natural habitat zone as an attempt to reproduce the fairly pure, dense stands of wild cereals on similar terrain; in the marginal zone, cultivation would have been restricted to favorable microenvironments (Flannery 1973). Sherratt (1980) contends that early farmers would have engaged in small-field horticulture near predictable water sources rather than clear large expanses for unpredictable rainfall agriculture. He suggests that the location of many preagricultural and early agricultural settlements in areas with a high water table represents such a farming practice. The locational evidence is persuasive, although the archaeobotanical evidence used to support the argument is less so. The presence of sedge seeds mixed with grain is said to show that grain fields at Ali Kosh were located directly on the edge of marshy areas (Helbaek 1969). But there may as easily have been a fortuitous conjunction of grain and sedge in an animal's diet.

Although pollen studies are the primary source of information about ancient vegetation, complementary information can be gleaned from the archaeobotanical record. Charcoal (or lack thereof) provides important clues to the vegetation around an ancient settlement. The vegetation closest to the settlement is that most directly influenced by human activities, such as cutting wood for fuel. In more arid and/or deforested areas, one would expect to find charred remains of shrubs or riverine woods. In richer environments, one would expect to find burned evidence of climax vegetation. For example, oak charcoal seems to predominate at Çayönü in southeastern Turkey (van Zeist 1972) and at Jarmo in northeastern Iraq, both of which are located in relatively rich environments (L.S. Braidwood et al. 1983:541). By contrast, high proportions of riverine types (poplar and tamarisk), which reflect overall dry conditions, are found at sites like Jericho (Western 1971) and Mureybit (van Zeist 1970), which have highly localized sources of moisture.

The value of archaeobotanical data for assessing aspects of animal husbandry has been alluded to in this chapter, but no such studies have been conducted. A broader, related issue concerning the nature of the archaeobotanical record, however, may be addressed.

Although it was perceptive of Helbaek to suggest that even in an agricultural setting wild plants could serve as human food, it is no longer acceptable to assume automatically that seeds in preagricultural situations are either food remains or accidental inclusions in the archaeobotanical record. In order to assess the relative importance of agricultural and wild plant products in an ancient economy, one has to understand the archaeological context and preservation circumstances of the assemblage. If seeds occur in fairly pure caches, they are likely to represent intentionally harvested food, which has been accidentally burned. If, as is more generally the case, they are sparsely distributed in very mixed assemblages, they may represent the remains of fuel, fodder, refuse, or accidental inclusions. It is clear that the simple identification of a plant as wild or domestic is insufficient to determine its role in the ancient economic system. Regardless of whether or not plants served as human food, plant remains do provide valuable economic evidence. For these reasons, the usefulness of archaeobotanical data is greatly enhanced when sample-by-sample inventories of assemblages and information about the nature of the archaeological deposits are provided in final reports. An understanding of depositional circumstances is clearly critical for testing theories of agricultural origins.

Conclusions

The archaeobotanical record of early plant cultivation and domestication in the Near East is frustratingly incomplete. Much important early work concentrated on identifying the wild ancestors of the cultigens and tracing the morphological changes resulting from domestication. As excavators became more confident that plant remains could be recovered through systematic sampling and flotation, the volume of archaeobotanical remains available for study increased. Our understanding of the environmental and cultural setting of early agriculture has grown tremendously since the publication of the first Jarmo report in 1960, but there is much archaeobotanical work still to be done.

First, there is simply not enough evidence from Epipaleolithic sites to provide a detailed picture of preagricultural plant use. The low density of charred material from these early sites is a real problem, but it is still important to seek these remains.

Second, we need material from more sites in each environmental zone in order to assess the typicality of the sites already excavated. A larger number of excavated sites would allow comparisons to be made between zones as well. For example, the early appearance of domesticated animals in the Zagros area could have established conditions for the development of agriculture there that differed from those in the Levant and Syria (see Hole 1984).

Third, comparative charcoal analyses would help establish the state of the local vegetation and enable us to trace land clearance and assess the degree of environmental disturbance caused by the new, permanent settlements.

Archaeobotanical work on the origins of plant cultivation in the Near East still suffers from an inadequate database. This situation is being ameliorated by continuing fieldwork at early village sites, as well as by necessary but underfunded work in the laboratory. Archaeobotanists can take advantage of a long tradition of interdisciplinary research by archaeologists, physical anthropologists, palynologists, and others to help interpret their plant remains. In turn, archaeobotanists can help put theories of the origins of cultivation on a firmer footing.

I have chosen to end this account in the seventh millennium B.C. By that time, the economic and dietary importance of food production was firmly established. The new subsistence base led to morphological changes in the plants, the creation of a new agricultural niche for both people and plants, and higher population densities supported by agricultural production. Agricultural development did not stand still. In subsequent millennia new crops were added; some of the older ones that were not stressed in this article became more important; and new technologies, such as irrigation, took on increasing importance (Miller 1991).

Although I provide no definitive conclusions about the origins of cultivation in the Near East, it is likely that by the time plants were domesticated human populations were already dependent on cereals for a substantial portion of their diet. Human groups, increasingly dependent on agricultural production for their livelihood and on larger, denser populations to meet their social needs, could no longer easily revert to a foraging way of life.

Acknowledgments

I would like to thank Patty Jo Watson, Mary Kennedy, Gail Wagner, and Stuart Fleming for reading and commenting on earlier drafts of this paper. I would also like to thank Jack Harlan, Willem van Zeist, and P. E. L. Smith for directing me to recently published sources.

Notes

1. Philip Smith (personal communication 1986) points out that Ganj Dareh is a significant exception for the Zagros Mountain region. At that site, evidence for animal control appears after the earliest evidence for the use of domesticated barley.

2. Jack Harlan (personal communication 1985) points out that growing conditions, chaff weight, and wide yearly fluctuations in yield of both wild and domesticated einkorn and emmer make generalizations difficult.

References Cited

Akroyd, W. R.
  1970   Wheat in Human Nutrition. FAO Nutritional Studies 23. FAO, Rome.

Aurenche, O., J. Évin, and J. Gascó
  1987   Une séquence chronologique dans la proche orient de 12,000 à 3,700 B.C. et sa relation avec les données du radiocarbon. In Chronologies du Proche Orient/Chronologies in the Near East, edited by O. Aurenche, J. Évin, and F. Hours, pp. 21–31. BAR International Series 379(i). British Archaeological Reports, Oxford.

Aurenche, O., J. Évin, and F. Hours
  1987   Chronologies du Proche Orient/Chronologies in the Near East. BAR International Series 379(i). British Archaeological Reports, Oxford.

Bar-Yosef, O., and M. E. Kislev
  1989   Early Farming Communities in the Jordan Valley. In Foraging and Farming, The Evolution of Plant Exploitation, edited by D. R. Harris and G. C. Hillman, pp. 632–42. Unwin Hyman, London.

Bar-Yosef, O., and J. C. Vogel
  1987   Relative and Absolute Chronology of the Epi-Paleolithic in the Southern Levant. In Chronologies du Proche Orient/Chronologies in the Near East, edited by O. Aurenche, J. Évin, and F. Hours, pp. 219–45. BAR International Series 379(i). British Archaeological Reports, Oxford.

Bender, B.
  1978   Gatherer-Hunter to Farmer: A Social Perspective. World Archaeology 10:204–22.

Binford, L. R.
  1968   Post-Pleistocene Adaptations. In New Perspectives in Archaeology, edited by S. R. Binford and L. R. Binford, pp. 313–41. Aldine, Chicago.

  1983   On the Origins of Agriculture. In In Pursuit of the Past, edited by L. R. Binford, pp. 195–213. Thames and Hudson, New York.

Bohrer, V.
  1972   On the Relationship of Harvest Methods to Early Agriculture in the Near East. Economic Botany 26:145–55.

Bottema, S., and W. van Zeist
  1981   Palynological Evidence for the Climate History of the Near East, 50,000–6000 B.P. In Préhistoire du Levant, edited by J. Cauvin and P. Sanlaville, pp. 111–32. Colloques Internationaux du C.N.R.S. 598. Editions du CNRS, Paris.

Braidwood, L. S., R. J. Braidwood, B. Howe, C. A. Reed, and P. J. Watson
  1983   Prehistoric Archeology along the Zagros Flanks. Oriental Institute Publications 105. University of Chicago, Chicago.

Braidwood, R. J., and B. Howe
  1960   Prehistoric Investigations in Iraqi Kurdistan. Studies in Ancient Oriental Civilization 31. University of Chicago Press, Chicago.

Cauvin, J.
  1978   Les premiers villages de Syrie-Palestine du IXème siècle au VIIème millénaire avant J.C. Collection de la Maison de l'Orient Méditerranéan 4, Série archéologique 3. Lyon.

1979 Les fouilles de Mureybet (1971–1974) et leurs signification pour les origines de la sédentarisation au Proche Orient. *In* Archeological Reports from the Tabqa Dam Project—Euphrates Valley, Syria, edited by D. N. Freedman, pp. 19–48. Annual of the American Schools of Oriental Research 44, Cambridge.

Childe, V. G.
1969 New Light on the Most Ancient East. W. W.
[1952] Norton, New York.

Cohen, M. N., and G. J. Armelagos
1984 Paleopathology at the Origins of Agriculture: Editors' Summation. *In* Paleopathology at the Origins of Agriculture, edited by M. N. Cohen and G. J. Armelagos, pp. 585–601. Academic Press, Orlando.

Dennell, R. W.
1974 Botanical Evidence for Prehistoric Crop Processing Activities. Journal of Archaeological Science 1:275–84.

1979 Prehistoric Diet and Nutrition: Some Food for Thought. World Archaeology 11:121–35.

Edwards, P. C.
1989 Revising the Broad Spectrum Revolution and its Role in the Origins of Southeast Asian Food Production. Antiquity 63:225–46.

Edwards, P. C., S. J. Bourke, S. M. Colledge, J. Head, and P. G. Macumber
1988 Late Pleistocene Prehistory in Wadi al-Hammeh, Jordan Valley. *In* The Prehistory of Jordan: The State of Research in 1986, edited by A. N. Garrard and H. G. Gebel, pp. 525–65. BAR International Series 396. British Archaeological Reports, Oxford.

FAO
1970 Amino-Acid Content of Foods. FAO Nutritional Studies 24. FAO, Rome.

Feldman, M.
1976 Wheats. *In* Evolution of Crop Plants, edited by N. W. Simmonds, pp. 120–28. Longman, London.

Fisher, W. B.
1978 The Middle East: A Physical, Social, and Regional Geography, 7th ed. Methuen, London.

Flannery, K. V.
1969 Origins and Ecological Effects of Early Domestication in Iran and the Near East. *In* The Domestication and Exploitation of Plants and Animals, edited by P. J. Ucko and G. W. Dimbleby, pp. 73–100. Aldine, Chicago.

1973 The Origins of Agriculture. Annual Review of Anthropology 2:271–310.

French, D. H., with G. C. Hillman, S. Payne, and R. J. Payne
1972 Excavations at Can Hasan III 1969–1970. *In* Papers in Economic Prehistory, edited by E. S. Higgs, pp. 181–90. Cambridge University Press, Cambridge.

Gill, N. T., and K. C. Vear
1980 Agricultural Botany, Monocotyledonous Crops, vol. 2. Duckworth, London.

Gowlett, J. A. J., and R. E. M. Hedges
1987 Radiocarbon Dating by Accelerator Mass Spectrometry: Applications to Archaeology in the Near East. *In* Chronologies du Proche Orient/Chronologies in the Near East, edited by O. Aurenche, J. Évin and F. Hours, pp. 121–44. BAR International Series 379(i). British Archaeological Reports, Oxford.

Harlan, J. R., and D. Zohary
1966 Distribution of Wild Wheats and Barley. Science 153: 1074–1080.

Hassan, F. A.
1976 Diet, Nutrition, and Agricultural Origins in the Near East. *In* Origine de l'élevage et de la domestication, edited by E. Higgs, pp. 227–47. IXème Congrès Union Internationale des Sciences Préhistoriques et Protohistoriques. Nice.

1981 Demographic Archaeology. Academic Press, New York.

Hayden, B.
1981 Research and Development in the Stone Age. Current Anthropology 22:519–48.

Helbaek, H.
1960 The Palaeoethnobotany of the Near East and Europe. *In* Prehistoric Investigations in Iraqi Kurdistan, by R. J. Braidwood and B. Howe, pp. 99–118. Studies in Oriental Civilization 31. University of Chicago Press, Chicago.

1966 Pre-Pottery Neolithic Farming at Beidha. Palestine Exploration Quarterly 98:61–66.

1969 Plant-Collecting, Dry-Farming, and Irrigation Agriculture in Prehistoric Deh Luran. *In* Prehistory and Human Ecology of the Deh Luran Plain, by F. Hole, K. V. Flannery, and J. A. Neely, pp. 383–426. Museum of Anthropology Memoir 1, University of Michigan, Ann Arbor.

1970 The Plant Husbandry of Hacilar. *In* Excavations at Hacilar, by James Mellaart, pp. 189–244. University Press, Edinburgh.

Henry, D. O.
1989 From Foraging to Agriculture. University of Pennsylvania Press, Philadelphia.

Hillman, G. C.
1978 On the Origins of Domestic Rye—*Secale cereale:* The Finds from Aceramic Can Hasan III in Turkey. Anatolian Studies 28:157–74.

1981 Reconstructing Crop Husbandry Practices from Charred Remains of Plants. *In* Farming Practice in British Prehistory, edited by R. Mercer, pp. 123–62. University Press, Edinburgh.

Hillman, G. C., S. M. Colledge, and D. R. Harris
1989 Plant-Food Economy during the Epipaleolithic Period at Tell Abu Hureyra, Syria: Dietary Diversity, Seasonality, and Modes of Exploitation. *In* Foraging and Farming, The Evolution of Plant Exploitation, edited by D. R. Harris and G. C. Hillman, pp. 240–68. Unwin Hyman, London.

Hillman, G. C., and M. S. Davies
1990 Measured Domestication Rates in Wild Wheats and Barley under Primitive Cultivation, and their Archaeological Implications. Journal of World Prehistory 4:157–222.

Hole, F.
1984 A Reassessment of the Neolithic Revolution. Paléorient 10(2):49–60.

Hole, F., K. V. Flannery, and J. A. Neely
1969 Prehistory and Human Ecology of the Deh Luran Plain. Museum of Anthropology Memoir 1. University of Michigan, Ann Arbor.

Hopf, M.
1969 Plant Remains and Early Farming at Jericho. *In* Domestication and Exploitation of Plants and Animals, edited by P. J. Ucko and G. W. Dimbleby, pp. 355–59. Aldine, Chicago.

1983 Jericho Plant Remains. *In* Excavations at Jericho. vol. 5, edited by K. M. Kenyon and T. A. Holland, pp. 576–621. British School of Archaeology in Jerusalem, London.

Hopf, M., and O. Bar-Yosef
1987 Plant Remains from Hayonim Cave, Western Galilee. Paléorient 13(1):117–20.

Kislev, M. E.
1984 Emergence of Wheat Agriculture. Paléorient 10(2):61–70.

1985 Early Neolithic Horsebean from Yiftah'el, Israel. Science 228:319–30.

1988 Nahal Hemar Cave, Desiccated Plant Remains: An Interim Report. 'Atiqot 18:76–81.

1989 Pre-Domesticated Cereals in the Pre-Pottery Neolithic A Period. *In* People and Culture Change, edited by I. Hershkovitz, pp. 147–51. BAR International Series 508(i). British Archaeological Reports, Oxford.

Kislev, M. E., O. Bar-Yosef, and A. Gopher
1986 Early Neolithic Domesticated and Wild Barley from the Netiv Hagdud Region in the Jordan Valley. Israel Journal of Botany 35:197–201.

Köhler-Rollefson, I.
1988 The Aftermath of the Levantine Neolithic Revolution in the Light of Ecological and Ethnographic Evidence. Paléorient 14(1):87–93.

Körber-Grohne, U.
1981 Distinguishing Prehistoric Cereal Grains of *Triticum* and *Secale* on the Basis of Their Surface Patterns Using the Scanning Electron Microscope. Journal of Archaeological Science 8:197–204.

Legge, A. J.
1972 Prehistoric Exploitation of the Gazelle in Palestine. *In* Papers in Economic Prehistory, edited by E. S. Higgs, pp. 119–24. Cambridge University Press, Cambridge.

Lisicyna, G.
1983 Die ältesten paläoethnobotanischen Funde in Nordmesopotamien. Zeitschrift für Archäologie 17:31–38.

Miller, N. F.
1984 The Use of Dung as Fuel: An Ethnographic Example and an Archaeological Application. Paléorient 10(2):71–79.

1991 The Near East. *In* Progress in Old World Palaeoethnobotany, edited by W. van Zeist, K. Wasylikowa, and K.-E. Behre, pp. 133-60. Balkema, Rotterdam.

Miller, N. F., and T. L. Smart
1984 Intentional Burning of Dung as Fuel: A Mechanism for the Incorporation of Charred Seeds into the Archeological Record. Journal of Ethnobiology 4:15–28.

Moore, A. M. T.
1982 Agricultural Origins in the Near East: A Model for the 1980s. World Archaeology 14:224–36.

1985 The Development of Neolithic Societies in the Near East. *In* Advances in World Archaeology, vol. 4, edited by F. Wendorf and A. E. Close, pp. 1–69. Academic Press, New York.

Moore, A. M. T., G. C. Hillman, and A. J. Legge
1975 The Excavation of Tell Abu Hureyra in Syria: A Preliminary Report. Proceedings of the Prehistoric Society 41:50–77.

Noy, T., A. J. Legge, and E. S. Higgs, with R. W. Dennell
1973 Recent Excavations at Nahal Oren, Israel. Proceedings of the Prehistoric Society 39:75–99.

Noy, T.
1988 Gilgal I—An Early Village in the Lower Jordan Valley, Preliminary Report of the 1987 Winter Season. Israel Museum Journal 7:113–14.

Rathbun, T. A.
1984 Skeletal Pathology from the Paleolithic through the Metal Ages in Iran and Iraq. *In* Paleopathology at the Origins of Agriculture, edited by M. N. Cohen and G. J. Armelagos, pp. 137–67. Academic Press, Orlando.

Renfrew, J. M.
1969 The Archaeological Evidence for the Domestication of Plants: Methods and Problems. *In* The Domestication and Exploitation of Plants and Animals, edited by P. J. Ucko and G. W. Dimbleby, pp. 149–72. Aldine, Chicago.

1973 Palaeoethnobotany. Columbia University Press, New York.

Rollefson, G. O., A. H. Simmons, M. L. Donaldson, W. Gillespie, Z. Kafafi, I. U. Köhler-Rollefson, E. McAdam, S. L. Rolston, and M. K. Tubb
1985 Excavation at the Pre-Pottery Neolithic B Village of 'Ain Ghazal (Jordan), 1983. Mitteilungen der Deutschen Orient-Gesellschaft zu Berlin 117:69–116.

Schick, T.
1986 Perishable Remains from the Nahal Hemar Cave. Journal of the Israel Prehistoric Society 10:95–97.

Schoeninger, M.
1981 The Agricultural "Revolution": Its Effect on Human Diet in Prehistoric Iran and Israel. Paléorient 7:73–91.

1982 Diet and the Evolution of Modern Human Form. American Journal of Physical Anthropology 58:37–52.

Sheratt, A.
1980 Water, Soil, and Seasonality in Early Cereal Cultivation. World Archaeology 11:313–30.

Sillen, A.
1981 Strontium and Diet at Hayonim Cave. American Journal of Physical Anthropology 56:131–37.

Smith, P.
1972 Diet and Attrition in the Natufians. American Journal of Physical Anthropology 37:233–38.

Smith, P., O. Bar-Yosef, and A. Sillen
1984 Archaeological and Skeletal Evidence for Dietary Change During the Late Pleistocene/Early Holocene in the Levant. *In* Paleopathology at the Origins of Agriculture, edited by M. N. Cohen and G. J. Armelagos, pp. 101–36. Academic Press, Orlando.

Smith, P. E. L., and T. C. Young, Jr.
1972 The Evolution of Early Agriculture and Culture in Greater Mesopotamia: A Trial Model. *In* Population Growth, edited by B. Spooner, pp. 1–59. MIT Press, Cambridge.

1983 The Force of Numbers: Population Pressure in the Central Western Zagros 12,000–4500 B.C. *In* The Hilly Flanks and Beyond, Essays on the Prehistory of Southwestern Asia presented to Robert J. Braidwood, edited by T. C. Young, P. E. L. Smith, and P. Mortensen, pp. 141–61. Oriental Institute, University of Chicago, Chicago.

Stevens, O. A.
1932 The Number and Weight of Seeds Produced by Weeds. American Journal of Botany 19:784–94.

Stewart, B.
1976 Paleoethnobotanical Report—Çayönü 1972. Economic Botany 30:219–25.

Stuiver, M., B. Kromer, B. Becker, and C. W. Ferguson
1986 Radiocarbon Age Calibration Back to 13,300 Years B.P. and the 14C Age Matching of the German Oak and U.S. Bristlecone Pine Chronologies. Radiocarbon 28(2B):969–79.

Sumerian Agriculture Group
1983 Bulletin on Sumerian Agriculture, vol. 3. Cambridge, England.

van Zeist, W.
1970 The Oriental Institute Excavations at Mureybit, Syria: Preliminary Report on the 1965 Campaign. Part III: The Paleobotany. Journal of Near Eastern Studies 29:167–76.

1972   Palaeobotanical Results of the 1970 Season at Çayönü, Turkey. Helinium 12:1–19.

1986   Some Aspects of Early Neolithic Plant Husbandry in the Near East. Anatolica 15: 49–67.

van Zeist, W., and J. A. H. Bakker-Heeres
1979   Some Economic and Ecological Aspects of the Plant Husbandry of Tell Aswad. Paléorient 5:161–69.

1985   Archaeobotanical Studies in the Levant 1. Neolithic Sites in the Damascus Basin: Aswad, Ghoraifé, Ramad. Palaeohistoria 24:165–256.

1986a   Archaeobotanical Studies in the Levant 2. Neolithic and Halaf Levels at Ras Shamra. Palaeohistoria 26:151–70.

1986b   Archaeobotanical Studies in the Levant 3. Late-Paleolithic Mureybit. Palaeohistoria 26: 171–99.

van Zeist, W., and S. Bottema
1966   Palaeobotanical Investigations at Ramad. Annales archéologiques arabes syriennes 16:179–80.

1982   Vegetational History of the Eastern Mediterranean. In Palaeoclimates, Palaeoenvironments and Human Communities in the Eastern Mediterranean Region in Later Prehistory, edited by J. L. Bintliffe and W. van Zeist, pp. 277–321. BAR International Series 133. British Archaeological Reports, Oxford.

van Zeist, W., and W. A. Casparie
1968   Wild Einkorn Wheat and Barley from Tell Mureybit in Northern Syria. Acta Botanica Neerlandica 17:44–53.

van Zeist, W., P. E. L. Smith, R. M. Palfenier-Vegter, M. Suwijn, and W. A. Casparie
1986   An Archaeobotanical Study of Ganj Dareh Tepe, Iran. Palaeohistoria 26:201–24.

van Zeist, W., and W. Waterbolk-van Rooijen
1985   The Palaeobotany of Tell Bouqras, Eastern Syria. Paléorient 11(2):131–47.

van Zeist, W., and H. Woldring
1980   Holocene Vegetation and Climate of Northwestern Syria. Palaeohistoria 22:111–25.

Watson, P. J.
1983   A Note on the Jarmo Plant Remains. In Prehistoric Archeology along the Zagros Flanks, edited by L. S. Braidwood, R. J. Braidwood, B. Howe, C. Reed, and P. J. Watson, pp. 501–503. Oriental Institute Publications 105. University of Chicago, Chicago.

Western, A. C.
1971   The Ecological Interpretation of Ancient Charcoals from Jericho. Levant 3:31–40.

Wright, G. A.
1978   Social Differentiation in the Early Natufian. In Social Archeology: Beyond Subsistence and Dating, edited by C. L. Redman, M. J. Berman, E. V. Curtin, W. J. Langhorne, N. M. Versaggi, and J. C. Wanser, pp. 201–33. Academic Press, New York.

Wright, H. E.
1977   Environmental Change and the Origin of Agriculture in the Old and New Worlds. In Origins of Agriculture, edited by C. A. Reed, pp. 281–318. Mouton, The Hague.

1980   Climatic Change and Plant Domestication in the Zagros Mountains. Iran 18:145–48.

1983   Climatic Change in the Zagros Mountains—Revisited. In Prehistoric Archeology along the Zagros Flanks, edited by L. S. Braidwood, R. J. Braidwood, B. Howe, C. A. Reed, and P. J. Watson, pp. 505–10. Oriental Institute Publications 105. University of Chicago, Chicago.

Zohary, D.
1989   Domestication of the Southwest Asian Neolithic Crop Assemblage of Cereals, Pulses, and Flax: The Evidence from the Living Plants. In Foraging and Farming, the Evolution of Plant Exploitation, edited by D. R. Harris and G. C. Hillman, pp. 358–73. Unwin Hyman, London.

Zohary, D., and M. Hopf
1973   Domestication of Pulses in the Old World. Science 182:887–94.

1988   Domestication of Plants in the Old World. Clarendon Press, Oxford.

Zohary, D., and P. Spiegel-Roy
1975   Beginnings of Fruit Growing in the Old World. Science 187:319–27.

Zohary, M.
1973   Geobotanical Foundations of the Middle East. Gustav Fischer, Stuttgart.

4

Indigenous African Agriculture

Studies on the origins of indigenous African agriculture have been much neglected. Investigations of agricultural origins and dispersals have been rather massive in Europe, the Near East, Japan, and North America, substantial in China and Mesoamerica, and just beginning to unfold in Southeast Asia, the South Pacific, and South America. Yet research in Africa remains inadequate and too unsystematic to provide more than a tenuous outline of events. We know that an indigenous African agriculture did emerge. The list of plants that were domesticated is impressive and includes all of the traditional classes, such as cereals, grain legumes, vegetables, oil seed crops, fibers, drugs, narcotics, fish poisons, and magic and ritual plants. The African inventory is short, however, on tree fruits and ornamentals compared to Asia, the South Pacific, and tropical America. The agricultural systems that evolved supported extensive village farming populations with local markets and regional trading centers and gave rise to the high cultures of Nok, Benin, Ghana, Mali, and other Sudanic and East African kingdoms.

This much we know, but when, where, and by what peoples agricultural systems developed remains obscure largely for want of systematic research on the problem. It is not so much that archaeological research has not been conducted in Africa, although the volume may be less for the area than on other continents. Rather, the Africanist archaeologists have had other foci and have been examining other time ranges. One substantial group has been searching for human origins, stimulated by the Leakeys, Dart, Broom, Howell, Bourliere, and others. Another major contingent has been preoccupied with Bantu Iron Age expansion. One group of studies remains too early and the other too late to help unravel the story of African agriculture.

It is possible that in Africa agriculture evolved so gradually that there was no significant change in the toolkit or other artifactual remains. The more or less "Mesolithic" artifact assemblages may have persisted unchanged while food procurement shifted from a primarily hunting-fishing-gathering economy to a food-producing one. For whatever reasons, the scarcity of archaeological evidence makes it necessary to rely more than we would like on evidence from the plants themselves. Because the several crops have different ecological adaptations and distinct geographical distributions, such evidence tells us something about where events took place, but the time range remains to be established. In this chapter I attempt to

JACK R. HARLAN

59

sketch the prehistoric setting in a general way, and then discuss the various agricultural complexes that evolved.

Prehistoric Setting

Evidence for past climatic changes is relatively conspicuous in Africa. The Sahara has expanded and contracted. Stabilized dunes may be found well into the broadleaved savanna zone, and pluvials occurred when most of the Sahara was inhabited by human populations. Lake Chad was once ten times its present size and Lake Rudolph was once so deep that water spilled into the Nile watershed (this requires a rise in lake level of some 80 m or more). The Nile itself has sometimes been much higher and sometimes much lower than it is now. Similar expansions and contractions of climatic regimes occurred in southern Africa, but not necessarily in synchrony or symmetry with northern Africa.

Intimate associations between plants and people may go back to the very roots of human origins. Our nearest surviving relatives are largely vegetarian. Grinding stones and sharp flakes with sheen characteristic of ancient sickle blades have been found in southern Egypt dating to some 18,000 years ago (Wendorf et al. 1979, 1980) (Fig. 4.1). The crop seeds associated with these implements have turned out to fall within the historical time range and can no longer be considered evidence for early agriculture, but the toolkit suggests an intensive kind of gathering and use of plant foods. Egypt was in a hyperarid phase at the time.

Grinding stones and sickle blades with sheen appear again along the Nile floodplain from about 14,500 to 11,000 years ago. These probably represent a later phase of the same lifestyle. By about 8,000 years ago a pluvial had set in and people ranged widely across the Sahara. Pollen samples and remnants of flora on the African massifs indicate a southward deflection of the Mediterranean winter rainfall regime. Shallow lakes that expand during the rainy season and shrink in the dry season were common throughout the Sahara at that time. People settled in camps or even in small villages near the shores of these lakes, fished, and hunted hippopotamus and crocodiles, as well as antelope, giraffe, etc., in the uplands. They also ground something, probably wild or domesticated cereal grains, with grinding stones.

By 7,000 years ago pastoralists with flocks of sheep, goats, and cattle roamed the Sahara. They also fished the playa lakes, hunted, and used grinding stones. Sickle blades with sheen are also found from this period. Precisely when people actually began to grow plants on purpose for food we do not know. The evidence is not yet available, but it is evident that an intimate association between people and plants was in effect for many thousands of years before significant changes in the mode of living took place.

It is suggestive that techniques of décrue agriculture are characteristic of indigenous African agriculture and are practiced widely in sub-Saharan zones (Harlan and Pasquereau 1969). It seems likely that the best place to learn décrue technique would be at the playa lakes of the Sahara that rose during the rains and sank in the dry season. Sowing seeds or transplanting seedlings in the moist soil as the waters recede is a technique easily learned, but only if the waters rise and fall with regularity.

It is also suggestive that the first farmers for whom we have archaeological evidence in Egypt arrived about 6,500 years ago equipped with sheep, goats, and cattle, and a cool-season complex of crops (e.g., barley, emmer, flax, lentil, chickpea, etc.), but their toolkits were African and not Near Eastern. They also knew the arts of décrue agriculture and how to exploit the yearly flood of the Nile.

By the middle of the third millennium B.C. (4,500 years ago), the effects of desiccation became noticeable and populations began to move out of the Sahara. The process accelerated about 2000 B.C. and much of the region was deserted by 1000 B.C. Presumably, many of the people moved southward toward more favorable rainfall regimes, but the rains in the south were summer rains and not appropriate for the crops of North Africa.

The prehistoric agriculture of the Nile Valley, however, consists of cool-season crops that are sown in the fall during the décrue of the Nile. These are not the tropical crops of indigenous sub-Saharan African agriculture. The archaeological record of these is extremely scanty, inadequate for tracing origins and dispersals, and generally too late to help much (Phillipson 1982). Camps (1969) identified two pollen grains from the Ameki site in the Hoggar as belonging to pearl millet (*Pennisetum glaucum*) dated to between 6000 and 4500 B.C. Grass pollen is notoriously difficult to identify and the validity of this evidence has been questioned. Clark (see Shaw 1976) has reported *Brachiaria* at 4000 B.C. and sorghum at 2000 B.C. at Adrar Bous in the Aïr region of the Sahara.

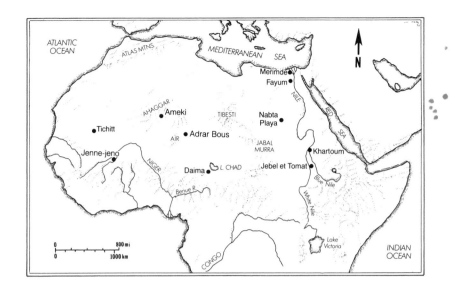

Fig. 4.1. Location of African archaeological sites mentioned in the text.

The *Brachiaria* may or may not have been domesticated.

Davies (1968) has also reported pearl millet from the Ntereso site in northern Ghana dated to about 1250 B.C., and Munson (1976) demonstrated a shift from gathering wild *Cenchrus,* or *kram-kram,* to the cultivation of pearl millet in the eleventh century B.C. in southern Mauritania. Otherwise, nearly all finds are dated after the first millennium A.D. A good review is provided by Shaw (1976).

In the bend of the Niger River, rice culture has been detected at Jenne-jeno dating to the first few centuries A.D., and foxtail (*Setaria*), sorghum, pearl millet, and purselane (*Portulaca*) were added shortly thereafter (McIntosh and McIntosh 1979). Bedaux et al. (1978) also report botanical finds representing a complete agricultural regime from the same general region dating to 1000 A.D. and later.

It seems almost certain that African plants had been domesticated well before the dates of the recovered remains. Sorghum, pearl millet, and possibly finger millet appear in the archaeological record in India by 1000 B.C. and Vishnu-Mittre (1968, 1974) has reported even earlier finds. His identification of finger millet, however, has been questioned (Hilu et al. 1979). It seems that African archaeology has not yet provided evidence for the times and places of initial domestication of indigenous African plants. In a general way, the archaeological evidence suggests that activities related to plant domestication took place in a broad belt between the Sahara and the equator, but

at this point archaeology does not tell us much more than that. It is necessary to trace the origins of each crop as best as we can from the botanical evidence.

Botanical Evidence for Crop History

Despite the current lack of early archaeological evidence, some conclusions about the history of crop plants may be reached based on botanical, genetic, ecological, and geographical evidence. The origin of sorghum has been studied in some detail by these methods (Harlan and Stemler 1976). The first step in the study was to collect and classify the wild races (see Figure 4.2).

Four distinct races of wild *Sorghum bicolor* were identified: (1) arundinaceum, adapted to the forest zone, found primarily along stream banks and clearings, (2) virgatum, confined almost entirely to the floodplains of the Nile, (3) aethiopicum, found in the Kassala region of Sudan and more sparsely toward the west along the fringes of the Sahara, and (4) verticilliflorum, widespread and abundant in the savanna zones of eastern and southern Africa but poorly represented west of Nigeria. The arundinaceum race flourishes best in areas that are too wet and humid for cultivated sorghum to grow well and thus can be eliminated on morphological grounds as being the primary progenitor. All of the races belong to the same biological species and are fully fertile when hybridized. It is impossible, therefore, to rule out some

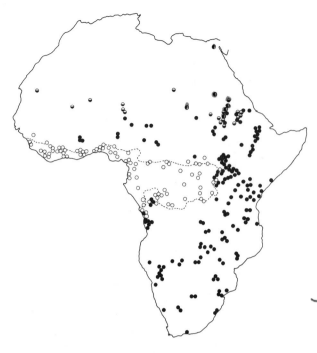

Fig. 4.2. Distribution of wild races of sorghum in Africa: clear circles, arundinaceum race; solid circles, verticilliflorum race; upper half clear, lower solid, aethiopicum race; right half clear, left solid, virgatum race; dotted line, approximate limits of tropical forest.

common type, but it is also found along the East African mountains where rainfall is high. This race is the best suited for high rainfall areas; it found its way to India where it is also grown under high rainfall conditions in the eastern and western Ghats. (3) Caudatum race is the dominant sorghum of eastern Nigeria, Chad, Sudan, and Uganda. It is quite distinctive and does not seem to have spread out of Africa until the nineteenth century. It is a major component of modern, high-yielding grain sorghums. (4) Kafir race is the dominant sorghum of southern Africa. It is a Bantu race and is sufficiently distinct to raise speculation that it might have been domesticated independently from local populations of verticilliflorum in southern Africa. It is also a major component of modern grain sorghums and reached India at a fairly early date. (5) Durra race is not especially abundant in Africa but is the dominant sorghum of the Near East and most of India and Pakistan. Morphologically, it is quite distinct. In Africa, it is largely confined to Islamicized cultivators along the fringes of the Sahara and in Ethiopia. For this reason, we have speculated that this race may have evolved in India from an early introduction of a

genetic contribution from any of the races, but the verticilliflorum race is clearly the primary progenitor of most sorghums.

If agriculture is relatively late in southern Africa compared to sub-Saharan regions, as the archaeological record tends to suggest, then the most logical region of sorghum domestication would be the northeastern quadrant of the continent.

The second step in the study was to collect and classify the cultivated sorghums of Africa. Figure 4.3 shows some of the results of that study, although there were too many collections to put them all on the map. Five basic races of cultivated sorghum plus races intermediate between any pair of them were identified. The basic races are quite clear-cut and distinct; the intermediate races blend characteristics of both parental types, as expected. The basic races are: (1) Bicolor race, a primitive kind that, of all the cultivated sorghums, most nearly resembles the wild forms. It is almost never the dominant sorghum of any district but is grown on a small scale almost anywhere that sorghum is cultivated. (2) Guinea race, basically a West African race where it is the most

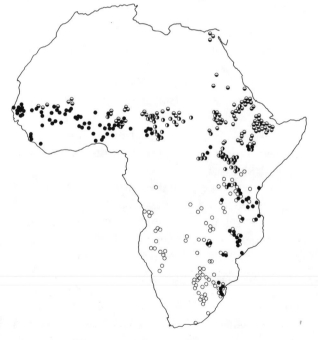

Fig. 4.3. Distribution of basic cultivated races of sorghum in Africa: clear circles, kafir race; solid circles, guinea race; upper half clear, lower solid, durra race; left half clear, right solid, caudatum race.

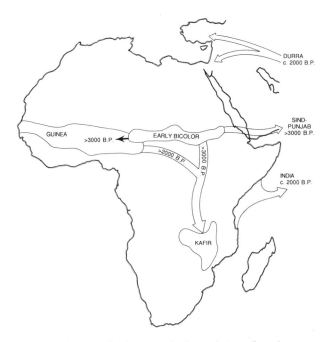

Fig. 4.4. Diagram of early events in the evolution of sorghum. Area labelled early bicolor is the probable area of initial domestication. The guinea race is basically West African, although it occurs in East Africa and India; the kafir race is characteristic of southern Africa. (After Harlan and Stemler 1976)

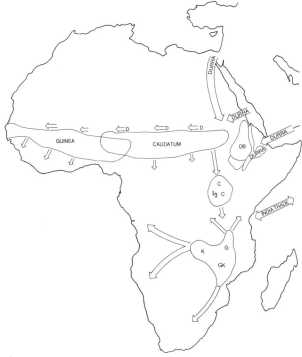

Fig. 4.5. Diagram of later events in the evolution of cultivated sorghum. (After Harlan and Stemler 1976)

bicolor and that it returned to Africa during the Arab conquests.

Based on this evidence, we have postulated an early domestication in the Chad-Sudan savanna, producing a primitive bicolor type. The early bicolor spread to West Africa where the guinea race evolved and to southern Africa where the kafir race evolved. It also went on to India where the durra race evolved. Figure 4.4 shows this early primary distribution of the primitive race.

A later distribution can also be postulated, as illustrated in Figure 4.5, which shows the late evolution of the caudatum race, the expansion of the kafir race, and the entry of the durra race from India and the Near East. While this history is speculative, it does fit the biological evidence (Harlan 1982).

The time of dispersal is still a problem. It is proposed that the early bicolor must have been widely dispersed before 3000 B.P. (see Figure 4.4). This date was selected as the most recent date possible because sorghum is reported in India by that time. How much time before 3000 B.P., however, we are as yet unable to say, but one thousand years earlier would not be

unreasonable. Only more archaeological research will resolve the matter. The sorghum from the Daima site in Nigeria dated to the ninth or tenth centuries A.D. is a caudatum type very much like that grown in the region today. The sorghum of Jebel el-Tomat, dated to the fourth century A.D., however, is a primitive bicolor. This is compatible with the proposed theory, but much more evidence is needed.

The present distributions of wild progenitors of other crops can provide useful clues, but changes in climate and intensive disturbance of habitats complicate the picture. Figure 4.6 shows the known distribution of wild races of pearl millet, together with the northern pearl millet belt. Within the darkly shaded area, pearl millet is the staff of life of the people. The crop is grown in the lightly shaded area, but it is less important there than sorghum. The wild forms actually grow in the Sahara and the source of the crop's drought resistance is evident. The distribution appears to be somewhat truncated, suggesting a wider distribution during the last pluvial followed by a retreat with the desiccation of the Sahara. It seems likely that the crop was originally domesticated in what is now a desert region (Brunken et al. 1977).

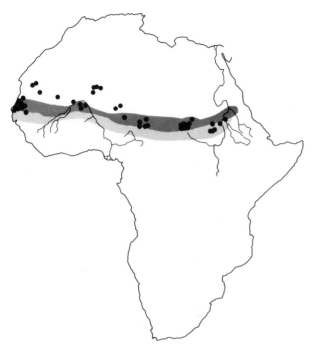

Fig. 4.6. Distribution of pearl millets. Circles, location of wild pearl millet, *Pennisetum violaceum;* dark shading, northern pearl millet belt; light shading, area in which pearl millet is grown, but in which sorghum is the dominant crop. The crop is also associated with the drier zones of East and southern Africa.

Fig. 4.7. Distribution of wild *Oryza barthii,* the progenitor of African rice.

Figure 4.7 shows the known distribution of the wild annual rice, which is the progenitor of the African rice domesticate. The species ranges much more widely than the crop ever did. Figure 4.8 shows the area where African rice is, or was, cultivated, and the yam belt where yam is the primary food of the people. These distributions tell us something about where these plants were domesticated but very little about when the process took place (Harlan 1982).

The Savanna Complex

The most widespread and characteristic African agricultural complex is one adapted to the savanna zones (Table 4.1). In the thorn bush savanna of the Sahel and at the margins of agriculture near the Kalahari Desert, pearl millet becomes the most important cultivated plant because of its great drought resistance. Here, too, the watermelon assumes a special role not only as a food plant but also as a source of water. Marginal farming is often integrated with livestock herding.

In the broadleaved savanna where rainfall is greater and more crops can be grown, sorghum is usually the most important cereal. In some parts of West Africa, African rice is the leading cereal and the primary source of calories. This rice is a different species from the more familiar Asian rice.

Several trees native to the savanna are heavily exploited. Although they may not be actually planted or cultivated, they are encouraged in the sense that they are seldom cut down, while other native trees are used for fuel and building materials. Thus, extensive subspontaneous stands of karite or white acacia may develop. The fruit of the karite yields an edible fat that is appreciated by local tribes, and the tree has a semi-sacred status. Among many tribes, it is possible to sell land and at the same time preserve ownership of the trees. The white acacia (*Acacia albida*) has the unique feature of shedding its leaves at the beginning of the rains and of leafing out at the start of the dry season. It does not, therefore, compete very much with cultivated plants, and the cultivators are convinced that crops grow better with acacia trees than without them. There is some research that indicates the cultivators are correct. As a legume, the acacia provides some nitrogen to the soil. The leaf litter is probably beneficial but, more often than not,

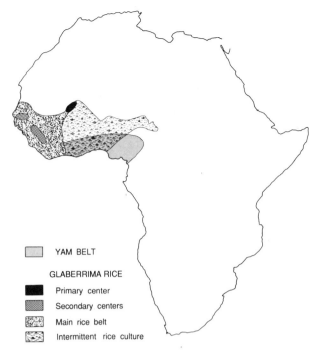

YAM BELT

GLABERRIMA RICE

Primary center

Secondary centers

Main rice belt

Intermittent rice culture

Fig. 4.8. Location of yam and rice belts in West Africa. (Adapted in part from Portères 1955)

the leaves are consumed by goats and sheep whose droppings are also good for the soil. Finally, soil moisture relations appear to be better on lands where the acacias grow. The locust bean tree (*Parkia*) be-

longs to the same family, and the sweet pods are edible and choice.

The baobab is another popular savanna tree that is encouraged and protected. It is usually found near villages, but it is not clear whether the trees are present because of the villages, or whether the villages were situated because of the trees. Its fruit and leaves may be eaten, fiber is stripped from the bark, and large, old trees, which are often hollow, can be used as cisterns to store water during the dry season. Because of the lack of surface water in the dry season, some villages would have to be abandoned if it were not for baobab cisterns.

Although rice is usually associated with the wet tropics, the African rice is clearly of savanna origin. The wild progenitor is an annual adapted to waterholes that fill up during the rains and dry out in the dry season. The life cycle fits this particular niche. Along more permanent streams and lakes a perennial species is more common. After the plant was domesticated, probably somewhere near the bend of the Niger River, people moved it into the forest where it could be grown in the uplands during the rainy season. A number of West African tribes became heavily dependent on rice as a staple. To these people, rice is the staff of life and the crop takes on a sacred character.

Fonio is also a plant revered by some tribes and

Table 4.1. Cultivated Plants of the Savanna Complex

Latin Name	Common Name and Use
Acacia albida	White acacia, subspontaneous, companion crop
Adansonia digitata	Baobab, multiple use, subspontaneous
Butyrospermum paradoxum	Karite, oil from fruit, encouraged
Colocynthis citrullus	Watermelon, edible fruits, potherb, cultivated, weedy
Corchorus olitorius	Tossa jute, pot herb, leaves and seedlings, cultivated
Digitaria exilis	Fonio, a cereal, cultivated and weedy
D. iburua	Black fonio, limited cultivation
Hibiscus cannabinus	Kenaf, a pot herb in Africa, cultivated
H. sabdarifa	Roselle, leaves and calyxes, cultivated
Lagenaria siceraria	Bottle gourd, widely used, cultivated
Oryza glaberrima	African rice, cereal, cultivated and weed races
Parkia biglobosa	Tree with sweet pods, locust bean, subspontaneous
Pennisetum glaucum	Pearl millet, cereal, the drier zones, cultivated and weedy
Polygala butyracea	Black beniseed, oil in seeds, West Africa
Solanum aethiopicum	African tomato, edible fruits
S. macrocarpon	A nightshade, fruits and leaves, weedy
Sorghum bicolor	Sorghum, cereal, sweet stalks, fodder, cultivated, weedy
Voandzeia subterranea	Earthpea, Bambara groundnut, pulse cultivated

incorporated into their religions and mythologies. It is a low-yielding cereal that requires considerable labor to cultivate because of its very small seeds, but it is a prestige crop, a chief's food, and is greatly appreciated for its culinary value. It is said to make the best couscous of all the cereals.

Kenaf, roselle, and tossa jute (*Corchorus olitorius*) are grown for fiber in Asia but are pot herbs in the hands of African farmers. The calyxes of roselle have a very pleasant acid taste and are frequently used in soups. The jute is harvested quite young and is highly mucilaginous when cooked. The Bambara groundnut (*Voandzeia subterranea*) is widely grown in the African savannas as a garden plant, not as a field crop. Seeds are produced underground, as in the peanut, but they are much larger. Its wild progenitor is known only from the Cameroon-Nigeria border area in broadleaved savannas.

The Forest Margin Complex

Crops of the forest margin complex are listed in Table 4.2. Some of these are true forest plants (e.g., the colas, grains of paradise, akee apples, guinea millet, and piasa), but there are reasons to believe that the agricultural complex evolved along the savanna-forest ecotone and then invaded the forest. One piece of evidence for this is the importance of the "annual" yams in the system. These tubers are propagated veg-

etatively from year to year and so are not truly annuals, but they produce tubers on an annual basis. At the start of the rains, a vine grows very rapidly from the tuber. The food reserves are mobilized and poured into the vine. Toward the end of the rains, the process is reversed and the food reserves of the vine are directed into new tubers underground. This growth cycle is an adaptation to savanna climates with pronounced dry seasons. The dormant tuber persists through the dry months and is protected underground from fires characteristic of the savanna environment. The true forest yams have perennial tubers, which are usually woody and not easily consumed.

Thus, while the yam-growing tribes of West Africa are forest people, their staple crop has clear ties to the savanna environment. The African oil palm is another example. Under plantation culture, it thrives in climatic regimes with an equitable distribution of rainfall and a minimal dry season, yet it cannot tolerate the dense shade of the tropical forest. As a wild plant, it favors the edge of the forest. Rice, as we have seen, is also a savanna plant that has moved into the wet forest zone.

Cowpea, on the other hand, has a somewhat reverse history. It is primarily a crop of the savanna, but the wild races are viny climbers of the forest belt. It is probable that there were two steps in the evolution of the cowpea, especially if a weedy race that colonizes disturbed habitats in the savanna was the

Table 4.2. Cultivated Plants of the Forest Margin Complex

Latin Name	Common Name and Use
Afromomum melegueta	Grains of paradise, spice, West Africa
Blighia sapida	Akee apple, aril eaten, forest, West Africa
Brachiaria deflexa	Guinea millet, cereal, Guinea only
Coffea canephora	Robusta coffee, forest zones
Cola acuminata	Cola nut, forest, West Africa
C. nitida et alia	Cola nut, forest, West Africa
Dioscorea bulbifera	Air potato, aerial tubers, wide distribution
D. rotundata	White guinea yam, forest, West Africa
Elaeis guineensis	Oil palm, forest, West Africa to Angola
Hibiscus esculentus	Okra, garden vegetable, common in West Africa
Kerstingiella geocarpa	A groundnut, limited culture, West Africa
Lablab niger	Hyacinth bean, now widespread, East Africa
Plectranthus exculentus	Hausa potato, tuber, West Africa
Solenostemon rotundfolius	Piasa, becoming rare, West Africa
Sphenostylis stenocarpa	Yampea, tuberous legume, West Africa
Telfaira occidentalis	Fluted gourd, fruit and seeds, West Africa
Vigna unguiculata	Cowpea, pulse, West Africa

race that was domesticated. If this is the case, one may visualize a degradation of the forest-savanna ecotone by human activities resulting in derived savanna formations. During the process, weedy cowpeas evolved from the wild forest race. In the second step, people began to cultivate the weedy forms and domesticated races evolved from them.

The cola nut is very important in the culture of many West African tribes. It is a mild narcotic and relieves pain and hunger. It is also a stimulant containing caffeine. The nut is often featured in welcoming ceremonies. When an honored guest enters a home and is seated on a stool, cola nuts are brought in a special ceremonial wooden bowl. (These carved cola nut bowls are prized by some collectors of African art.)

Among forest tribes, however, the most intense religious feelings are reserved for the yam (Coursey and Coursey 1971). Although there are usually several yam festivals in a year, the festival of the new yams is generally the most important and the most sacred. It is held at the first harvest, and digging yams is strictly forbidden until the ceremonies are properly completed. This is a resurrection and renewal festival with much ritual cleansing. The village is purified and stools are ritually washed and sprinkled with the blood of sacrificed animals. Libations of palm oil and palm wine are poured, and there is special music and dancing of a ceremonial nature. At the conclusion of the sacred rituals, the chief or a priest ceremonially digs the first yam of the season with a digging stick that must not be tipped with iron. After this, yams may be dug for food.

The division between rice cultivators and yam cultivators in West Africa is very sharp (Miege 1954). People on the left bank of the Bandama River in Ivory Coast grow yams and people on the right bank grow rice (Fig. 4.8). The rice-growing tribes, of course, have rice festivals and ceremonies and to them rice is a sacred plant. The reason for the sharply drawn line between the two groups is probably to be found in history. As we have seen, African rice is a savanna plant and had to be adapted to upland culture in the forest. Rice-growing groups were expansive and spread through the West African forests. At the Bandama River, however, they met the highly organized yam-growing tribes already well established and spread no further in that direction. Of course, the rice-growers cultivate some yams and the yam-growers may plant some rice, but the exogenous crop never approaches the cultural and religious position of the basic staple.

The African oil palm is another plant of great cultural significance. In traditional agriculture, it was never cultivated in plantations as is the case today. It was, rather, a useful plant that was seldom cut down even in slash-and-burn agricultural systems. In such systems, which are widely practiced in the West African forest zone, a patch of forest is cut during the dry season, the wood is allowed to dry, and then it is burned before the start of the rainy season. After one or two crops are grown in the cleared area, the patch is abandoned for the forest to rejuvenate. Ideally, a field will be abandoned to the forest for 15 to 20 years before another cycle of slash and burn is begun. With current population pressures, however, this chronology is no longer possible, and the period of bush fallow is now dangerously short. At any rate, slash-and-burn systems eventually result in degraded and derived savannas, and if oil palms are spared, they gradually become the dominant tree. Large, extensive stands of subspontaneous oil palms develop in this way.

The palm is popular not only because the edible oil is used in cooking but also because it yields a sweet juice that can be fermented into wine. Encouragement of the tree is reinforced by native beliefs and superstitions. It is said by some that if an oil palm tree is cut down, someone in the village will die. Another superstition helps to disseminate seeds. Fruits are beaten off the bunches in the forest because if this were done in the village, the flying fruits would symbolize people leaving the village. The two beliefs together assure that the oil palm will prosper even if no one ever plants a seed or establishes a plantation.

The Ethiopian Complex

A small group of crops was evidently domesticated in Ethiopia and adjacent highlands (Table 4.3). Vavilov includes Ethiopia as one of his eight centers of origin, but most of the crops he lists actually originated elsewhere and had secondary centers of diversity in Ethiopia (Harlan 1969). Six of the seven species in Table 4.3 are native to Africa and are local domesticates. The oats are part of an agricultural complex introduced from the Near East.

The relationship between oats and other cereals is of some interest in attempting to understand the processes of domestication. The wild forms of the com-

Table 4.3. Cultivated Plants of the Ethiopian Complex

Latin Name	Common Name and Use
Avena abyssinica	Tetraploid oats, weeds in barley and emmer fields
Catha edulis	Chat, a mild narcotic, chewed fresh
Coffea arabica	Primary coffee of commerce
Eleusine coracana	Finger millet, perhaps domesticated in Uganda
Ensete ventricosum	Enset, a relative of banana, stem base eaten
Eragrostis tef	Tef, the principal cereal of Ethiopia
Guizotia abyssinica	Noog, the main edible oil crop of Ethiopia

mon European oat are found around the Mediterrannean and in the Near East, but the plant seems to have been domesticated in northern Europe. Wild oats are very weedy in nature and readily take advantage of agricultural practices designed for the production of other cereal grains. Weed oats spread to northern Europe by infesting barley and emmer fields. Weed oats normally shatter their seeds at maturity and infest the soil. The seeds must then lie dormant until the next season, when they germinate with the planted crop. Another strategy is to suppress shattering genetically so that the seed is harvested with the crop. If there is no good method for separating the seeds, both crop and weed will be sown together at planting time. A mixture of barley and oats called dredge was sown for many centuries in Europe.

Ethiopian oats have repeated the same history. They are a different species from the European oat. The Ethiopian variety is a tetraploid with 14 pairs of chromosomes while the common oat is a hexaploid with 21 pairs. Ethiopian oats occur in mixtures with barley and emmer wheat. I have never seen a pure stand of oats in Ethiopia except under experimental conditions. The oats have evolved races that are semishattering and some that are nonshattering. In the first case, some of the seed shatters and infests the soil; the rest is harvested with the crop and sown the following season. The nonshattering forms are treated like a crop. The cultivators do not object to having oats in their barley and wheat and make no effort to eliminate them. If the Ethiopian oats are domesticated, it is a case of domestication by indifference. Some plants insert themselves into the domestic fold without deliberate human effort.

To most Ethiopians, the most important native crop is tef. It is considered a noble, a royal, grain. The seeds are minute and the crop is labor intensive because the seed bed must be very carefully prepared and the fields meticulously weeded. Unlike fonio, however, tef yields well and the Ethiopian farmers find the crop rewarding to grow. Tef is used to prepare the national bread called enjera. The seeds are ground to flour on a household grinding stone. The flour is moistened and allowed to ferment for two or three days. The dough is then baked on a large clay plate. The result is a highly nutritious and palatable bread. Tef holds a very high place in Ethiopian culture (Huffnagel 1961).

Enset is the staple of several tribes in south-central Ethiopia. The stem base yields a product that is nearly pure starch, although this is fermented for some time and the microflora probably add some vitamins and other nutritive factors. Noog is the primary edible oil of the country but is grown very little elsewhere. Finger millet is now used more for beer making than as a food cereal. Chat is one of the major exports. Since it must be chewed fresh, its distribution is restricted more or less to lands around the Red Sea. Most of it is now exported by air. Coffee is another story, since the coffee of commerce traces to the Ethiopian highlands and is one of the most important agricultural items in world trade.

Agriculture on the Ethiopian plateau is based primarily on a Near Eastern crop complex that is not indigenous to tropical Africa. It includes cool-season plants, such as barley, emmer, and other tetraploid wheats, chickpea, lentil, flax, fava bean, pea, and cabbage, together with a few warm-season exotics like safflower and sesame. The date of arrival of this complex on the plateau has not been established, but it is thought that an indigenous agriculture was already in place at the time. At least it seems unlikely that wheat and barley growers would bother to domesticate tef and finger millet, or that farmers with a complete crop complex would take up enset growing.

Summary

African agriculture is complex, diverse, innovative, and adaptable to a wide range of situations. Slash-and-burn systems are practiced in the forest; décrue techniques are used where water regimes are suitable. Swamp rice and upland rice are grown in the west. Pearl millet cultivation is pushed to the limits of ag-

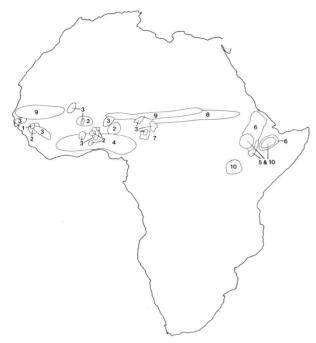

Fig. 4.9. Probable areas of domestication of selected African crops. 1, *Brachiaria diflexa*; 2, *Digitaria exilis* and *Digitaria iburua*; 3, *Oryza glaberrima*; 4, *Dioscorea rotundata*; 5, *Musa ensete* and *Guizotia abyssinica*; 6, *Eragrostis tef*; 7, *Voandzeia* and *Kerstengiella*; 8, *Sorghum bicolor*; 9, *Pennisetum americanum*; 10, *Eleusine coracana*. (Adapted from Harlan 1971)

riculture on the desert margins, and the deserts themselves are exploited by nomadic herdsmen. Yam growers of the forest modify landforms by mounding and ridging. Native trees are encouraged and exploited. The diversity and flexibility of the systems are impressive in their adaptability to difficult environmental problems.

The distribution of the crops and their wild progenitors does not reveal a pattern indicating a center of origin. Plants were domesticated throughout sub-Saharan Africa, from one side of the continent to the other (Fig. 4.9). If there were a center, the evidence for it has not been found. The time range is equally elusive: indigenous African agriculture could be as old as any or it could be younger than most. There are many discoveries yet to be made and answers to our questions must await further research.

References Cited

Bedaux, R. M. A., T. S. Constandse-Westermann, L. Hacquebord, A. G. Lange, and J. D. Van der Waals
   1978   Recherches archeologiques dans le delta interieur du Niger. Palaeohistoria 20:91–220.

Brunken, J., J. M. J. de Wet, and J. R. Harlan
   1977   The Morpholology and Domestication of Pearl Millet. Economic Botany 31:163–74.

Camps, G.
   1969   Ameki: neolithique ancien du Hoggar. Memoire du Centre de Recherches Archeologiques, Préhistoriques et Ethnologiques 10. Algiers.

Clark, J. D.
   1970   The Prehistory of Africa. Praeger Publishers, London.

Coursey, D. G., and C. K. Coursey
   1971   The New Yam Festivals of West Africa. Anthropos 66:444–84.

Davies, O.
   1968   The Origins of Agriculture in West Africa. Current Anthropology 9:479–82.

Davies, O., and K. Gordon-Gray
   1977   Tropical African Cultigens from Shongweni Excavations, Natal. Journal of Archaeological Science 4:153–62.

Harlan, J. R.
   1969   Ethiopia: A Center of Diversity. Economic Botany 23:309–14.

   1982   The Origins of Indigenous African Agriculture. *In* Cambridge History of Africa, vol. 1: From Earliest Times to c. 500 B.C., edited by J. Desmond Clark, pp. 624–57. Cambridge University Press, Cambridge.

Harlan, J. R., and J. Pasquereau
   1969   Décrue Agriculture in Mali. Economic Botany 23:70–74.

Harlan, J. R., and A. B. L. Stemler
   1976   The Races of Sorghum in Africa. *In* Origins of African Plant Domestication, edited by J. R. Harlan, J. M. J. de Wet, and A. B. L. Stemler, pp. 465–78. Mouton Publishers, The Hague.

Hilu, K. W., J. M. J. de Wet, and J. R. Harlan
   1979   Archaeobotanical Studies of *Eleusine coracana* ssp. *coracana* (Finger Millet). American Journal of Botany 66:330–33.

Huffnagel, H. P.
   1961   Agriculture in Ethiopia. FAO, Rome.

McIntosh, Susan K., and Roderick J. McIntosh
1979   Initial Perspectives on Prehistoric Subsistence in the Inland Niger Delta (Mali). World Archaeology 2:227–43.

Miege, J.
1954   Les cultures vivrieres en Afrique occidentale. Cahiers d'Outre-Mer 7:25–50.

Munson, P. J.
1976   Archaeological Data on the Origin of Cultivation in the Southwestern Sahara and their Implications for West Africa. *In* Origins of African Plant Domestication, edited by J. R. Harlan, J. M. J. de Wet, and A. Stemler, pp. 197–209. Mouton Publishers, The Hague.

Phillipson, D. W.
1977   The Excavation of Gobedra Rock-shelter, Axum: An Early Occurrence of Cultivated Finger Millet in Northern Ethiopia. Azania 12:53–82.

1982   Early Food Production in Sub-Saharan Africa. *In* Cambridge History of Africa, vol. 1: From Earliest Times to c. 500 B. C., edited by J. Desmond Clark, pp. 770–829. Cambridge University Press, Cambridge.

1985   African Archaeology. Cambridge University Press, Cambridge.

Portères, R.
1955   Les céreales mineures du genre *Digitaria* en Afrique et en Europe. Journal d'Agriculture Tropicale et de Botanique Appliquée 2:349–675.

Shaw, T.
1976   Early Crops in Africa: A Review of the Evidence. *In* Origins of African Plant Domestication, edited by J. R. Harlan, J. M. J. de Wet, and A. Stemler, pp. 107–153. Mouton Publishers, The Hague.

Vishnu-Mittre
1968   Protohistoric Records of Agriculture in India. Transactions of the Bose Research Institute 31:87–106. Calcutta.

1974   Palaeobotanical Evidence in India. Evolutionary Studies in World Crops: Diversity and Change in the Indian Subcontinent, edited by J. B. Hutchinson, pp. 3–30. Cambridge University Press, Cambridge.

Wendorf, F., R. Schild, N. El Hadidi, A. Close, M. Kobusiewicz, H. Wieckowska, B. Issawi, and H. Haas
1979   Use of Barley in the Egyptian Late Paleolithic. Science 205:1341–1347.

Wendorf, F., R. Schild, and A. Close (editors)
1980   Prehistory of the Eastern Sahara. Academic Press, New York.

5

The Origins of
Crop Agriculture
in Europe

ROBIN W. DENNELL

My main aims in this chapter are to review the current archaeobotanical evidence for the earliest crop agriculture in Europe and to show how the origins of European crop agriculture have been, and can be, explained. Because archaeological investigations of crop agriculture involve more than merely noting when cultigens first appeared in an area, a large part of this chapter is concerned with assessing their importance within the overall subsistence strategies of prehistoric communities in different parts of Europe. The outcome of this investigation suggests that it is misleading to study the first appearance of cultigens in isolation from other types of evidence: in many parts of Europe, crop agriculture initially functioned as a minor adjunct of well-established hunting-and-gathering strategies.

In many ways, prehistorians studying early farming in Europe are fortunate compared with their colleagues working in other areas. Agricultural origins have been studied in Europe longer than anywhere else, and the quantity of data now available—at first sight—is extremely impressive. There is also a wealth of supporting data on the chronology, material culture, architecture, animal domesticates, and environmental background that is usually unavailable for other regions. Despite this, both the quantity and quality of archaeobotanical data for the first crop agriculture in Europe are still far poorer than one would wish. Serious attempts to rectify this situation have been made only in the last twenty years, and in many areas of Europe the evidence for early cultigens is still very inadequate.

These inadequacies can often be blamed on the scarcity of serious attempts to recover archaeobotanical data from prehistoric sites. Over much of Europe, however, early cultigens will continue to be poorly evidenced simply because they were of little *initial* value to those who used them. For that reason, such evidence as large carbonized grain and seed samples, crop processing and storage areas in settlements, and ancient field systems may not occur in many regions of Europe until long after cultigens were first used.

Three problems in particular complicate the study of early crop agriculture in Europe. The first is that it is difficult to treat early European plant domestication in isolation from animal husbandry, because crop yields depended heavily on animal manure as fertilizers, while livestock—including oxen for plowing—often relied on crops for much of their winter feed. Animal feeding requirements were probably one of the main demands on any prehistoric crop-growing community (see, e.g., Dennell 1978; Barker 1985 for discussion). The second problem is the need to consider the local (Mesolithic) hunting-gathering background, as well as contemporaneous developments in the Near East where agriculture originated at an early date. As discussed below, the role of European Mesolithic societies in the expansion of farming across Neolithic Europe may have been seriously underestimated. The third problem is that the diversity of European prehistoric studies is as great as that of the societies under study. This is immediately apparent in the number of languages in which excavation reports are published: those working in Europe have to digest publications in languages as diverse as Finnish and French. An additional but less obvious problem is the effect of different academic traditions, whereby similar types of material are often studied and published in entirely different ways to the detriment of interregional studies. In archaeobotanical studies there are important differences in emphasis between continental and British researchers. The former—particularly in Germany and Switzerland—tend to work within a tightly defined multidisciplinary framework, often with a major interest in vegetational history. British workers, especially in recent years, have tended to favor a holistic approach with a greater emphasis on the interactions between populations, their plant and animal resources, and the environment.

Europe: Preliminary Considerations

The diversity of the European landscape often comes as a surprise to those from larger and more homogenous areas, such as North America or Australia. Dividing Europe into logical, homogenous units is difficult, and most who do so resort to a variety of criteria, whether geographic, climatic, political, or historical. If we bear in mind the state of the current database on early European plant domesticates, we can use the following units: Southeast Europe, the northern margins of the Mediterranean, central Europe, Atlantic Europe, the Alps, and East and Northeast Europe.

Southeast Europe

Southeast Europe encompasses Greece, Bulgaria, Yugoslavia, and southern Romania. The archaeological record for early farming in this region is very similar to that of much of the Near East. In both regions, the evidence for early farming includes "tell" or multiperiod settlements, geometrically painted ceramics, and an emphasis on cereal and legume cultivation and sheep/goat husbandry. There is also little that is immediately pre-Neolithic, except for a few cave sites at some distance from what seem to have been the main areas for early farming.

Geologically, Southeast Europe is complex. It comprises several mountainous areas that contain low-lying and often extensive valley systems, many of which were cultivated by the fifth millennium B.C.[1] The drainage of this region is dominated by the Danube and its tributaries. Historically, and doubtless in prehistory too, these river systems also provided the main routes of communication within the region. The climate of Southeast Europe reflects first the effects of distance from the Mediterranean, and second marked differences in altitude over small distances. In general terms, winters become colder and longer, and summers cooler and wetter from south to north. Local altitudinal differences are superimposed upon this large-scale gradient. In intermontane valleys, winters are less severe and summers are hotter and drier than in the mountains. These upland areas are poor for crop cultivation, but often provide valuable summer grazing for sheep and goat. One of the main themes of agrarian land use in this region, as well as in the Mediterranean, in historic times has been that of transhumance, with valley-based farming communities grazing their sheep in the hills in the summer and returning them in autumn to be fed through the winter on stubble and whatever natural fodder was available in the valley.

Northern Margin of the Mediterranean

This unit is easy to demarcate because it is so precisely defined by its warm, dry summers and generally frost-free winters and by the distribution of the classic Mediterranean crop, the olive. As in Southeast Europe, this region is marked by major altitudinal differences over often short distances between coastal plains and inland mountains. Not surprisingly, transhumance between the two areas has been an important feature of many areas in the Mediterranean basin for at least three millennia.

Central Europe

For our purposes this notoriously ill-defined area can be described as covering the distribution of the Linienbandkeramik (LBK) culture (ca. 4500–4000 B.C.). This is a remarkably uniform phenomenon that is centered on Germany, much of Poland, Austria, and Czechoslovakia and extends into eastern France and southern Holland. In terms of prehistoric agriculture the main features of this region are areas of loess and other well-drained soils that are extensive, discontinuous, and usually located in river valleys lying between upland areas of only moderate relief. The Rhine, Elbe, and their tributaries would have provided the main communication routes over much of this region. Climatically, the region experiences cool and often wet summers and moderately cold winters with abundant frost and snow.

Atlantic Europe

Atlantic Europe comprises Portugal, northern Spain, western and northern France, the British Isles, Belgium, coastal Netherlands, northern Germany, Denmark, Norway, and southern Sweden. Two themes help unify this otherwise highly diverse region. One is the sea, which presents opportunities for fishing, commerce, and other maritime activities. The other is climate, which again reflects the influence of the Atlantic with its prevalent westerly winds, high rainfall, and ameliorating influence on winter temperatures. The northern part of this region, comprising the British Isles and the Norwegian coast, can be treated as a subunit characterized by rainfall, allegedly temperate climate, and notoriously fickle weather.

The potential for crop agriculture varies enormously across this region. In the northern and highland parts of Britain, as well as Norway and much of Sweden and Denmark, the severity of the winters and the shortness of the growing season greatly constrain crop production, which often is based on barley and oats. Further south, wheat, as well as a variety of legumes and vegetables, tend to be the favored crops. The overwintering of livestock is an important aspect of many crop economies in this region. In southerly areas this can be done by grazing livestock on straw and stubble, but further north leaf fodder and hay increase in importance.

The Alps

This area encompasses Switzerland, southwestern Germany, western Austria, northern Italy, and parts of eastern France. This tiny area has an archaeobo-tanical significance that is entirely out of proportion to its size, being the area where this type of research began following the discovery of the so-called "lake dwellings" in the 1850s (see the next section). As might be expected, this region is characterized by extensive areas of high mountains, many over 4000 m high; substantial areas of summer grazing; and only limited lowland areas, usually located around lake margins. Farming systems in this region have long emphasized dairy products, the use of summer pasture, and the production of fodder in the lowlands for overwintering livestock.

East and Northeast Europe

This region, which includes Sweden, southern Finland, northwestern Russia, and northern Poland, is marked by long cold winters and short but often hot summers and by a generally low relief (excluding such areas as the Polish Tatra range). An important aspect of early farming evidence in this region is that hunting, gathering, and foraging were often a better alternative to farming. Indeed, in some areas agriculture was adopted only within the last millennium or so.

History of Archaeobotanical Research in Europe

Archaeobotanical research in Europe can be traced back to the beginning of systematic prehistoric research in the 1850s. Until recently, European archaeobotany has been dominated by studies of the origin and diffusion of cultivated plants. It began with the brilliant work of Oswald Heer (1866) on the plant remains from the so-called "lake dwellings" of Switzerland. Even now, this work still ranks among the most innovative and thoughtful ever undertaken in European archaeobotany. In addition to identifying the material from these sites, Heer made many perceptive comments about the type of crop processing activities likely to have been performed there and about the way samples varied in composition within single horizons—topics that did not receive serious attention until the 1970s. After his work, little comparable research was done until after World War I. De Candolle's (1884) *Origin of Cultivated Plants* was a useful synthesis of what had been learned about the history of various cultigens from literary and linguistic sources, but it touched only lightly on archaeological discoveries. By the 1930s, however, enough had been learned for Bertsch and Bertsch (1949) to draft their *Geschichte unserer Kulturpflanzen,* which contained numerous maps showing the distribution of various cultigens throughout Europe during different prehistoric periods.

Until recently, the number of people active in European archaeobotany was very small, both absolutely and in proportion to those undertaking pollen analysis. In this context it is interesting to note that suggestions that slash-and-burn cultivation was the earliest type of crop agriculture in northwestern Europe were based on inferences from pollen diagrams (Iversen 1941) and not on carbonized plant remains. Because the basic interest in archaeobotanical research concerned the origin and dispersal of different cultigens, only minimal collaboration between archaeologists and archaeobotanists was needed. Archaeologists tended to be more interested in material culture than in subsistence, and archaeobotanists usually showed more interest in vegetational history than in agriculture. Neither party had much interest in retrieving evidence of plant foods from archaeological sites in a large-scale or systematic manner. When attempts were made to estimate the importance of a crop, an estimate of its abundance relative to other plants usually was considered adequate. In addition, the first appearance of a cultigen often was assumed to indicate the adoption of crop agriculture as a major component of the economy. This seemed logical in that crop agriculture supposedly spread through the colonizing activities of farmers (see section below on origins of crop cultivation). Equally logical was the supposition that the first appearance of pottery was also a reasonably accurate indicator of early farming because early farmers used pottery, whereas Mesolithic hunter-gatherers by and large did not. As will be seen later, these assumptions may be true for some areas of Europe but are certainly not so for others, and their uniform application has probably seriously distorted our understanding of early European crop agriculture.

Major developments in European archaeobotany occurred in the 1970s. By then, the New Archaeology was already underway and encouraged a change in emphasis from regional diachronic studies to local synchronic ones. Many prehistorians began to show a greater interest in how the components of a prehistoric cultural system—the physical environment, subsistence, social organization, technology, ideology, and so forth—interacted with each other while a

site and its neighbors were occupied. One important consequence of this focus was that both archaeologists and archaeobotanists began to study prehistoric subsistence as a topic in its own right. As a result, the development of large-scale flotation techniques (e.g., Jarman et al. 1972) led to a dramatic increase in the quantity of plant remains retrieved from excavations. Much attention has also been paid since the 1970s to establishing reliable methodologies for assessing the relative importance of each plant resource represented in an archaeological context, for determining which were cultigens and which were commensals of other crops, and for inferring the cropping system to which each crop belonged. Lastly, and importantly, much attention has been paid to evaluating the importance of crop agriculture relative to other food sources in prehistoric farming societies.

Research Goals in Archaeobotanical Studies of Early Farming

In order to investigate the early history of crop agriculture in Europe, the following questions must be asked:

1. When was a potential cultigen first used in an area?

In some parts of Europe, this question can often be answered fairly confidently because plant remains frequently occur in dense concentrations that are easily visible during excavation. Even so, it is worth noting that barley was not evidenced in early Neolithic Bulgaria until the use of large-scale flotation devices. In most areas of Europe, however, carbonized plant remains are present in only low densities and often are diffused throughout an archaeological horizon. If the number and size of samples are small, rarer cultigens may not be detected. In multicomponent sites (e.g., caves), seeds or grains may have percolated downward; in some cases, accelerator radiocarbon dating has indicated that cereal grains might have migrated downward through over 3 m of compacted archaeological deposits (Legge 1986). Consequently, claims for early or pre-Neolithic crop usage based on very small numbers of individual seeds or grains need to be regarded cautiously unless the remains are dated by radiocarbon.

In parts of northern Europe, the main evidence for the earliest crop cultivation is indirect and consists of pollen profiles indicating episodes of woodland clearance that are usually associated with the pollen of plants that can be weeds, such as *Plantago lanceolata,* of arable land. Cereal pollen is often, but not always, found as well. It was once thought that these clearances began in the Early Neolithic (ca. 3500–3000 B.C.), but recent British evidence indicates that woodland clearance also took place in the Mesolithic. This finding indicates that direct evidence for crop cultivation may not be preserved in archaeological settlements until some time after it was practiced.

Finally, the earliest evidence of plant cultigens in some parts of northern and eastern Europe sometimes consists of grain impressions in pottery. Although useful, such impressions need not indicate what was happening locally; for example, some of the pottery made by farmers could be present in the territories of hunter-gatherers as the result of gift-giving or exchange.

2. When was a potential cultigen first cultivated as a crop in its own right?

This question is difficult to answer if samples of carbonized plant remains contain several potential cultigens. Indeed, the commonest plant represented in a sample might be the least valued part of a crop if the sample represented the waste left after crop cleaning (Dennell 1974a). This point emphasizes the need for detailed contextual information on the circumstances under which plant material was preserved. Unfortunately, the absence of this information in most European archaeobotanical reports often makes it impossible to determine whether a plant was an actual or potential cultigen. Unless samples are both large and overwhelmingly composed of the remains of a cultigen, sample composition alone is a poor guide to when a potential cultigen was actually cropped.

Long-distance trade may also be significant in two circumstances. The first is when farming communities may have exchanged cereal produce for other goods with hunter-gatherers in adjacent territories, and the other is when they traded or exchanged with nomadic pastoralists. Hillman (1981) has suggested that imported produce should be discernible if the plant assemblages from a site represent fully processed crops with no evidence for such activities as grain threshing or winnowing. The absence of cereal pollen from or near sites that contain charred grain can also provide an indication that grain was imported. Until this factor is controlled, the scale of crop cultivation in Neolithic Europe may be overes-timated, especially in or near areas where hunter-gatherer populations were well established.

3. What was the relative importance of each crop?

Ranking prehistoric crops in terms of their importance has proved to be difficult, and so far two main approaches have been tried. The first is *quantitative.* This was initially done by counting the number of each type of grain or seed in samples from an archaeological horizon to arrive at a percentage (e.g., see Renfrew 1973; Hubbard 1976). There are two problems with this approach. The first is determining what the percentage actually refers to: the number of grains/seeds of that crop as a proportion of the total number in the annual plant diet; the amount by weight or the caloric/protein-value that a plant contributed to the total plant diet; the amount of time spent in cultivating it; or the proportion of arable land used for that crop. The second is that the method fails to take account of the biasing effect of different types of crop processing activities on the representation of material in archaeological samples of plant remains.

For these reasons, I proposed (Dennell 1976a) an alternative, *qualitative* approach that simply ranks the plant species represented archaeologically into three categories: staple, incidental, and casual. Ranking is effected by noting the domestic contexts in which each plant is represented. Thus, staple resources should figure primarily in cooking/storage contexts, while incidentals and casuals should be found mainly in refuse contexts. The main drawback to this approach is that it demands detailed, contextual information and high standards of sampling and recording. It is also inappropriate for sites that lack domestic contexts of crop processing, storage, and cooking.

In conclusion, therefore, it cannot be assumed that a plant was important economically simply because of its archaeobotanical abundance. Consequently, attempts to use the abundance of seeds or grains of particular crops to show changes in the importance of prehistoric crops must be treated circumspectly (Dennell 1977).

4. To what type of crop system did each cultigen belong?

This is one of the most interesting topics in archaeobotanical research, but also one of the least explored. Two approaches have been tried. The first is to look at the commensal plants associated with each crop to see if there is evidence for crop rotations (van Zeist

1968; Dennell 1978). An important prerequisite of this approach is the ability to decide whether commensals in fully processed crops represent contamination that occurred after crop cleaning or the relic proportion of commensals that grew in the crop itself (see Jones 1987). The second approach is to look at the weed flora associated with different types of crops and cropping systems. Modern crops (Hillman 1981; G. Jones 1984) have distinctive weed communities, and prehistoric crops (Groenman van Waateringe 1979; Wasylikowa 1981) were probably no different. The numbers and types of weed seeds that survive in archaeological samples of ancient crops, however, provide a very small and partial sample of the weeds that grew in fields. Consequently, multivariate analyses are probably needed to identify significant associations between the weed seeds and crop plants represented in archaeological samples.

5. What was the actual importance of crop cultivation?

Because the first appearance of cultigens is not the same as the first evidence of crop agriculture as a major part of the local economy, it is necessary to look at other sources of data. One is evidence for other types of plant resources. Another is information on local animal exploitation because changes in the importance of hunting or herding should provide some indication of how the subsistence strategy as a whole was changing. Changes in settlement patterns and in human skeletal remains can also provide useful information on dietary changes (see e.g., Price et al. 1985) and on the approximate importance of plant foods. Finally, environmental evidence in the form of woodland clearance and field system or drainage/irrigation ditch construction is also useful. All of these suggestions emphasize the need for an integrated, holistic approach to studies of early crop agriculture.

History of Early Crop Cultivation in Europe

In this section I concentrate on presenting the primary evidence for early crop cultivation in Europe. Because of space limitations, I emphasize only the major cereal and legume crops. The bibliography is selective rather than exhaustive, and I have utilized only the material that I consider to be both important and accessible. Wherever possible, I also favor re-

ports or syntheses that are in English, and I assume that the reader has a basic knowledge of the European Mesolithic and Neolithic. Background reading for those who do not is referenced at the end of the chapter.

In general terms, we can note an important distinction between the archaeobotanical record for early cultigens in Southeast and Central Europe and the Alps on the one hand and the rest of Europe on the other. In the former regions, there is often a substantial amount of archaeobotanical evidence from several early Neolithic sites. In addition, the first appearance of cereals and legumes usually coincides with that of sheep/goat and other domestic livestock, pottery, and substantial settlements. There is also little evidence for any of these features in local pre-Neolithic contexts. A different picture prevails over the rest of Europe. Cereals and legumes are much less well represented and can first appear in a wide variety of contexts, ranging from late Mesolithic to late Neolithic in date (in artifactual terms) and often without being associated with sheep/goat, pottery, and/or substantial settlements. In such areas, it is rarely possible to obtain more than an impressionistic view of what might have been happening in terms of early crop agriculture. As is discussed below, these differences cannot be wholly attributed to the intensive use of flotation techniques in some regions but not in others. With these points in mind, we can consider the "crisp" evidence from Southeast and Central Europe and the Alps.

"Crisp" Evidence

SOUTHEAST EUROPE
In this region, the earliest farming sites belong to the Early Neolithic, which begins ca. 6000 B.C. in Greece, 5000 B.C. in Bulgaria, and ca. 4700 B.C. in Yugoslavia. Early Neolithic sites in this region usually contain evidence for the earliest local use of pottery, polished stone artifacts, and domesticated crops and animals. The Neolithic of Southeast Europe is outlined by Barker (1985) and Whittle (1985) and regional accounts of the archaeobotanical evidence are provided by several authors, notably Lisitsina and Filipovich (1980), Renfrew (1969, 1979), and Dennell (1978).

Greece (excluding Franchthi Cave) There are two main sources of data on early crop agriculture in

Table 5.1. Cereals and Legumes from Early Neolithic (6200–5300 B.C.) Sites in Greece

	Gediki	Achilleion	Sesklo	Argissa	Soufli	Nea Nikomedia
Triticum monococcum	x	x	–	x	–	x
T. dicoccum	x	x	x	x	x	x
T. aestivum	–	–	–	–	–	–
Hordeum distichum	x	x	x	–	–	–
H. vulgare	–	–	–	x	–	x
Panicum miliaceum	–	–	–	x	–	–
Avena sp.	–	x	–	–	–	–
Pisum sp.	x	–	x	–	–	x
Vicia sp.	x	–	–	–	–	x
Lens sp.	x	x	–	x	x	x

Source: Renfrew 1979.

Greece (see Table 5.1 and Fig. 5.1). The first are the samples collected from Early Neolithic sites after 1958 and studied by Hopf (1962) and Renfrew (1966, 1969, 1979). These were predominantly "grab" samples, collected without the use of flotation techniques, and are summarized in Table 5.1. The small size of each sample and the lack of contextual data prohibit assessments about the importance of each cultigen, but they indicate the presence of three types of wheat and legumes and two types of barley. Recent material from the Early Neolithic of Prodromos can also be included (Halstead and Jones 1980).

The second source of evidence is from the site of Nea Nikomedia (Rodden 1962, 1965). The age of this site has been given as ca. 6230 B.C. on the basis of one radiocarbon date, but a younger one of ca. 5470 B.C. is more likely on the basis of two other radiocarbon dates of 5605 ± 90 and 5330 ± 75 B.C. Although flotation techniques were not used, botanical material was found dispersed throughout the cultural debris but not in storage or cooking contexts. A preliminary study of the botanical remains by van Zeist and Bottema (1971) showed a wide range of wild and domestic plants and considerable variation in sample composition. Unfortunately, the significance of this variability cannot be assessed without further contextual information.

Bulgaria The evidence for cultigens from early Neolithic Bulgarian sites is summarized in Table 5.2. The samples from Azmak and Karanovo I were collected without flotation and were studied by Renfrew (1969) and Hopf (1973). These samples indicated that

cereal agriculture was essentially wheat-based and supplemented by lentil and, perhaps, grass-pea.

The evidence from Chevdar and Kazanluk was recovered by large-scale flotation (Dennell 1978). Several types of samples were found. These included large, homogenous samples from ovens/hearths at Chevdar; heterogenous grain samples with large numbers of weed seeds and generally small grains on floor deposits at both Chevdar and Kazanluk; samples comprising numerous spikelet fragments, many weed seeds, and only a few grains from middens at Kazanluk; and samples comprising large cereal grains and small numbers of weeds, again from middens at Kazanluk. By considering sample composition, grain size, and context, I explained these differences as the result of different on-site crop processing activities, such as grain cleaning and dehusking (Dennell 1974a). These data were later used to suggest that emmer, barley, and legumes (mainly lentil) had been staple crops; others such as vetch, flax, and fruits had been much less important (Dennell 1976a). Samples of fully processed crops also allowed assessments of crop purity, which was the same as in later periods (Dennell 1974b). Some kind of crop rotation involving emmer, six-row barley, and pulses was suggested from the association of each crop with its weed flora and commensals (Dennell 1978). This suggestion was strengthened by the discovery of nematode remains often associated with intensive cropping systems (Webley and Dennell 1978). Supplementary data on the catchment areas of Neolithic and later sites in Bulgaria are given in Dennell and

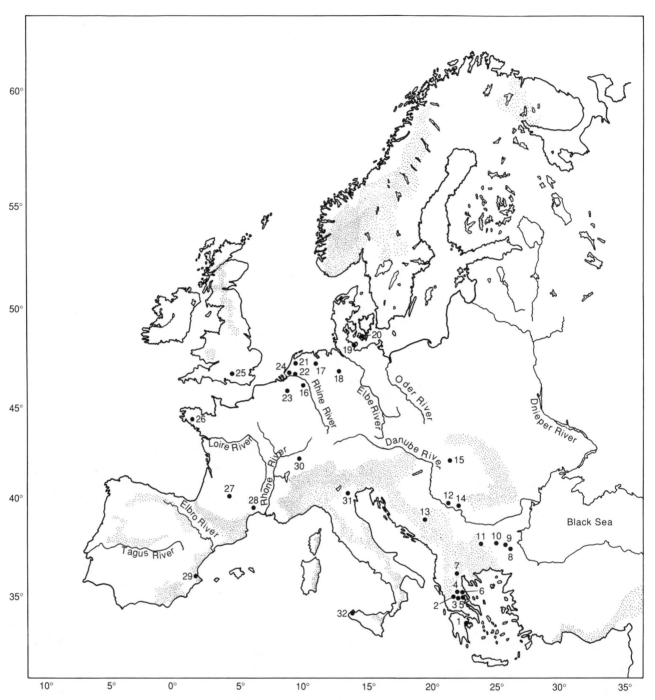

Fig. 5.1. Location of European sites mentioned in the text. **Greece:** 1, Franchthi Cave; 2, Prodromos; 3, Achilleon; 4, Argissa; 5, Sesklo; 6, Ghediki; 7, Nea Nikomedia. **Bulgaria:** 8, Azmak; 9, Karanovo; 10, Kazanluk; 11, Chevdar. **Yugoslavia:** 12, Starčevo; 13, Obre. **Romania:** 14, Icoana. **Poland:** 15, Korlat. **Germany:** 16, Alderhoven Plateau; 17, Dummer; 18, Eitzun; 19, Siggenben-Sud. **Denmark:** 20, Lidsč. **Netherlands:** 21, Swifterband; 22, Hazendonk; 23, Esloo, Stein; 24, Vlaardingen. **Britain:** 25, Windmill Hill; **France:** 26, Tevièc, Hödièc; 27, Roucadour; 28, Abeurador. **Spain:** 29, Coveta de l'Or. **Switzerland:** 30, Cortaillod (and other nearby sites). **Italy:** 31, Molino Casarotto; 32, Grotta dell'Uzzo.

Table 5.2. Cereals and Legumes from Early Neolithic (5000–4600 B.C.) Sites in Bulgaria

	Azmak	Karanovo I	Chevdar	Kazanluk
Triticum monococcum	x	x	x	x
T. dicoccum	x	x	x	x
T. aestivum	x	–	x	x
Hordeum vulgare var. *nudum*				
H. vulgare	–	–	x	–
Vicia sp.	x	–	–	x
Lathyrus cicera	x	–	–	–
Lens sp.	x	–	x	x
Pisum sp.	–	–	x	x

Source: Renfew 1979.

Webley (1975) and on the likely productivity of subsistence agriculture at Chevdar in Dennell (1978:99–112).

Yugoslavia The archaeobotanical record for Yugoslavia is still poor, considering its size and archaeological wealth. The main data from early Neolithic sites were studied by Hopf (1974) and Renfrew (1979) and are summarized in Table 5.3. As in Bulgaria, wheat, barley, pea, and lentil were probably the main cultigens. In one of the few studies of the local surroundings of early Neolithic sites in Yugoslavia, Barker (1975) indicated considerable variation in the location of such sites and, presumably, in their crop regimes. Clearly, much more work needs to be done before there is an adequate account of the earliest crop husbandry in Yugoslavia.

CENTRAL EUROPE

There are several useful English syntheses of the evidence from Linienbandkeramik (LBK) settlements (ca. 4500–4300 B.C.), including those by Barker (1985:139–47), Bogucki (1988), Bogucki and Grygiel (1983), Hammond (1981), and Whittle (1985:76–95). Because this culture appears suddenly and marks the first local use of pottery, domestic crops and animals, and substantial settlements, it often has been assumed to represent an intrusive tradition, best explained as the result of agricultural colonization from areas further south. Its origins, however, are still uncertain. Local late Mesolithic populations may have played a greater part in this early Neolithic culture than has commonly been supposed (see Whittle 1985:94). A long-noted feature is a marked preference for settlement on loess, presumably because it is well-drained, fertile, and easily cultivated (Sielmann 1976). On the other hand, sites such as Esloo and Sittard in the Netherlands are on river terraces, and loess areas on the western and eastern edge of the LBK distribution were not always occupied. In terms of material culture, LBK sites tend to be very uniform and are primarily characterized by timber-framed long-houses

Table 5.3. Cereals and Legumes from Early Neolithic (5000–4600 B.C.) Sites in Yugoslavia

	Starcevo	Vrsnik III	Anza I–III	Obre I	Kakanj	Dani
Triticum monococcum	x	x	x	x	x	x
T. dicoccum	x	x	x	x	x	?
T. aestivum	–	x	x	–	–	–
T. compactum	–	x	–	x	x	–
Hordeum vulgare	x	x	x	x	x	–
Secale sp.	–	–	–	–	–	x
Pisum sativum	x	–	x	x	x	–

Source: Renfrew 1979

(up to 45 m in length and 6 m wide) and handmade, unpainted, and incised pottery. Sites are often large with up to 100 structures. The actual settlements may have been far smaller: at Esloo perhaps only 11 to 17 huts were used simultaneously, and on the Alderhoven Plateau in western Germany settlements may have comprised small clusters of widely separated houses (Hammond 1981; Lüning 1982; Whittle 1985:88). Likewise, estimates of the amount of land in use at any one time for crop agriculture have dwindled in recent years. Earlier suggestions that the inhabitants of LBK settlements practiced slash-and-burn cultivation, relocated their settlements once crop yields fell, and thus colonized large areas quickly have been heavily criticized in recent years (e.g., Rowley-Conwy 1981). Instead, the amount of land under crops may have been only 12 to 30 ha (Milisauskas 1984), and that land may have been cropped for several years in succession. This latter suggestion is supported by Willerding's (1980) comment that some of the weeds represented on LBK sites indicate the presence of hedgerows and fixed fields.

The basic archaeobotanical text on LBK plant husbandry is by Willerding (1980), who has summarized the data from almost 100 sites. Emmer, barley, and pulses are the most common cultigens, while other plants, such as flax, opium poppy, and fruits, are also indicated. One point to stress is that plant remains are not generally well preserved or common on LBK sites, and samples tend to be small. Their archaeological surfaces often have been eroded, hindering attempts to study on-site crop processing and other domestic activities, as can be done in Southeast Europe. Much of the botanical evidence also comes from impressions in pottery and daub, and these are unlikely to indicate the range of species evident in samples of carbonized remains. It is difficult to assess at present the type of variation in crop regimes over the area of LBK settlement beyond noting that barley is represented as carbonized remains only in eastern France (Bakels 1984) and in areas north and west of the Harz Mountains, although barley impressions in pottery are found over a much larger area (see Whittle 1985:87). This may reflect the movement of pottery rather than the actual extent of barley. It is, however, probably significant that emmer is the most commonly represented cereal, and legume remains are sporadic.

The literature on the LBK is vast and only a small selection is cited here, in addition to those already mentioned. Key settlement data are contained in Bakels (1978) and Modderman (1970, 1977) for the Netherlands; Kuper et al. (1974, 1977) for Germany; Kruk (1980) for Poland; and Ilett (1983) for France. Additional archaeobotanical reading includes Knörzer (1968, 1972, 1973, 1977) for Germany; Hajnalová (1973, 1976) for Czechoslovakia; Hartyányi and Novaki (1975) and Tempír (1964, 1973) for Hungary; and Kamieńska and Kulczycka-Leciejewiczowa (1970) and Wasyilikowa (1984) for Poland.

THE ALPS

Agricultural settlements associated with pottery, domestic livestock, and a wide range of cultivated plants (see Table 5.4) were established along the major lake systems of Switzerland and in the upper valleys and tributaries of the Rhine and Rhone after ca. 3500 B.C. Four major cultural groups are recognized by their pottery and other artifacts: the Cortaillod group on the western plateau and in the Jura Mountains; the Pfyn group around the lakes of Zürichsee and Bodensee; the Egozwil group around the lakes of Burgaschisee and Wauwilermoos; and the St. Leonard group in the upper Rhine and Rhone valleys. The quality of preservation of organic material is often stunning, and a detailed picture has been built up of animal and crop husbandry, as well as of the types of textiles and wooden artifacts that were used. Most settlements were small and many may have been the equivalent of individual farms or hamlets. The largest contained perhaps 150 to 180 people and was occupied for several decades. Barley, emmer and bread wheat, flax, and peas have been found in storage contexts, and einkorn, lentil, and millet, as well as a wide range of wild plants, have also been found.

Although the subsistence and dating of the local late Mesolithic are poorly understood, it may overlap with the early Neolithic (Gregg 1988:13–14). If so, foragers and farmers could have coexisted for several generations. The origins of the Swiss Neolithic are still obscure: many have argued that it represents colonists from areas of Bandkeramik settlements, but it seems at least as likely that some local populations acquired the necessary resources and developed agriculture locally (see Barker 1985:124).

Convenient English summaries of the Swiss data are provided by Barker (1985:118–24), Gregg (1988:10–15), and Sakellardis (1979), who also includes summary tables of the botanical and faunal data. Further botanical data on early alpine settlements can be found in Heitz et al. (1981), Jacomet-

Table 5.4. Cereals and Legumes from Early Neolithic (3500–3000 B.C.) Sites in Switzerland

	Chavannes	Cortaillod	Lüscherz	Neuenstadt	Port	St. Blaise	Thun	Wangen
Triticum monococcum							x	x
T. dicoccum							x	x
T. aestivum	x		x	x				S
T. compactum	x		X		S	x	S	x
Hordeum distichum								S
H. hexastichum		x	x		S	x	x	S
Panicum sp.			x			x		x
Lens sp.						x		
Pisum sp.			x		S	x	x	x

Source: Sakellaridis 1979: 344–45, 358–59.
Key: x = present; X = frequent; S = storage context

Engel (1980, 1981, 1986, 1987), Jacomet and Schibler (1985), Jøorgensen (1975), Küster (1984), Baudais-Lundstrom (1978, 1984), Schlichtherle (1985), and van Zeist and Casparie (1973).

Data from subalpine Italy show that foragers adopted pottery, cereals, and sheep/goat around 4000 B.C. without otherwise changing their lifestyle to any significant degree. At Molino Casarotto, for example, the remains of red deer and boar dominated the faunal sample, while cattle, sheep, and goats comprised only 3 percent of the total. Cereals were represented, but in very small amounts, and their remains were far rarer than those of water chestnuts. Farming does not seem to have become well established in this region until after 3000 B.C. (see Barker 1985:124–26).

"Diffuse" Evidence

THE MEDITERRANEAN BASIN

With a few notable exceptions, research into agricultural origins in this region is still poorly developed. The archaeobotanical record is very weak for the Adriatic coastline of Yugoslavia, much of central and northern Italy, southern Spain, and many of the Mediterranean islands. In addition, a substantial number of sites in the northern Adriatic and the Bay of Languedoc in the Marseilles area have probably been drowned by rising sea levels since the sixth millennium B.C. A further bias is caused by the lack of open-air sites outside southern Italy, as almost all the key data on early crop agriculture are from caves. These are often located in areas of low arable productivity, and many may have been used seasonally in connection with animal herding or hunting. For these reasons, discussion has to focus on those areas where useful results have been obtained, namely Franchthi Cave in Greece, Grotta dell'Uzzo in Sicily, southern France, and, to a lesser extent, coastal Yugoslavia and southern Italy.

Franchthi Cave This cave site in southern Greece is tremendously important for studies of agricultural origins in the Mediterranean because of its long cultural sequence from the late Upper Pleistocene to beyond the Neolithic. Its excavation included one of the most intensive sieving and flotation operations ever mounted (summarized in Diamant 1975). Preliminary reports are provided by Jacobsen (1976, 1981), by Payne on the fauna (1975), and by Hansen and Renfrew (1978) on the botanical data. The last mentioned are summarized in Table 5.5. This information is provocative in showing that wild barley and lentil were utilized in the late Pleistocene and early Holocene, and it strengthens arguments that at least some cultigens were present in Southeast Europe before the Neolithic. Plants such as pear and pistachio also seem to have been used throughout the Franchthi sequence. Evidence published so far indicates that domestic forms of wheat, barley, and possibly lentil do not appear until the Neolithic layers (ca. 6000 B.C.), which also contain the first indications of pottery and sheep/goat. This evidence for discontinuity between the Mesolithic and Neolithic might suggest that agriculture was introduced in a developed form from elsewhere.

Grotta dell'Uzzo This cave site in Sicily rivals Franchthi in having a long and carefully researched sequence that spans the Mesolithic and Neolithic (Table 5.4). In the Mesolithic layers (8500–6000 B.C.), remains of grass pea (*Lathyrus*), pea (*Pisum*), wild strawberry, wild olive, and wild grape were found. The early Neolithic layers postdating 6000 B.C. contained remains of *Lathyrus/Pisum* and wild strawberry, but also einkorn, emmer, bread wheat, barley, and lentil, along with the remains of domestic animals. Pottery, however, seems to have appeared after the first usage of these domestic resources (Constantini 1989).

Southern France General summaries in English of the Mesolithic–Neolithic transition in southern France are provided by Lewthwaite (1986), Mills (1983), Phillips (1975), and Trump (1980). The evidence from this area is fascinating in showing a gradual transition to farming between 6000 and 4000 B.C. within the context of highly stable Mesolithic foraging strategies. Sheep appear to have been the first major local introduction. Geddes (1985) has argued convincingly that these sheep were probably Asiatic in origin and were probably derived from Italy and, ultimately, Greece. Whether they arrived of their own accord or were acquired through contacts with adjacent herders and farmers is currently unknown.

Local Mesolithic traditions in material culture and site location persisted well into the Neolithic, the beginning of which is defined by the first appearance of pottery in the early fifth millennium B.C. Mesolithic plant usage is evidenced by the sporadic appearance of dwarf chickling (*Lathyrus cicera*) and bitter vetch

Table 5.5. Cereals and Legumes from Franchthi Cave, Greece, and Grotta dell'Uzzo, Italy

	Franchthi Cave			Grotta dell'Uzzo	
	Late Palaeolithic	Mesolithic	Early Neolithic	Mesolithic	Early Neolithic
Triticum monococcum					■
T. dicoccum					■
Hordeum spontaneum	■	■	■		
H. distichum	■		■		
Avena sp.	■	■	■		
Lens sp.	■	■	■	■	■
Pisum sp. [a]		■			
Vicia sp.	■	■			

Sources: Hansen 1978; Constantini 1989.
[a] This group includes *Lathyus* at Grotta dell'Uzzo.

(*Vicia ervilia*), for which a local origin does not raise any problematic issues. More controversial are the claims by Vacquer et al. (1986) of domestic and Asiatic forms of chickpea, lentil, and pea from a Mesolithic context at Abeurador, dated to ca. 6790 ± 90 B.C. Further information is required on their identification as domestic and Asiatic, and confirmation of their context by accelerator radiocarbon dating would also be needed if these identifications are confirmed. When cereals first appear in southern France is unclear, but small quantities of the remains of emmer (*Triticum dicoccum*), bread wheat (*T. aestivum*), and/or barley (*Hordeum* sp.) have been found in contexts at such sites as Chateauneuf-les-Martigues and Grotte des Eglises from the late fifth millennium B.C. (Courtin and Erroux 1974).

If we exclude the controversial finds from Abeurador, crop agriculture appeared after pottery and sheep and probably developed on a large scale only after 4000 B.C. Crop agriculture appears to have begun, therefore, in several ways in the Mediterranean basin: by a process of sudden change, along with the adoption of sheep/goat and pottery, as at Franchthi; suddenly, with the adoption of both cereals and sheep/goat but not pottery, as at Grotta dell'Uzzo; and as a process of gradual and probably minor change by foragers, who also acquired sheep and pottery, as in southern France.

One area of the Mediterranean that shows a major discontinuity between the Mesolithic and Neolithic is southern Italy and eastern Sicily. Enormous (up to 500 m by 750 m) ditched enclosures, associated with pottery and probably domestic cereals and livestock, appear between 5000 and 3000 B.C. (see Barker 1985:65–67; Whittle 1985:103). These sites often are located on areas of well-drained and easily tilled soils, suggesting that crop cultivation may have been important (Jarman and Webley 1975). Although these settlements might represent direct colonization from the Aegean, the evidence is still ambiguous, and a local origin cannot be ruled out. The same is also true of coastal Yugoslavia, where recent data (Chapman and Müller 1990) show that pottery, cereals, and domestic livestock were used after 5000 B.C.

Elsewhere, however, the dominant impression is one of a predominantly foraging way of life that continues well into the Neolithic and gradually involves the acquisition of pottery, cereals, legumes, and sheep/goat. Italian data are summarized by Barker (1985:65–67), who suggests that between 4500 and 3000 B.C. there was a variety of subsistence strategies in and south of the Po Valley. These included year-round foraging, seasonal foraging by herders, year-round herding, and herding-cum-cultivation. All these are recognized formally as Neolithic because of their pottery, but farming as a general way of life is not in evidence until after 3000 B.C.

Data from Spain are sparse. Pottery was probably in use by the early fifth millennium B.C., and einkorn, emmer, bread wheat, and barley were used ca. 4500 B.C. at the Coveta de l'Or site (Hopf and Schubart 1965). Other botanical reports (e.g., Hopf and Catalan 1970; Hopf and Muñoz 1974) add little to this picture. In southern Spain, there seems to have been a considerable degree of continuity between the Epipalaeolithic and Bronze Age with only a minor effect on previous traditions resulting from the introduction of pottery, cereals, and sheep (Guilaine et al. 1982). Most of this evidence comes from seasonally used caves, however, and a different picture might emerge from examinations of open-air sites.

ATLANTIC SEABOARD

There has been some excellent recent work on the Mesolithic and/or Neolithic in the Netherlands, Denmark, southern Sweden, Britain, and Ireland. Unfortunately, there has been little comparable work in Portugal, western and northern France, Belgium, and Norway; research in northern Spain has been primarily a by-product of cave and midden-oriented Palaeolithic and Mesolithic investigations; and work in southwestern France on the Mesolithic and Neolithic has always been overshadowed by the richer Palaeolithic sequences of the Dordogne. Nevertheless, the archaeological record of the Atlantic seaboard and the Baltic/North Sea region contains some important regularities.

The essential background to understanding the origins of crop agriculture in these areas is the local Mesolithic. Although most of our knowledge of this is from northwestern Europe, subsistence before 4000 B.C. was primarily based on a broad range of "traditional" resources, notably red and roe deer, pig, and aurochs; plants (such as hazel and acorn); and in coastal regions, sea mammals (such as seal), fish, and shellfish. The last were often collected in large numbers, as evidenced by the number of shell middens, which are archaeologically one of the most conspicuous types of Mesolithic sites in many coastal regions of northwestern Europe. In terms of regional productivity, population densities were probably higher than those of Central Europe. Some (e.g., Rowley-Conwy 1983; Zvelebil 1986) have argued that many of these groups were largely sedentary and often socially differentiated (see also Price and Brown 1985). The beginning of the Neolithic in these parts of Europe, as elsewhere, is defined by the first appearance of pottery. In many cases, previous traditions of resource procurement continued unchanged. In short, Mesolithic hunter-gatherer-fishers often became simply Neolithic pottery-using hunter-gatherer-fishers. Sadly, this point has often been ignored in studies of Mesolithic and hunter-gatherer behavior since pottery can provide useful information on exchange networks and social differentiation. In areas such as southern England, however, the first pottery seems to have appeared considerably later than the earliest evidence for cereals (see below), and so no hard and fast rule can be drawn.

Otherwise, the transition from foraging to farming often spanned several centuries, extending from the time when domestic resources first appeared to the time when farming became a dominant part of local subsistence. During this transitional period, neither cereals nor domestic livestock, of which sheep are the least ambiguous, appear to have been more than minor additions to existing subsistence strategies. In this light, it is not surprising to find occasional carbonized cereal grains, palynological evidence for woodland clearance and cereal cultivation, or sheep bones in otherwise Mesolithic contexts or to find few major changes in subsistence until well into the Neolithic. Although more data are needed, the following evidence provides some indication of how protracted the transition to farming was along the Atlantic seaboard.

COASTAL NETHERLANDS, NORTHERN GERMANY, AND SCANDINAVIA

The first crop agriculture in these areas occurred within the context of well-established foraging (Mesolithic) communities. After 4000 B.C., many of these in the Netherlands, northern Germany, and southern Scandinavia began to use pottery, and the necessary techniques may have been acquired independently or from adjacent (LBK) farming groups. At some sites dated to between 4000 and 3000 B.C., there is some evidence for the consumption and, sometimes, for the local cultivation of cereals, along with the herding of domestic livestock. These, however, were generally minor additions to an otherwise foraging way of life until well after 3000 B.C.

Coastal Netherlands Two features of the general background to early farming in this area are worth noting. One is that there were LBK farming sites in southern Holland by 4000 B.C. from which agricultural resources could have been obtained; the other is that the amount of arable land decreased because of rising sea levels in the fourth millennium B.C.

The Mesolithic settlements in the Drenthe area are described by Barker (1985:164–66). Unfortunately, there are no data on the subsistence, although it is likely that a wide range of fish, waterfowl, and terrestrial mammals were exploited. Pottery may have been used as early as 4300 B.C., although the dating is insecure, and it might not have been in general use until well after 4000 B.C. (Zvelebil and Rowley-Conwy 1986:77). The key evidence on early farming in this area comes from the site of Swifterband 3, dated to ca. 3400–3300 B.C. This site was located on one of several low clay levees in an area of tidal flats that were flooded in winter. The presence of chaff fragments, as well as grains of six-row barley and emmer, shows that these were grown locally and not imported (Casparie et al. 1977). Cereal growing must have been on a small scale and would probably have taken place during seasonal visits between March/April and late September (Barker 1985:171). Otherwise, fishing, fowling, and hunting seem to have been the main sources of food. The same conditions prevailed at the site of Hazendonk (ca. 3370 B.C.), which contained numerous remains of fish and some grains and chaff fragments of einkorn and barley (Looue Kooijmans 1976). The persistence of this way of life is evidenced by the sites of the Vlaardingen culture (ca. 2500–2000 B.C.), which have produced much evidence of hunting and fowling and only a little evidence for cereal cultivation (van Zeist 1968:55–65).

Northern Germany Little is known of developments towards farming in this area, but they seem to be similar to those in the coastal Netherlands. The earliest pottery at the site of Dummer dates from 4110 B.C. and is only a little later than the LBK site of Eitzun, which is dated to 4530 ± 210 B.C. and is located 100 km to the southeast (see Zvelebil and Rowley-Conwy 1986:78–79). There is no evidence at all, however, for any domestic crops. Moreover, they are not evidenced in later layers (ca. 3670 B.C.) at Dummer (Zvelebil and Rowley-Conwy 1986:78). It seems likely that foraging provided the main source of food in this area throughout most of the Neolithic.

Scandinavia Relevant data on southern Scandinavia are summarized by Barker (1985:232–39). The appearance of cereal cultivation and stock rearing in this area coincides with the first appearance of Trichterbecker (TRB or "funnel-necked beaker") pottery between 3300 and 2700 B.C. On stylistic grounds, it now seems doubtful that the appearance of pottery represents the immigration of farming groups. Subsistence data confirm this view. According to Madsen (1982), the same types of coastal and lakeside hunting sites that were used in the (pottery-using) Ertebølle phase of the late Mesolithic were also used in the early TRB, and they were used to take the same type of fish, fowl, and game. These sites may have been used seasonally by groups operating from larger residential sites, which were situated inland and had some potential for crop cultivation. Faunal data indicate that, although some sheep/goat were kept, they were less important than cattle and pig, both of which were important in the preceding Mesolithic. Evidence for cereal cultivation is scarce until well after 3000 B.C. At Lidsč, for example, a pit yielded only 9 grains of barley, 2 of emmer, and 2 of einkorn, but they produced 58 seeds of *Rubus* and over 8,500 seeds of *Chenopodium* (Jøorgensen and Fredskild 1978). Cereals are only sparsely represented in other TRB sites elsewhere in Denmark (see Jøorgensen 1976, 1981).

Data from southern Sweden also show that early crop cultivation was incorporated on a small scale into existing foraging strategies (e.g., see Hultén and Welinder 1981). As in Denmark, crop cultivation does not appear before the TRB culture, which some (e.g., Welinder 1982) regard as an intrusive tradition, although others (e.g., Barker 1985:234; Zvelebil and Rowley-Conwy 1986:81) are skeptical. The Swedish evidence is interesting in indicating that early farming was even more small-scale than in Denmark and was then abandoned for several centuries. Subsistence data (Welinder 1982) indicate that hunting continued as the main subsistence activity at TRB sites, although there is some evidence for the cultivation of emmer and barley. There is, however, no evidence for any cereal cultivation in central Sweden between 2700 and 2300 B.C., and it seems likely that this was abandoned in favor of sealing, fishing, and hunting (Zvelebil and Rowley-Conwy 1986:81).

Data from Norway on early farming are extremely sparse, and such carbonized plant remains as have been found are summarized by Griffin (1981). According to Berglund (1985), developments might have been like those in central Sweden: a little crop cultivation at ca. 2700 B.C. and then none until ca. 2000 B.C. As elsewhere in Scandinavia, the first evidence of pottery (at ca. 3000 B.C. in southern Norway) has little economic significance.

BRITISH ISLES

Britain can be treated as a case in its own right, in view of its size, diversity of environments, and amount of relevant data. As elsewhere, the advent of farming was initially assumed to be marked by the synchronous appearance in the Neolithic of cereal cultivation, animal herding, and pottery. For this reason, supposedly "early" Neolithic sites, such as Windmill Hill, were often cited as showing the earliest farming communities in Britain at ca. 2900 B.C.

A very different picture has emerged over the last ten years through radiocarbon dating and palynological studies. It now seems that the Mesolithic and Neolithic overlapped in mainland Britain for at least 300 years and in Ireland for at least 800 years (Williams 1989; Green and Zvelebil 1990). Pollen evidence indicates that cereals may have been used before the Neolithic. Edwards and Hirons (1984) concluded that the earliest reliable indicators of cereals in Britain extend back to ca. 4000 B.C. in contexts that are otherwise indistinguishable from securely Mesolithic ones.

Interpretation of this evidence is still confused. The traditional model envisaged a period of agricultural colonization from the continent (e.g., Case 1969) on the grounds that foragers were incapable or unwilling to acquire these resources themselves. An alternative is that indigenous Mesolithic communities acquired cereals, sheep, and pottery of their own accord from the continent, and incorporated these into existing practices (Dennell 1983:184–87). Which process was the more important is perhaps subsidiary to the point that foragers and farmers coexisted in Britain and Ireland during much of the fourth millennium B.C., before farming became of major regional importance.

Undisputed evidence for cereal cultivation in the form of cereal grain remains or impressions is not found in England until after 3000 B.C. The main evidence for this is still the grain impressions on the pottery from Windmill Hill (Helbaek 1952). Because almost all these were of emmer, Helbaek (1952) concluded that this had been the most important crop. I revised (Dennell 1976b) this view by showing that most of the pottery sherds with emmer impressions had been imported from an area over 50 miles away, which was well suited for wheat cultivation, whereas the pottery made locally contained mostly barley impressions. This conclusion has since found general (e.g., Whittle 1985:221) and partial (e.g., Monk 1986) acceptance.

COASTAL FRANCE, NORTHERN SPAIN, AND PORTUGAL

A useful discussion of what little is known about the transition to agriculture in these areas can be found in Zvelebil and Rowley-Conwy (1986:68–73). As elsewhere, the advent of "the Neolithic" has usually been defined in ceramic terms only and does not appear to indicate more than the acquisition of pottery by well-established foragers. In western France, much of the record for coastal exploitation before the end of the sixth millennium B.C. has been lost due to rising sea levels. The earliest sites with pottery date to ca. 4500–4000 B.C. and seem to indicate a basic continuity with the preceding "Mesolithic" in terms of site location, lithic technology, and resource exploitation. A few also contain faunal assemblages that are dominated by the remains of wild animals but also contain a few sheep or goat bones. The only evidence for cereal cropping comes from the inland site of Roucadour. Despite Scarre's (1983:267) assertion that "even the earliest of the known pottery-using sites had an agricultural base," there is no clear evidence that agriculture was ever more than a minor component of a foraging lifestyle until well into the Neolithic. The same is true of the Pyrenees. As Bahn (1983) points out, if there was a "Neolithic revolution" in this area, it did not happen until the Bronze Age and, even then, pastoralism was far more important than crop agriculture.

A slightly different situation may have prevailed in Brittany in northwestern France, although more data on Neolithic and Mesolithic plant usage are needed. Palynological studies indicate the presence of ruderals and cereals in profiles dating back to the fifth millennium B.C., and so some form of agricultural clearance may have occurred by that time. Data from such sites as the Mesolithic middens of Téviec and Hödièc (ca. 4500–4000 B.C.), however, show an overwhelming predominance of coastal and marine resources, albeit with some very slight evidence for dog, sheep, and domestic cattle. These sites may, of course, represent only part of the annual activities of Mesolithic groups in the area, and crop cultivation and livestock herding may have been more important inland. There is no evidence as yet, however, that crop cultivation was important in this area until late in the third mil-

lennium B.C., long after the beginning of the Neolithic, as defined by the first appearance of pottery (see Hibbs 1983).

BALTIC REGION AND WESTERN RUSSIA

As in northern Europe, the history of the earliest crop cultivation in the Baltic and western Russia is essentially one of a minor component being added to hunter-gatherer subsistence strategies, long after the first appearance of pottery and the ensuing advent of the Neolithic (see Fig. 5.2).

The onset of agriculture in southern Finland coincided with the appearance of the Boat Axe culture at ca. 2500 B.C. Crop agriculture, however, did not take place on a large scale until the mid-first millennium A.D. and even then, as now, wild resources continued to be important (Zvelebil 1978). Much the same picture emerges from western Russia (Dolukhanov 1979, 1986). In the forest zone in the northern part of western Russia, pottery-using foragers persisted from the mid-fifth millennium B.C. until well into the first millennium B.C. Further south in the forest-steppe zone, crop cultivation was practiced on a large scale between ca. 4000 and 2500 B.C. at sites with Triploye assemblages, most of which are found between the Dneister River and the Romanian border. This may indicate a process of agricultural expansion similar to that of the LBK culture into central Europe; whether it supplanted or assimilated local foraging populations is still unclear. Thereafter, pastoralism seems to have been of major importance, and large-scale crop agriculture is not evidenced until the end of the second millennium B.C. Botanical data from southern Russia (Janushevich 1984) indicate a gradual expansion eastward of emmer from the Romanian border areas toward the Don River, but there is no clear indication that crop cultivation was of major importance in these easterly regions until historic times.

Explanations for the Origins of Crop Cultivation in Europe

Despite the immense variation in the material culture of early farming groups and in the quality, quantity, and type of evidence for early agriculture from different parts of Europe, the archaeological evidence indicates three types of patterns (see Figure 5.3). These are:

1. Areas where farming communities appear suddenly.

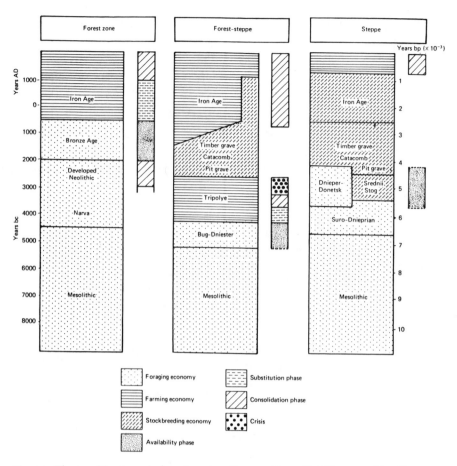

Fig. 5.2. The transition to agriculture in eastern Europe. (From Dolukhanov 1986:118)

In some areas, notably much of Southeast and Central Europe and the Alps, the early Neolithic appears as an intrusive phenomenon into a local Mesolithic foraging context. In those areas where the Mesolithic has been researched, the Mesolithic and Neolithic are spatially distinct and often coexist as distinct entities for up to several centuries before the former disappears.

The remains of villages associated with pottery and domestic crops and animals are a conspicuous feature of the archaeological record in these areas. Some of these settlements, especially in Southeast Europe, were occupied for several centuries; elsewhere, they were occupied for perhaps a few decades. Most were also small by comparison with later examples and rarely contained more than 200 inhabitants. As discussed above, current estimates favor numbers of between 150 and 200 for the largest tell settlements in Bulgaria (Dennell 1978), perhaps 20 to 60 people for most LBK settlements (Hammond 1981), and around

100 for the first farming settlements in Switzerland (Barker 1985:127). The density of early farming communities in these areas was also low, at least at a regional level. Many areas remained unoccupied long after farming first appeared. Examples are large parts of mainland Greece and the Mediterranean islands (Halstead 1981; Cherry 1981); other areas were only lightly settled, such as large parts of Bulgaria (Dennell 1983:156) and Germany (e.g., Hammond 1981).

Overall, the areas where farming appeared suddenly tend to be those with the best archaeobotanical data. Some of the best evidence has come from the tell or occupation mounds of Southeast Europe where large amounts of plant material were processed and cooked near hearths and ovens; sometimes it has come from settlements like Chevdar, Azmak, and Ezero in Bulgaria, which were gratifyingly combustible (see Dennell 1978). Swiss lakeside settlements often were exemplary traps for archaeobotanical material, although LBK sites are generally poor for such

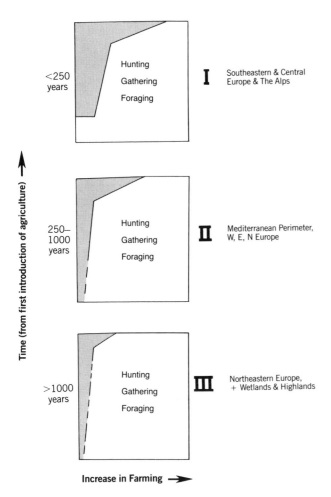

Fig. 5.3. Three archaeological patterns for the transition to crop agriculture in Europe. The vertical axis represents the length of time taken from the first introduction of cereals and legumes to the establishment of a predominantly agricultural way of life. The horizontal axis represents the importance of farming relative to hunting and gathering during this transition.

I, cereals appeared suddenly in the Early Neolithic, along with domestic livestock and pottery. Although there may be a slight overlap with the local Mesolithic, communities entirely dependent upon foraging seem to have disappeared within 250 years. This pattern seems to apply to large areas of Southeast and Central Europe and the Alps.

II, there is no clear point at which domestic crops and animals and pottery were first used in the Mesolithic. Sites used specifically for farming are very rare, and foraging seems to have provided most of the food for several hundred years. In areas such as much of the Mediterranean perimeter and western, eastern, and northern Europe, agriculture was of minor importance for up to 1,000 years after the first appearance of its components.

III, crop agriculture is never wholly successful and was often tried intermittently during prehistory. This pattern predominates in much of northeastern Europe and in many highland and wetland areas. In each case, three questions can be asked: (1) where did the crop and animal resources come from? (2) what factors influenced their subsequent use and importance? and (3) what processes (usually in the Late Neolithic or Early Bronze Age) resulted in a predominantly agricultural lifestyle?

remains, especially when their surfaces have been eroded.

The main questions to be clarified in these areas are: (1) where did the farmers and their resources come from? (2) why was farming sufficiently advantageous that farming communities could be established? and (3) why were foraging settlements later discontinued?

2. Areas where farming occurs gradually, and long before farming settlements appear.

In western Europe, the British Isles, Scandinavia, and along most of the Mediterranean littoral, there is no sharp distinction between the Mesolithic and Neolithic in terms of material culture, site histories, and resource usage. Instead, cereals, legumes, and domestic livestock (especially sheep and goat) appear gradually and in no fixed order. Independent farming settlements are rare, and it is often several centuries before a distinct farming culture appears. The density of communities using agricultural resources is usually low. For example, over much of southern France and lowland Britain, large areas of viable farming land were not used for agriculture for several centuries (see Mills 1983; Bradley 1978).

Foraging sites seem to have been unsuitable for the preservation of significant amounts of plant remains, and this may help explain the paucity of evidence from northern and western Europe in the Neolithic. Cave sites, such as those in southern France, were probably used on a seasonal basis only, and plant material retrieved from these may have been grown elsewhere. Open sites, such as middens, are by their nature inappropriate for the preservation of large amounts of carbonized plant remains in much the same way as farming sites are not likely to yield evidence of such activities as fish-drying.

The main questions to be answered here are: (1) where did the crops and domestic livestock come from? (2) why did foraging continue to form the main resource base? and (3) why did it take so long for agriculture to become the primary resource base?

3. Areas where farming was unsuccessful as a long-term resource base.

In Sweden, Finland, and probably in many highland areas of temperate Europe, the record for early farming is similar to type 2 above, but is discontinuous. Additional questions here are: (1) why was it unsuccessful? and (2) why eventually did it become viable in some areas, but not others?

Traditional Explanations

Traditional explanations were heavily influenced by the nineteenth-century agricultural colonization of North America and Australia by Europeans. This colonization model was based on the supposition that early Neolithic farmers possessed an overwhelming demographic and economic superiority over local Mesolithic hunter-foragers and thus could appropriate their lands and dispossess them. The expansion of the Neolithic across Europe could thus be envisaged as a "wave of advance" (Fig. 5.4), as suggested by Ammerman and Cavalli-Sforza (1984) and Renfrew (1987). The model indicates the progressive expansion of farming communities into Mesolithic hunter-gatherer territories, much in the manner suggested by Childe (1958).

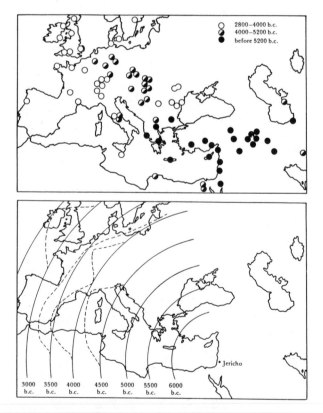

Fig. 5.4. Radiocarbon dating and early European agriculture. Top: the spread of agriculture (implied largely by the presence of pottery) across Europe according to the radiocarbon dates available in the early 1960s (after Clark 1965, fig. 2); and below, the "wave of advance" of farmers, according to Ammerman and Cavalli-Sforza (1984). The broken curved lines denote regional variations in the rate at which farming was adopted. (From Barker 1985:6)

This model has been criticized on several grounds as an all-embracing explanation for the spread of early farming across Europe (e.g., see Dennell 1983:152–89; Barker 1985:250–56; Zvelebil 1986:176–80). First, it is often unclear whether the expansion of the Neolithic indicates the first usage of pottery or cultigens; if the latter, it is often unclear whether the first evidence for cultigens indicates that crop agriculture was of major importance. Second, as seen already, the size and density of early farming populations do not seem to have been sufficiently high to cause or require high rates of emigration into new areas. Third, the "wave of advance" was often static for longer than would be expected if agriculture had expanded inexorably across Europe because of its inherent superiority. For example, early crop farming developed in pockets along the Mediterranean perimeter at an early date; however, in eastern Greece, the eastern parts of southern Italy, probably Sicily, and in small areas of southern France, it made little significant expansion from those areas for at least a millennium afterward. Likewise, in northern Europe, the early farming (LBK) settlements of Korlat and Eitzun are virtually contemporaneous (ca. 4500 B.C.), and yet are 1300 km apart. By contrast, the earliest site with evidence for agriculture that is north of Eitzun is Siggenben-Sud, which is only 200 km away but 1,300 radiocarbon years later (see Zvelebil and Rowley-Conwy 1986:79). Another instance is Sweden, where the agricultural frontier expanded but then retreated. A fourth weakness of the "wave of advance" model is the protracted nature of the transition from foraging to farming in many areas. As noted above, the transition to agriculture often spanned at least several centuries; what the evidence suggests is less the rapid adoption of agriculture than its slow and very gradual assimilation.

A second traditional explanation that has been tried in the last twenty years favors local domestication of cereals and legumes. Mesolithic populations in Southeast Europe and along the Mediterranean perimeter were probably able to domesticate some of these plants locally. As noted above, wild barley grains have been found in late Pleistocene and early Holocene deposits at Franchthi Cave in southern Greece, and similar finds might be expected from elsewhere along the Mediterranean perimeter. (As seen above, however, this is not true of Grotta dell'Uzzo.) One should also note the pollen grains of what seem to be cereals from the Mesolithic site of Icoana in Romania (Cârciumaru 1973). In this case, the identification seems to have been careful and the context secure. On the other hand, it is highly unlikely that wild einkorn and barley grew in temperate Europe before the Neolithic, so the domestic forms and derivatives of these must have been introduced. Because domestic forms of rye are known from the aceramic Neolithic of Turkey (Hillman 1978), it is possible that they, too, were introduced into Europe. The only major cereal crop that might have originated in Europe is probably oats, but this seems to have been unimportant until late in prehistory.

The early Holocene distribution of those legumes that were later domesticated is still problematic, as is the relationship between present-day wild and domestic forms. In the absence of information on pre-Neolithic legumes in Europe, domestic pea, lentil, and vetch are usually assumed to have originated in the Near East (Zohary and Hopf 1988). If the claims of Vacquer et al. (1986) for Mesolithic domestic legumes at Abeurador in France are confirmed, however, a European origin for some domestic legumes seems reasonable.

Explanations of the origins of European crop agriculture in terms of local domestication have only a limited applicability, however. First, there is no reason to suppose that this could have occurred in temperate Europe where cultigens must have been introduced. Secondly, models of local domestication do not explain why farming appeared later as one moves from southeastern to northwestern Europe.

Toward an Alternative Explanation

Early farming in Europe always occurred in areas where there were already hunter-forager communities. These cannot be regarded as irrelevant to the pattern of agricultural expansion. First, it is likely that they were capable of domesticating or, at least, of carefully husbanding plant and animal resources if they saw it was in their interests to do so. Numerous ethnographic studies over the last twenty years have shown that modern and recent hunter-gatherers can manage their environment in deliberate and productive ways. Australian data have been particularly important in this context over the last two decades. Allen (1974), for example, documented the deliberate reaping, threshing, storage, and sowing of grasses by some aboriginal groups; Lourandos (1980) described the planned construction of channels to trap eels; and Jones (1969) discussed how Tasmanian aborigines deliberately used fire to clear woodlands as a way of increasing the productivity of game and plant foods.

Ethnographic data also indicate that hunter-gatherers are not innately conservative and unwilling to change their lifestyles. As Schrire (1980) has shown, the San bushmen are far from being examples of unchanging hunter-gatherers. Instead, they have alternated between hunting/gathering and herding their own or others' livestock since they were first encountered by Europeans in the sixteenth century. Shrire's (1985) edited volume stresses the flexibility of hunter-gatherers and their ability to innovate and adopt new strategies if they perceive it is in their interest to do so. The realization that recent hunter-gatherers can turn to herding and crop cultivation if they perceive this to be advantageous has major implications for studies of agricultural origins in Europe.

These and similar studies have blurred a formerly crisp distinction between hunting/gathering and farming and imply that European Mesolithic groups may have practiced some form of food production. Strong interest in this possibility has come from several British workers, especially those derived from the Cambridge-based palaeoeconomic school of Higgs and his associates. As was pointed out by some researchers (e.g., Jarman 1972; Jarman and Wilkinson 1972), if the criteria for recognizing animal and plant domestication were applied consistently, one would have to conclude that domestication occurred before the Neolithic and involved "wild" resources, such as red deer, hazel, and wild barley. If this were so, the question to be asked would not be "When and where did domestication first occur?" but "Which resources may have been domesticated but were later discarded in favor of others?" Pertinent examples here might be red deer, which were husbanded and perhaps herded in the Mesolithic (e.g., Jarman 1971) but then discarded in favor of sheep, which could be stocked at higher rates, were easily herded, and, at least in later periods, could be used for their wool and milk as well as their meat. On the plant side, acorns and hazel nuts may have been key Mesolithic plant foods that were later replaced by other storable, protein-rich plants, such as legumes and cereals, which had higher yields, were easier to process on a large-scale, and could be more easily harvested in greater abundance should the occasion demand it.

An alternative approach is to study early farming in terms of the ways that foragers and farmers interacted at a regional level. The "frontier" between the two could have taken a variety of forms—from mobile to static—and both forms could have been either porous or impervious (see Dennell 1983, 1985). As indicated in Figure 5.5, there are several ways that foraging populations could have acquired the necessary techniques or resources for developing agriculture. Zvelebil and Rowley-Conwy (1986) have recently developed the notion of the "frontier" as a spatial concept by suggesting how it may have evolved through time. According to their model, hunter-gatherer populations with access to farming communities would have passed through three phases—availability, substitution, and consolidation—in their own transition to farming. The earliest evidence for cultigens should occur in the first phase; in the second phase, cultigens would gradually become more important but not significantly disrupt existing practices; and in the third phase, they would become one of the mainstays of the economy. As Zvelebil and Rowley-Conwy (1986) indicate in general terms and as suggested here archaeobotanically, the length of time involved in this process is often considerable.

The three types of archaeological patterns identified in this chapter can be explained in terms of different types of interactions between foragers and farmers. In areas where farming appears as an intrusive phenomenon, it is likely that the initial impetus for agriculture came from the outside through colonization. Thereafter, it is likely that the inhabitants of these communities developed symbiotic relationships with neighboring foraging groups, as I have suggested for the LBK (Dennell 1985) and Zvelebil (1986) and Gregg (1988) have suggested for the Alpine forelands. It is debatable whether the success of farming in these areas stemmed from the stability of these relationships as much as from its own strengths. In areas where farming appeared gradually within the matrix of local Mesolithic traditions, it is more probable that the resources were acquired from neighboring groups and then developed indigenously, with only minor modifications to existing subsistence strategies and without the need to establish independent farming groups.

The reasons why crop cultivation was first adopted as a minor addition to existing practices probably varied from area to area. In areas such as Jutland and Finland, the eventual adoption of crop cultivation may be explained by the failure of traditional resources, such as the oyster and seal, respectively (e.g., Zvelebil and Rowley-Conwy 1986:88). Elsewhere, the critical factor may have been the need for a storable resource for winter usage, or the desirability of a reliable resource, or one whose production could be expanded at short notice, or one that used otherwise unproductive members of a group, such as the very young or infirm, to produce food by weeding, crop processing, etc. Other reasons might include the attraction of a storable resource that could be used to generate surpluses and thus enhance the status of individuals or groups.

Zvelebil (1986) has suggested that the adoption of farming proceeded through the three phases of availability, substitution, and consolidation, irrespective of how it began. In general terms, the process was accomplished far more rapidly in those areas where it appeared suddenly than it did along the Mediterranean and in northern and western Europe. What is interesting, however, is that even in these areas agriculture was firmly consolidated by ca. 3000 B.C. To that extent, there was an agricultural revolution in temperate Europe, even though it occurred long after farming was first practiced.

Conclusions

The concept of the "Neolithic" as signifying the appearance of agriculture probably has done more to obscure than to illuminate the nature of the processes involved. This is because there is a very rich record of hunter-gatherers buried under the general rubric of "the Neolithic." Such is the weight of tradition,

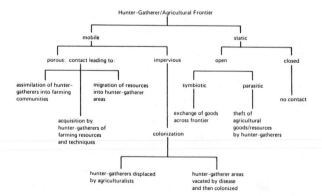

Fig. 5.5. Some examples of the types of frontiers that could have existed in Europe between early farmers and hunter-gatherers at the time of contact. (From Dennell 1985, fig. 6.4)

however, that studies of Holocene hunter-gatherers usually end with the first appearance of pottery. Conversely, prehistorians of the early Neolithic (and, in many areas of Europe, the remainder of the Neolithic and often much of the Bronze Age) have tended to focus their attention on the presence of pottery and on the generally scant evidence for agriculture, and have overlooked the often impressive evidence for the continuity of previous foraging strategies.

Recommendations for Future Research

The following recommendations are made for future studies of early crop agriculture in Europe:

1. More data are needed on plant usage before and after the Mesolithic-Neolithic transition. Detailed contextual data are also needed on the provenience of archaeobotanical samples from early farming settlements in order to clarify the importance and use of each cultigen.

2. Because plant remains are not generally common in early Neolithic sites outside Southeast and Central Europe, flotation techniques should be obligatory on excavations of Neolithic (and later) sites. In some cases, accelerator radiocarbon dating may also be necessary to confirm that cereal grains and seeds and domestic legumes are contemporaneous with the deposits in which they are found rather than the result of percolation from later contexts.

3. At present most European archaeobotanists are based in northwestern Europe, and few experts in this field are resident in Portugal, Spain, France, Italy, and Greece. More funding is required both for training archaeobotanists and for post-excavation analysis.

4. Scanning electron microscopy (SEM) work is needed to verify critical identifications, particularly of the wild and domestic forms of oats and many of the legumes. Data on the last-mentioned seem particularly useful if, as suggested above, some may have been domesticated locally in Europe.

5. Major changes in our understanding of the transition from foraging to farming in Europe are likely to occur through chemical studies of human skeletal remains. Techniques that have been, or are being, developed in recent years include measuring the ratio of $N14$ to $N15$, and $C12$ to $C13$ (e.g., see Keegan 1989) and the amounts of trace elements in bone tissue (e.g., Aufderheide 1989). These techniques do not provide a direct indication of which foods were eaten but may indicate which groups of foods with similar chemical "signatures" were consumed. Analyses of skeletons through time may therefore show gross shifts in diet from one food group to another, and these in turn may be linked to the conventional types of archaeobotanical and archaeozoological evidence cited in this chapter. Chemical analyses may also indicate differences in protein intake between males and females, between adults and children, and/or between high- and low-ranking individuals.

Another exciting development arises from the identification of DNA in ancient human bone and tissue (see, for example, Hagelberg and Clegg 1991). If it can be routinely identified in European Mesolithic to Bronze Age skeletal remains, DNA data may become vital to key debates over whether agriculture spread over most of Europe through ethnic movements—as suggested by Childe and many others—or through the movement of resources but not people, as argued here.

6. The teaching of European agricultural origins should take into account the new view of the Neolithic that is now emerging. Traditionally, a rigid distinction has been drawn between the Palaeolithic and Mesolithic as the study of hunter-gatherers on the one hand, and the Neolithic and Bronze Age as the study of farmers on the other. As I have argued in this chapter, the Neolithic over much of Europe is primarily about the coexistence of farmers and hunter-gatherers, many of whom gradually acquired agricultural resources and developed them on their own accord.

Suggested Reading

Several syntheses of European prehistory provide useful background reading for both the general reader and for those wishing to study the subject in depth. In order to understand present debates over the respective roles of colonization, diffusion, and acculturation, Gordon Childe's works are still essential starting points, particularly his best-known work, *The Dawn of European Civilization* (1925; revised until 1957) and what is arguably his finest general summary, *The Prehistory of European Society* (1958). Graham Clark and Stuart Piggott's (1965) *Prehistoric*

Societies and Piggott's (1965) *Ancient Europe* provide similarly well-written and magisterial overviews. Colin Renfrew's (1973) *Before Civilisation* shows how radiocarbon 14 dating undermined previous views on the importance of the Near East as a source of innovation. More recent accounts of the European Neolithic can be found in Sarunas Milisauskas's (1978) *Prehistoric Europe;* Patricia Phillips's (1981) *European Prehistory;* Tim Champion's (1984) *Prehistoric Europe;* and Alistair Whittle's (1985) *The Neolithic of Europe.*

Three works specifically on early European agriculture should be mentioned. One is the volume from the Cambridge palaeoeconomic school of the late Eric Higgs, *Early European Agriculture,* edited by Michael Jarman et al. (1982), and the other is a more rounded work from the same tradition, *Prehistoric Farming in Europe* (1985) by Graeme Barker. *Hunters in Transition,* edited by Marek Zvelebil (1986), provides an important perspective on early farming in Europe from the viewpoint of the indigenous Mesolithic communities.

Acknowledgments

I would like to thank Wes Cowan and Patty Jo Watson for their patience while I compiled this chapter, and for arranging my participation at the AAAS symposium in Los Angeles in 1985. Glynis Jones and an anonymous critic wielded very hefty sticks, and Paul Halstead, Linda Hurcombe, and Marek Zvelebil offered some worthwhile carrots. Mavis Torrey is thanked for helping with the preparation of the manuscript.

Notes

1. All dates cited in this chapter are uncalibrated radiocarbon dates, unless otherwise stated.

References Cited

Allen, H.
  1974   The Bagundji of the Darling Basin: Cereal Gatherers in an Uncertain Environment. World Archaeology 5 (3):309–22.

Ammerman, A. J., and L. L. Cavalli-Sforza
  1984   The Neolithic Transition and the Genetics of Population in Europe. Princeton University Press, Princeton.

Aufderheide, A. C.
  1989   Chemical Analysis of Skeletal Remains. *In* Reconstruction of Life from the Skeleton, edited by M. H. Íscan and K. A. R. Kennedy, pp. 236–60. Alan Liss, New York.

Bahn, P.
  1983   The Neolithic of the French Pyrenees. *In* Ancient France: 6000–2000 B.C., edited by C. Scarre, pp. 184–222. University Press, Edinburgh.

Bakels, C. C.
  1978   Four Linearbandkeramik Settlements and Their Environments. Analecta Praehistorica Leidensia 11:1–248.

  1984   Carbonised Seeds from Northern France. Analecta Praehistorica Leidensia 17:1–25.

Barker, G. W.
  1975   Early Neolithic Land Use in Yugoslavia. Proceedings of the Prehistoric Society 41:85–104.

  1985   Prehistoric Farming in Europe. Cambridge University Press, Cambridge.

Baudais-Lundstrom, K.
  1978   Plant Remains from a Swiss Neolithic Lakeshore Site: Brise-Lames, Auvernier. Berichte der deutschen botanischen Gesellschaft 91:67–83.

Berglund, B. E.
  1985   Early Agriculture in Scandinavia: Research Problems Related to Pollen-Analytical Studies. Norwegian Archaeological Review 18 (1–2):77–105.

Bertsch, K., and F. Bertsch
  1949   Geschichte unserer Kulturpflanzen. Stuttgart.

Bogucki, P.
  1988   Forest Farmers and Stockherders: Early Agriculture and its Consequences in North-Central Europe. Cambridge University Press, Cambridge.

Bogucki, P., and R. Grygiel
  1983   Early Farmers of the North European Plain. Scientific American 248 (4):96–104.

Bradley, R.
  1978   The Prehistoric Settlement of Britain. Routledge and Kegan Paul, London.

Candolle, A. de
  1884   The Origin of Cultivated Plants. Kegan Paul, London.

Cârciumaru, M.
  1973   Analyse pollinique des coprolites livrés par quelque stations archéologiques des deux bords du Danube dans la zone des "Portes de Fer." Dacia 17:53–60.

Case, H.
  1969   Neolithic Explanations. Antiquity 43:176–86.

Casparie, W. A., B. Mook Kamps, B. Palfenier-Vegter, P. C. Struijk, and W. van Zeist
  1977   The Palaeobotany of Swifterband. Helinium 17:28–55.

Champion, T. (editor)
  1984   Prehistoric Europe. Academic Press, London.

Chapman, J., and J. Muller
  1990   Early Farmers in Dalmatia. Antiquity 64:127–34.

Cherry, J.
  1981   Pattern and Process in the Earliest Colonisation of the Greek Islands. Proceedings of the Prehistoric Society 47:41–68.

Childe, V. G.
  1925   The Dawn of European Civilisation. Routledge, Kegan and Paul, London.

  1958   The Prehistory of European Society. Penguin, London.

Clark, J. G. D.
  1965   Radiocarbon Dating and the Expansion of Farming from the Near East over Europe. Proceedings of the Prehistoric Society 21:58–73.

Clark, J. G. D., and S. Piggott
  1965   Prehistoric Societies. Hutchinson, London.

Constantini, L.
  1989   Plant Exploitation at Grotta dell'Uzzo, Sicily: New Evidence for the Transition from Mesolithic to the Neolithic Subsistence in Southern Europe. In Foraging and Farming: The Evolution of Plant Domestication, pp. edited by D. R. Harris and G. C. Hillman, pp. 197–206. Unwin Hyman, London.

Courtin, J., and J. Erroux
  1974   Aperçu sur agriculture préhistorique dans le Sud-Est de la France. Bulletin de la Societé préhistorique française 71:321–34.

Dennell, R. W.
  1974a   Botanical Evidence for Prehistoric Crop Processing Activities. Journal of Archaeological Science 1:275–84.

  1974b   The Purity of Prehistoric Crops. Proceedings of the Prehistoric Society 40:132–35.

  1976a   The Economic Importance of Plant Resources Represented on Archaeological Sites. Journal of Archaeological Science 3:229–47.

  1976b   Prehistoric Crop Cultivation in Southern England: A Reconsideration. Antiquaries Journal 56 (1):11–23.

  1977   On the Problems of Studying Prehistoric Climate and Crop Agriculture. Proceedings of the Prehistoric Society 43:361–69.

  1978   Early Farming in South Bulgaria from the VIth to the IIIrd Millennia B.C. BAR International Series 45. British Archaeological Reports, Oxford.

  1983   European Economic Prehistory: A New Approach. Academic Press, London.

  1985   The Hunter-Gatherer; Agricultural Frontier in Prehistoric Temperate Europe. In The Archaeology of Frontiers and Boundaries, edited by S. Green and S. Perlman, pp. 113–39. Academic Press, London.

Dennell, R. W., and D. Webley
  1975   Prehistoric Settlement and Land Use in Southern Bulgaria. In Palaeoeconomy, edited by E. S. Higgs, pp. 97–109. Cambridge University Press, Cambridge.

Diamant, S.
  1979   A Short History of Archaeological Sieving at Franchthi Cave, Greece. Journal of Field Archaeology 6:203–17.

Dolukhanov, P. M.
  1979   Ecology and Economy in Neolithic Eastern Europe. Duckworth, London.

  1986   The Late Mesolithic and the Transition to Food Production in Eastern Europe. In Hunters in Transition, edited by M. Zvelebil, pp. 109–19. Cambridge University Press, Cambridge.

Edwards, K. J., and K. R. Hirons
  1984   Cereal Pollen Grains in Pre-Elm Decline Deposits: Implications for the Earliest Agriculture in Britain and Ireland. Journal of Archaeological Science 11:71–80.

Geddes, D.
  1985   Mesolithic Domestic Sheep in West Mediterranean Europe. Journal of Archaeological Science 12:25–48.

Green, S., and M. Zvelebil
  1990   The Mesolithic Colonisation and Agricultural Transition of Southeast Ireland. Proceedings of the Prehistoric Society 65:57–88.

Gregg, S. A.
1988   Foragers and Farmers. Chicago University Press, Chicago.

Griffin, K.
1981   Plant Remains from Archaeological Sites in Norway: A Review. Zeitschift für Archaölogie 15:163–76.

Groenman van Waateringe, W.
1979   The Origin of Crop Weed Communities Composed of Summer Annuals. Vegetatio 41 (2):57–59.

Guilaine, J., M. Barbaza, D. Geddes, and J.-L. Vernet
1982   Prehistoric Human Adaptations in Catalonia (Spain). Journal of Field Archaeology 9:407–16.

Hagelberg, E., and J. B. Clegg
1991   Isolation and Characterization of DNA from Archaeological Bone. Proceedings of the Royal Society of London Series B. 244:45–50.

Hajnalová, E.
1973   Príspevok k štúdiu, anal'ýze a interpretácii nálezov kultúrnych rastlín na Slovensku. Slovenská Archeólogia 21 (1):211–20.

1976   Odtlačky kultúrnych rastlín z neolitu na východnom Slovensku. Archaeologia Rozhledy 29:121–36.

Halstead, P.
1981   Counting Sheep in Neolithic and Bronze Age Greece. In Pattern of the Past: Studies in Memory of David Clarke, edited by I. Hodder, G. Isaac, and N. Hammond, pp. 307–39. Cambridge University Press, Cambridge.

Halstead, P., and G. Jones
1980   Early Neolithic Economy in Thessaly—Some Evidence from Excavations at Prodomos. Anthropologika 1:4–5.

Hammond, F.
1981   The Colonisation of Europe: The Analysis of Settlement Process. In Pattern of the Past: Studies in Memory of David Clarke, edited by I. Hodder, G. Isaac and N. Hammond, pp. 211–48. Cambridge University Press, Cambridge.

Hansen, J.
1978   The Earliest Seed Remains from Greece: Palaeolithic through Neolithic at Franchthi Cave. Berichte der deutschen botanischen Gesellshaft 91:39–46.

Hansen, J., and J. M. Renfrew
1978   Palaeolithic-Neolithic Seed Remains at Franchthi Cave, Greece. Nature 271:349–52.

Hartyányi, B. P., and G. Novaki
1975   Samen—und Fruchtfunde in Ungarn von der Neusteinzeit bis zum 18. Jahrhundert. Agrártörténeti Szemle 17:1–88.

Heer, O.
1866   Die Pflanzen de Pfahlbauten. Neujahrsblatt der Naturforschenden Gesellschaft in Zurich 68:1–54.

Heitz, A., S. Jacomet, and H. Zoller
1981   Vegetation, Sammelwirtschaft, und Ackerbau im Zurichseegebiet zur Zeit der neolithichen und spätbronzezeitlichen Ufersiedlungen. Helvetia Archaeologia 45–48:139–52.

Helbaek, H.
1952   Early Crops in Southern England. Proceedings of the Prehistoric Society 12:194–223.

Hibbs, J.
1983   The Neolithic of Brittany. In Ancient France 6000–2000 B.C., edited by C. Scarre, pp. 271–323. University Press, Edinburgh.

Hillman, G.
1978   On the Origins of Domestic Rye—Secale Cereale: The Finds from Aceramic Can Hasan III in Turkey. Anatolian Studies 28:157–74.

1981   Crop Husbandry Practices from Charred Remains of Crops. In Farming Practices in British Prehistory, edited by R. Mercer, pp. 123–62. University Press, Edinburgh.

Hopf, M.
1962   Bericht über die Untersuchung von Samen und Holzkohlenresten von der Argissa-Magula aus den präkeramischen bis mittelbronzezeitlichen Schichten. In Die Deutschen Ausgrabungen auf der Argissa-Magula in Thessalien, edited by V. Milojčič, J. Boesneck, and M. Hopf, pp. 101–10. Rudolf Habelt Verlag, Bonn.

1973   Frühe Kulturpflanzen aus Bulgarien. Jahrbuch des Römanisches-Germanisches Zentralmuseums, Mainz 20:1–47.

1974   Pflanzenreste aus Siedlungen der Vinča-Kulture in Jugoslawien. Jahrbuch des Römanisches-Germanisches Zentralmuseums, Mainz 21:1–11.

Hopf, M., and M. P. Catalan
1970   Neolithische Getreidefunde in der Höhle von Nerja (Prov. Málaga). Madrider Mitteilungen 11:18–34.

Hopf, M., and A. M. Muñoz
1974   Neolithische Pflanzenreste aus der Höhle Los Murciélagos bei Zuheros (Prov. Córdoba). Madrider Mitteilungen 15:9–27.

Hopf, M., and H. Schubart
1965   Getreidefunde aus der Coveta de L'Or (Prov. Alicante). Madrider Mitteilungen 6:20–38.

Hubbard, R. N. L. B.
1976   Crops and Climate in Prehistoric Europe. World Archaeology 8 (2):159–68.

Hultén, B., and S. Welinder
1981   A Stone Age Economy. Thesis and Papers in North-European Archaeology 11. Lund.

Ilett, M.
1983   The Early Neolithic of North-Eastern France. *In* Ancient France 6000–2000 B.C., edited by C. Scarre, pp. 6–33. University Press, Edinburgh.

Iversen, J.
1941   Land Occupation in Denmark's Stone Age. Danmarks Geologische Undersgelse 2 (66):1–68.

Jacobsen, T. W.
1976   17,000 Years of Greek Prehistory. Scientific American 234 (6):76–87.

1981   Franchthi Cave and the Beginning of Settled Village Life in Greece. Hesperia 50:303–18.

Jacomet, S.
1981   Neue Untersuchungen botanischer Grossreste an jungsteinzeitlichen Seeufersiedlungen im Gebiet der Stadt Zürich (Schweiz). Zeitschrift für Archäologie 15:125–40.

1986   Kulturpflanzenfunde aus der neolithischen Seeufersiedlung Cham-St. Andreas. Jahrbuch der Schweizerischen Gesellschaft für Ur- und Frühgeschichte 69:55–62.

1987   Ackerbau, Sammelwirtschaft und Umwelt der Egozwiler und Cortaillodsiedlungen. Ergebnisse, samenanalytischer Unutersuchungen. *In* Zürich 'Kleiner Hafner', edited by P.J. Suter. Berichte der Zürcher Denkmalpflege 3:144–66.

Jacomet, S., and J. Schibler
1985   Die Hahrungsversorgung eines jungsteinzeitlichen Pfyndorfes am unteren Zürichsee. Archäologie der Schweiz 8:125–41.

Jacomet-Engel, S.
1980   Botanische Makroreste aus der neolithichen Seeufersiedlungen des Areals "Pressehaus Ringier" in Zürich (Schweiz). Stratigraphische und vegetationskundliche Auswertung. Viertieljahresschrift der Naturforschenden Gesellschaft in Zürich 125(2):73–175.

Janushevich, Z. V.
1984   The Specific Composition of Wheat Finds from Ancient Agricultural Centres in the USSR. *In* Plants and Ancient Man, edited by W. van Zeist and W. A. Casparie, pp. 267–73. Balkema, Rotterdam.

Jarman, H. N.
1972   The Origins of Wheat and Barley Cultivation. *In* Papers in Economic Prehistory, edited by E. S. Higgs, pp. 15–26. Cambridge University Press, Cambridge.

Jarman, H. N., A. J. Legge, and J. A. Charles
1972   Retrieval of Plant Remains from Archaeological Sites by Froth Flotation. *In* Papers in Economic Prehistory, edited by E. S. Higgs, pp. 15–26. Cambridge University Press, Cambridge.

Jarman, M. R.
1971   Culture and Economy in the North Italian Neolithic. World Archaeology 3:255–65.

Jarman, M. R., and D. Webley
1975   Settlement and Land Use in Capitanata, Italy. *In* Palaeoeconomy, edited by E. S. Higgs, pp. 177–221. Cambridge University Press, Cambridge.

Jarman, M. R., G. N. Bailey, and H. N. Jarman
1982   Early European Agriculture: Its Foundation and Development. Cambridge University Press, Cambridge.

Jarman, M. R., and P. F. Wilkinson
1972   Criteria of Animal Domestication. *In* Papers in Economic Prehistory, edited by E. S. Higgs, pp. 83–96. Cambridge University Press, Cambridge.

Jones, G.
1984   Interpretation of Archaeological Plant Remains: Ethnographic Models from Greece. *In* Plants and Ancient Man, edited by W. van Zeist and W. A. Casparie, pp. 43–61. Balkema, Rotterdam.

1987   A Statistical Approach to the Archaeological Identification of Crop Processing. Journal of Archaeological Science 14:311–23.

Jones, R.
1969   Firestick Farming. Australian Natural History 16:224–28.

Jøorgensen, G.
1975   *Triticum aestivum* s.l. from the Neolithic site of Weier in Switzerland. Folia Quaternaria (Cracow) 46:7–21.

1976   Et kornfund fra Sarup. Kuml 47–64.

1981   Cereals from Sarup. With Some Remarks on Plant Husbandry in Neolithic Denmark. Kuml 221–31.

Jøorgensen, G., and B. Fredskild
1978   Plant Remains from the TRB Culture, Period MN V. *In* The Final TRB Culture in Denmark, edited by K. Davidsen, pp. 189–92. Arkaeologiske Studier 5. Copenhagen.

Kamieńska, J., and A. Kulczycka-Leciejewiczowa
1970   The Neolithic and Early Bronze Age Settlement at Samborzec in the Sandomierz District. Archaeologià Polona 12:223–46.

Keegan, W.
1989   Stable Isotope Analysis of Prehistoric Diet. *In* Reconstruction of Life from the Skeleton, edited by M. H. Ícan and K. A. R. Kennedy, pp. 223–36. Alan Liss, New York.

Knörzer, K.-H.
1968   6000 jährige Geschichte der Getreidenahrung im Rheinland. Decheniana 119:113–24.

1972   Subfossile Pflanzenreste aus der bandkeramischen Siedlung Langweiler 3 und 6, Kreis Julich, und ein urnenfelderzeitlicher Getreidefund innerhalb dieser Siedlung. Bonner Jahrbuch 172:395–403.

1973   Der bandkeramische Siedlungsplatz Langweiler 2: Pflanzliche Grossreste. Rheinische Ausgrabungen 13:139–52.

1977   Pflanzliche Grosssreste des bandkeramischen Siedlungsplatzes Langweiler 9. Rheinische Ausgrabungen 18:279–303.

Kruk, J.
1980   The Neolithic Settlement of Southern Poland. BAR International Series 93. British Archaeological Reports, Oxford.

Kuper, R., J. Löhr, J. Lüning, and P. Stehli
1974   Untersuchungen zur neolithischen Besiedlung der Aldenhovener Platte IV. Bonner Jahrbuch 174:424–508.

Kuper, R., H. Löhr, J. Lüning, P. Stehli, and A. Zimmerman
1977   Der Bandkeramische Siedlungsplatz Langweiler 9, Gem. Aldenhoven, Kr. Duren. Rheinische Ausgrabungen 18.

Küster, H.
1984   Neolithic Plant Remains from Eberdingen-Hochdorf, Southern Germany. *In* Plants and Ancient Man, edited by W. van Zeist and W. A. Casparie, pp. 307–21. Balkema, Rotterdam.

Legge, A. J.
1986   Seeds of Discontent: Accelerator Dates on Some Charred Plant Remains from the Kebaran and Natufian Cultures. *In* Archaeological Results from Accelerator Dating, edited by J. Gowlett and R. Hedges, pp. 13–21. Oxford University Committee for Archaeology, Monograph 1.

Lewthwaite, J.
1986   The Transition to Food Production: A Mediterranean Perspective. *In* Hunters in Transition, edited by M. Zvelebil, pp. 53–66. Cambridge University Press, Cambridge.

Lisitsina, G. N., and L. A. Filipovich
1980   Paleoetnobotanischeske naxodki na Balkanskom ployostrove. Studia Praehistorica (Sofia) 4:5–90.

Looue Kooijmans, L. P.
1976   Local Developments in a Borderland. A Survey of the Neolithic of the Lower Rhine. Oudheidkundige Mededelingen 57:227–97.

Lourandos, H.
1980   Change or Stability?: Hydraulics, Hunter-Gatherers and Population in Temperate Australia. World Archaeology 11 (3):245–64.

Lundstrom-Baudais, K.
1984   Palaeo-ethnobotanical Investigation of Plant Remains from a Neolithic Lakeshore Site in France. *In* Plants and Ancient Man, edited by W. Van Zeist and W. A. Casparie, pp. 293–305. Balkema, Rotterdam.

Lüning, J.
1982   1982 Research into the Bandkeramik Settlement of the Aldenhovener Platte. Analaecta Praehistorica Leidensia 15:1–29.

Madsen, T.
1982   Settlement Systems of Early Agricultural Societies in East Jutland, Denmark: A Regional Study of Change. Journal of Anthropological Archaeology 1:197–236.

Milisauskas, S.

1978   European Prehistory. Academic Press, London.

1984   Settlement Organization and the Appearance of Low Level Hierarchical Societies during the Neolithic in the Bronocice Microregion, Southeastern Poland. Germania 62:1–30.

Mills, N.

1983   The Neolithic of Southern France. *In* Ancient France 6000–2000 B.C., edited by C. Scarre, pp. 91–145. University Press, Edinburgh.

Modderman, P. J. R.

1970   Linearbandkeramik aus Esloo und Stein. Analecta Praehistorica Leidensia 3.

1977   Die neolithische Besiedlung bei Hienheim, Ldkr. Kelheim. Analecta Praehistorica Leidensia 10.

Monk, M.

1986   Evidence from Macroscopic Plant Remains for Crop Husbandry in Prehistoric and Early Historic Ireland: A Review. Journal of Irish Archaeology 3:31–36.

Payne, S.

1975   Faunal Changes at Franchthi Cave from 20,000 B.C. to 3000 B.C. *In* Archaeozoological Studies, edited by A. T. Clason, pp. 120–31. North Holland Publlishing Company, Amsterdam.

Phillips, P.

1975   Early Farmers of West Mediterranean Europe. Hutchinson, London.

1981   European Prehistory. Hutchinson, London.

Piggott, S.

1965   Ancient Europe. Aldine, Chicago.

Price, T. D., and J. A. Brown

1985   Prehistoric Hunter-Gatherers: The Emergence of Cultural Complexity. Academic Press, London.

Price, T. D., M. J. Schoeninger, and G. J. Armelagos

1985   Bone Chemistry and Past Behaviour: An Overview. Journal of Human Evolution 14:419–47.

Renfrew, C.

1973   Before Civilisation. Jonathon Cape, London.

1987   Archaeology and Language. Jonathan Cape, London.

Renfrew, J.

1966   A Report of Recent Finds of Carbonised Cereal Grains and Seeds from Prehistoric Thessaly. Thessalika 5:21–36.

1969   The Archaeological Evidence for the Domestication of Plants: Methods and Problems. *In* The Domestication and Exploitation of Plants and Animals, edited by P. J. Ucko and G. W. Dimbleby, pp. 149–72. Duckworth, London.

1973   Palaeoethnobotany. Methuen, London.

1979   The First Farmers in South East Europe. Archaeo-Physika 9:243–65.

Rodden, R. J.

1962   Excavations at the Early Neolithic Site at Nea Nikomedia, Greek Macedonia. Proceedings of the Prehistoric Society 28:267–88.

1965   The Early Neolithic Village in Greece. Scientific American 212(4):83–91.

Rowley-Conwy, P.

1981   Slash and Burn in the Temperate European Neolithic. *In* Farming Practice in British Prehistory, edited by R. Mercer, pp. 85–96. University Press, Edinburgh.

1983   Sedentary Hunters: The Ertebolle Example. *In* Hunter-Gatherer Economy in Prehistory, edited by G. Bailey, pp. 111–26. Cambridge University Press, Cambridge.

Sakellardis, M.

1979   The Mesolithic and Neolithic of the Swiss Area. BAR International Series 67. British Archaeological Reports, Oxford.

Scarre, C.

1983   The Neolithic of West-Central France. *In* Ancient France: 6000–2000 B.C., edited by C. Scarre, pp. 223–70. University Press, Edinburgh.

Schlichtherle, H.

1985   Samen und Früchte. *In* Quantitative Untersuchungen an einem Profilsockel in Yverdon, Av. des Sports, edited by C. Strahm and H.-P. Uerpmann, pp. 7–43. Freiburg Im Bresgau.

Schrire, C.

1980   An Inquiry into the Evolutionary Status and Apparent Identity of San Hunter-Gatherers. Human Ecology 8 (1):9–32.

1985   Past and Present in Hunter-Gatherer Studies. Academic Press, London.

Sielmann, B.

1976   Der Einfluss der geographischen Umwelt auf die linien- und stichbandkeramische Besiedlung des Mittelelbe-Saale-Gebeites. Jahresschrift für mitteldeutsche Vorgeschichte 60:305–29.

Tempír, Z.

1964   Beiträge zur ältesten Geschichte des Pflanzenbaus in Ungarn. Acta Archaeologica Academiae Scientiarum Hungaricae 16:65–98.

1973   Finds of Prehistoric and Early Historic Remains of Food Plants and Weeds in Some Sites in Bohemia and Moravia. Vedecké Pràce 13:19–47.

Trump, D.

1980   The Prehistory of the Mediterranean. Allen Lane, London.

van Zeist, W.

1968   Prehistoric and Early Historic Food Plants in the Netherlands. Palaeohistoria 14:41–173.

van Zeist, W., and S. Bottema

1971   Plant Husbandry in Early Neolithic Nea Nikomedia, Greece. Acta Botanica Neerlandica 20(5):524–38.

van Zeist, W., and W. A. Casparie

1973   Niederwil, a Palaeobotanical Study of a Swiss Neolithic Lake Shore Settlement. Geologie en Mijnbouw 53:415–28.

Vacquer, J., D. Geddes, M. Barbaza, and J. Erroux

1986   Mesolithic Plant Exploitation at the Balma Abeurador (France). Oxford Journal of Archaeology 5(1):1–18.

Wasylikowa, K.

1981   The Role of Fossil Weeds for the Study of Former Agriculture. Zeitschrift für Archäologie 15:11–23.

1984   Fossil Evidence for Ancient Food Plants in Poland. In Plants and Ancient Man, edited by W. van Zeist and W. A. Casparie, pp. 257–66. Balkema, Rotterdam.

Webley, D., and R. W. Dennell

1978   Palaeonematology: Some Recent Evidence from Neolithic Bulgaria. Antiquity 52:136–37.

Welinder, S.

1982   The Hunter-Gathering Component of the Central Swedish Neolithic Funnel-Beaker (TRB) Culture. Fornvannen 77:154–66.

Whittle, A.

1985   Neolithic Europe: A Survey. Cambridge University Press, Cambridge.

Willerding, U.

1980   Zum Ackerbau der Bandkeramiker. Materialhefte Ur-und Fruhgeschichte Niedersachsens 16:421–56.

Williams, E.

1989   Dating the Introduction of Food Production into Britain and Ireland. Antiquity 63:510–21.

Zohary, D., and M. Hopf

1988   Domestication of Plants in the Old World. Clarendon Press, Oxford.

Zvelebil, M.

1978   Subsistence and Settlement in the North-Eastern Baltic. In The Early Post-Glacial Settlement of Northern Europe, edited by P. M. Mellars, pp. 205–41. Duckworth, London.

Zvelebil, M. (editor)

1986   Hunters in Transition. Cambridge University Press, Cambridge.

Zvelebil, M., and P. Rowley-Conwy

1986   Foragers and Farmers in Atlantic Europe. In Hunters in Transition, edited by M. Zvelebil, pp. 67–93. Cambridge University Press, Cambridge.

6

Prehistoric Plant Husbandry in Eastern North America

BRUCE D. SMITH

From their first radiation into the vast woodlands of eastern North America at the end of the Pleistocene, human populations depended to a substantial degree upon a variety of different plant species as food sources. Spanning more than ten millennia, the ensuing prehistoric period in the Eastern Woodlands witnessed dramatic changes in the nature, strength, and diversity of the relationships between human populations and their plant food sources: wild and cultivated, indigenous and introduced.

As outlined in this summary chapter, current understanding of this long and complex coevolutionary relationship between prehistoric human and plant populations of the Eastern Woodlands rests both on an archaeobotanical database that has substantially increased over the past two decades (Yarnell and Black 1985) and associated significant advances in the technology and methodology of archaeobotanical analysis (Watson 1976; Smith 1985a).

To facilitate discussion of evolving plant husbandry systems in the Eastern Woodlands, I have divided what was an unbroken, seamless developmental trajectory of 10,000 years into a temporal sequence of six periods of varying length that bracket ever-increasing levels of human dependence upon, and intervention in, the life cycle of plant species:

1. Early and Middle Holocene foragers prior to 7000 B.P. (5050 B.C.)

2. Middle Holocene collectors, 7000–4000 B.P. (5050–2050 B.C.)

3. The initial domestication of eastern seed plants, 4000–3000 B.P. (2050–1050 B.C.)

4. The development of farming economies, 3000–1700 B.P. (1050 B.C.–A.D. 250)

5. The expansion of field agriculture, 1700–800 B.P. (A.D. 250–A.D. 1150)

6. Maize-centered field agriculture after 800 B.P. (A.D. 1150)

While the descriptive headings assigned to each of these temporal periods indicate the most important aspects of the leading edge of the ongoing coevolutionary process, it is important to emphasize that the nature and timing of this developmental process was not uniform across the Eastern Woodlands. Some areas of the East developed more slowly and in directions different from those emphasized in this chapter.

Early and Middle Holocene Foragers prior to 7000 B.P. (5050 B.C.)

Prior to 7000 B.P. the bands of hunting and gathering groups that occupied the interior Eastern Woodlands of North America appear to have rather uniformly exploited plant and animal resources distributed fairly evenly along river valley corridors. Although the small, short-term seasonal camps of these forest foragers have yielded few carbonized plant materials, the impoverished archaeobotanical assemblages (along with faunal materials) suggest a broad-based utilization of both closed-canopy climax forest plant resources (particularly oak mast and hickory nuts) and early successional species more commonly associated with edge areas and more disturbed situations. The seeds, berries, and nuts of twenty plant species are represented in the archaeobotanical assemblages recovered from nine Early Holocene sites (Smith 1986:11; Meltzer and Smith 1986): *Quercus, Carya, Fagus, Corylus, Juglans, Castanea, Celtis, Diospyros, Acalypha, Amaranthus, Ampelopsis, Chenopodium, Galium, Phalaris, Phytolacca, Polygonum, Portulaca, Rhus, Vitis.* These Early Holocene foraging populations filled the role of opportunistic dispersal agents for plant propagules, and their small camps created ephemeral, disturbed soil opportunities for some pioneer species. There is no evidence for any greater degree of intervention by human populations in the life cycle of eastern plant species prior to 7000 B.P.

Middle Holocene Collectors: 7000–4000 B.P. (5050–2050 B.C.)

During the Middle Holocene period, the opportunistic response by human populations in the Eastern Woodlands to changes in river valley habitats resulted in both the creation of more permanently disturbed anthropogenic habitat patches, and increased human intervention in the life cycles of a number of plants of economic importance.

The change to zonal atmospheric flow across the Midwest during the Middle Holocene hypsithermal climatic episode (Knox 1983) resulted in a shift of mid-latitude fluvial systems from the earlier pattern of episodic pulses of sediment removal and river incision, to a phase of river aggradation and stabilization. This change in stream flow characteristics in turn resulted in the expansion of both backwater (backswamp and oxbow lake) and active stream shal-

low water and shoal area aquatic habitats along segments of some mid-latitude river systems (Smith 1986, 1987a). These aquatic habitats supported an increasingly abundant and easily accessible variety of aquatic resources. At the same time, the more geomorphologically active main channel portions of the floodplain environment continued to provide varied and abundant disturbed habitat opportunities for floodplain species of pioneer plants, including three of particular interest: *Chenopodium berlandieri, Iva annua,* and a wild *Cucurbita* gourd (Smith 1987a; Smith et al 1992; Cowan and Smith 1992). In response to the establishment of biotically richer floodplain segments, the widely scattered short-term occupational lenses of the antecedent Early Holocene and initial Middle Holocene were replaced by deep shell mound and midden mound settlements. Located on suitable topographic highs in the floodplain close to areas of enriched aquatic resources, and reflecting a narrowing of site preference, these midden mounds and shell mounds were reoccupied annually and reused over a long period of time, during the low-water, late-spring to early-winter growing season. This annual reoccupation and reuse of specific floodplain locations through the growing season represented the initial emergence of continually disturbed "anthropogenic" habitats in the interior river valleys of the eastern United States. A wide range of human activities would have resulted in frequent disturbance of the soil of these localities (construction of houses, wind breaks, storage and refuse pits, drying racks, earth ovens, hearths). In addition, floodwater sediments, along with the accumulation of plant and animal debris and human excrement, particularly at site edges, would have both altered the ground surface and improved the soil chemistry of such midden and shell mounds. These continually disturbed and fertilized "anthropogenic" habitat patches provided an excellent "dump heap" colonization opportunity (Anderson 1956; Fowler 1957, 1971; Smith 1987a) for a variety of early successional floodplain plant species, with seeds being introduced each year both by floodwaters and by human collectors.

During the 3,000-year-period from 7000 to 4000 B.P. these anthropogenic habitat localities witnessed increasing human intervention, to varying degrees, in the life cycle of a number of the plant species growing in their disturbed and enriched soils. The degree of human intervention in the life cycle of a plant species, and the extent to which the species in question depends upon humans for its continued survival, can be measured along a continuum extending from wild-status plants through eradicated, tolerated, and encouraged weed categories, to cultivation and domestication (Smith 1985a, fig. 1). Accurately placing prehistoric human-plant relationships along this continuum is often difficult, however, in the absence of any morphological changes reflecting domesticated status. Such morphological changes represent the single unequivocal indicator of cultivation and domestication that can be observed directly in the archaeological record. As discussed in the following section of this chapter, distinctive changes in the morphology of *Cucurbita*, goosefoot, sumpweed, and sunflower (*Helianthus annuus*) seeds recovered from fourth millennium archaeobotanical assemblages document their domestication by 4000–3000 B.P., if not earlier.

Although not reflected in the archaeological record of 7000 to 4000 B.P., the general sequence of coevolutionary development leading to the fourth millennium domestication of these three plants can be reconstructed with a reasonable level of confidence. Introduced from wild floodplain stands into midden mound and shell mound settlements each year either by floodwaters or human collectors, the seeds of *Cucurbita,* sumpweed and goosefoot would have quickly colonized these disturbed habitat patches where they would have thrived as weedy invaders. While the sunflower (*H. annuus*) also initially colonized these "domestilocalities" (Smith 1987a) as a weedy invader, it was apparently adventive to the Eastern Woodlands during the Middle Holocene, rather than being indigenous to the region.

Quickly established, and initially neither requiring nor receiving any human assistance other than inadvertent soil disturbance and enrichment, these colonizing weedy stands of *Cucurbita,* goosefoot, sumpweed, and sunflower, through simple toleration, provided a supplement to "natural" floodplain stands. The next level of human intervention in the life cycle of these weedy colonizers—a transition from simple toleration to inadvertent, and then active encouragement—was critical in the coevolutionary trajectory leading to domestication. When colonizing plant species were separated into "valuable" or "quasi-cultigen" versus "no value" species, with such "no value" plants actively discouraged, then the element of deliberate management or husbandry was introduced, to whatever minimal degree, and domestilocalities were transformed from inadvertent or incidental gardens into managed, or "true" gardens.

From the eradication of competing "weeds," it would have been a small, but very important escalation of the level of human intervention in the life cycle of the quasi-cultigens for their stands to have been expanded by planting within the domestilocality—likely by women (Watson and Kennedy 1991). It is this critical step of planting, even on a very small scale, that, if sustained over the long term, marks both the beginning of cultivation and the onset of automatic selection within affected domestilocality plant populations for interrelated adaptation syndromes associated with domestication (Heiser 1988). From here on, "the initial establishment of a domesticated plant can proceed through automatic selection alone" (Harlan et al. 1973:314). The nature of these automatic selection processes and the morphological changes in plants related to the adaptation syndromes they cause are described in the next section, with specific reference to goosefoot, sumpweed, and sunflower.

While fourth millennium morphological changes provide the basis for assigning domesticated status to these three species, *Cucurbita* gourds represented in Middle Holocene archaeobotanical assemblages had until recently also been identified as a domesticate on quite different grounds. Middle Holocene archaeobotanical assemblages from sites in west-central Illinois (Koster, Napoleon Hollow, Kuhlman, Lagoon), Missouri (Phillips Spring), Kentucky (Carlston Annis, Bowles, Cloudsplitter), and Tennessee (Hayes, Bacon Bend, Anderson) have yielded rind fragments assignable to the genus *Cucurbita* (Fig. 6.1, Table 6.1). Eight direct accelerator dates on rind and seed fragments leave no doubt as to the presence of gourd-like *Cucurbita* populations in the middle-latitude, riverine areas of eastern North America between 7000 and 4000 B.P.

Because it was thought that no indigenous species of wild *Cucurbita* gourd had ever been present in the East, a number of researchers (Chomko and Crawford 1978; Conard et al. 1984; Asch and Asch 1985c) concluded that the Middle Holocene *Cucurbita* material of the Eastern Woodlands was a variety of the tropical domesticate *C. pepo* that had been introduced, along with the concept of agriculture, from Mesoamerica by 7000 B.P.

More recently, however, both Decker (1986, 1988), Smith (1987a; Smith et al 1992), and Cowan and Smith (1992) have proposed a much more plausible explanation for this early *Cucurbita* rind material—that it reflects a more widespread prehistoric

Fig. 6.1. Location of archaeological sites and regions discussed in the text that have provided information about the development of farming economies in eastern North America.

range of distribution of an indigenous wild gourd that still grows today in the East. All of the Middle Holocene *Cucurbita* rind fragments recovered to date are thin, with reported thickness values falling below 2.0 mm, well within the range of published rind thickness values for present-day indigenous wild gourds of eastern North America (Cowan and Smith 1992; Decker et al 1992). Rind fragments thicker than 2.0 mm, probably representing domesticated plants, do not, in fact, appear in archaeological contexts in the East until after 3000 B.P. (at Salts Cave and Cloudsplitter rock shelter in Kentucky).

Along with rind fragments, *Cucurbita* seeds recov-

ered from Middle Holocene contexts should indicate when domesticated forms of *Cucurbita* first appeared in the East. Closely related wild and cultivated varieties of *Cucurbita* can be differentiated on the basis of seed morphology (Decker and Wilson 1986), and temporal changes in seed size and shape should mark the transition of *Cucurbita* from wild to cultivated and domesticated status along a developmental trajectory of incipient plant husbandry. Unfortunately, morphological analyses of Middle Holocene *Cucurbita* seed assemblages from the Eastern Woodlands—employing the Decker guidelines—have not yet been conducted. In addition, with the exception of the

Table 6.1. *Cucurbita* Remains from Eastern North America prior to 2000 B.P.

Site	Temporal/Cultural Context	Rind Fragments (N) Weight	Seeds (N) Size	References
Koster	Horizon 8b	(7) 27 mg		Asch and Asch 1985c
	7100 ± 300 (Accelerator—Rind)			Conard et al. 1984
	7000 ± 80			Conard et al. 1984
	6960 ± 80			Conard et al. 1984
	6820 ± 240 (Accelerator—Rind)			Ford 1985: 345
	Horizon 8a	(1)		Asch and Asch 1985c
	6860 ± 80			
Napoleon Hollow	Lower Helton	(3)		
	7000 ± 250 (Accelerator—Rind)			
	6800 ± 80			
	6730 ± 70			
	6630 ± 100			
	Upper Helton	(1)		
	6130 ± 110			
	5670 ± 90			
	5350 ± 70			
Anderson	6990 ± 120 (Accelerator—Seed)		(1)	Anna Dixon, pers.
	AA—1182			comm. 1987
Hayes	Unit III Levels E 20, 22, 23	(6) 0.4 g		Crites 1985
	5660 ± 190 B.P.			
	5525 ± 290 B.P.			
	Unit IV Level A 25	(16) .01 g		Crites 1985
	5340 ± 120 (Accelerator—Rind)			
	Unit III Level 017	(2)		Crites 1985
	5140 ± 185			
	4390 ± 170			
Carlston Annis	5730 ± 640 (Accelerator—Rind)			Watson 1985: 112
	<4500 ± 60 B.P.			
	>4250 ± 80 B.P.			
	>4040 ± 180 B.P.			
Phillips Spring	4310 ± 70		(65) length x = 10.5	King 1985
	4240 ± 80		width x = 7.0	King 1985
	4222 ± 57			King 1885
	3928 ± 41		(50) fragments	King 1985
Cloudsplitter	4700 ± 250 (Accelerator—seed)		(2) length x = 8.7	Cowan 1990
			width x = 5.3	
Bacon Bend	4390 ± 155	(4)		Chapman 1981:35–36
Lagoon	Feature 102a	(1)		Asch and Asch 1985c
	4300 ± 600			
	Titterington	(1)		Asch and Asch 1985c
	4030 ± 75			
Napoleon Hollow	Titterington	(1)		Asch and Asch 1985c
	4500 ± 500			
	4060 ± 75			
Bowles	4060 ± 220 (Accelerator—Rind)			Watson 1985:113
	>3440 ± 80			
Kuhlman	Titterington	(1)		Asch and Asch 1985c
	4010 ± 130			
Cloudsplitter	3620–3060	(9)		Cowan 1990
Iddens	3655 ± 135	(129+)	(140+)	Chapman 1981:132–34
	3470 ± 75			
	3205 ± 145			
Peter Cave	>3415 ± 105 B.P.		(2)	Watson 1985:113
Riverton	Feature 8a	(1)		Yarnell 1976
	ca. 3200			
Salts Cave	ca. 2500		length: x = 11.3	Yarnell 1969:51
			width: x = 7.3	
Cloudsplitter	2800–2300 B.P.	(29)	(5) length: x = 12.7	Cowan 1985, 1990
			width: x = 7.3	King 1985

Cloudsplitter seeds dating to 4700 B.P and a single possible *Cucurbita* seed from the Anderson site dating to 6990 ± 120 B.P. (AA-1182) (Table 6.1), the 65 seeds securely dated to ca. 4300 B.P. from the Phillips Spring site in west-central Missouri represent the only *Cucurbita* seed material recovered from Middle Holocene contexts in the East. When shape analysis of these early seed assemblages is carried out, however, I think that the results will support the temporal trend already documented in the rind assemblages—a transition to domesticated forms of *Cucurbita* after 3000 B.P. Currently there is no evidence to support the proposition that a domesticated variety of *C. pepo* was present in the East prior to 3000 B.P.

It has yet to be determined whether the cultivated, fleshier squash varieties of *C. pepo* that first appear in the East after 3000 B.P. (Salts Cave, Yarnell 1969:51; Cloudsplitter, Cowan 1990) were introduced into the East, or brought under domestication locally from an indigenous wild progenitor. An independent trajectory of domestication for *C. pepo* in the East, however, is the "much more likely" of the two alternatives (Heiser 1989:474), particularly in light of recent research by Decker-Walters et al (1992), and within the context of the first domesticated *Cucurbita* following on the heels of the preceding (4000–3500 B.P.) domestication of other indigenous seed plants. In addition, the recent direct dating of several small *Lagenaria* gourds recovered from the Windover site on the east coast of Florida (Brevard County) at 7300 B.P. (Doran et al. 1990) suggests that wild fruits may have washed ashore and become established during the Middle Holocene or earlier. Thus a mechanism of human-facilitated diffusion of domesticated forms from Mesoamerica need not be invoked to explain the presence of either the bottle gourd or *Cucurbita* gourd in the East during the Middle Holocene.

Another plant species of considerable economic importance may also have been encouraged by Middle Holocene groups, to their great benefit, without the species in question ever being brought under strict cultivation. Munson (1986) has proposed that the dramatic increase in representation of hickory nuts (*Carya*) in archaeobotanical assemblages of the Midwest riverine region after about 7500 B.P. reflects both improved methods for processing nuts (hide-lined and rock-heated boiling pits for separating nut meats from hulls) and active management of hickory trees to increase the quantity of hickory nuts available for human collectors. By simply girdling (and thereby killing) all nonpro-

ductive trees, human foragers could have created large stands of widely spaced hickories with increased long-term nut production and lower squirrel populations. Such stands would have required little maintenance other than annual, controlled burning of undergrowth. Alternatively, the increased representation of hickory nuts in Middle Holocene archaeobotanical assemblages may simply reflect the increasing abundance of *Carya* in midwestern forests (Delcourt and Delcourt 1981).

While the degree to which Middle Holocene groups managed and intervened in the life cycles of hickory trees and camp-follower gourds are topics of active and continuing debate, there is unequivocal evidence for the independent domestication of seed plants in the Eastern Woodlands during the fourth millennium B.P.

Initial Domestication of Eastern Seed Plants: 4000–3000 B.P. (2050–1050 B.C.)

The ten-century span from 4000 to 3000 B.P. brackets the earliest evidence of morphological changes reflecting domesticated status of the three, and perhaps four, annual seed crops brought under domestication prior to the first appearance of maize in the East at about A.D. 200. For sumpweed (*Iva annua*) and sunflower (*Helianthus annuus*) the morphological change indicating domestication is an increase in achene size, while a reduction in seed coat (testa) thickness reflects the transition to domesticated status in goosefoot (*Chenopodium berlandieri*). Both of these morphological changes (increase in seed size, reduction in testa thickness) reflect the automatic response by cultivated seed plants to the selective, if inadvertent, pressures inherent in the deliberate planting of seed stock.

Within domestilocality spring seed beds there would have been extremely intense competition between young plants: "The first seed to sprout and the most vigorous seedlings to sprout are more likely to contribute to the next generation than the slow or weak seedlings" (Harlan et al. 1973:318). As a result, strong selective pressure within seed beds would favor those seeds that would both sprout quickly because of a reduced germination dormancy (as reflected in a thinner or absent seed coat or testa), and grow quickly because of greater endosperm food reserves, as reflected by an increase in seed size.

With a mean achene length value of 4.2 mm, the

ca. 4000 B.P. Napoleon Hollow *Iva* assemblage, generally acknowledged to represent a domesticated crop (Yarnell 1983; Asch and Asch 1985c:160; Ford 1985:347), reflects a marked (31 percent) size increase over modern wild *Iva* populations, which have mean achene length values of 2.5 to 3.2 mm.

Yarnell (1983) and Ford (1985:348) both identify the ca. 2850 B.P. Higgs site sunflower achenes, which have a mean length value of 7.8 mm (as opposed to modern wild populations that range in mean length from 4.0 to 5.5 mm) as representing the earliest evidence for domestication of this species in the East. In addition, Yarnell (1978:291) suggests that the Higgs site sunflowers were "in early, but not initial stages of domestication." Ford (1985:348) suggests that such initial stages of domestication may have taken place at about 3500 B.P.

This proposed 3500–3400 B.P. transition to domesticated status for *Helianthus annuus* corresponds with the earliest available evidence for the presence of a thin-testa domesticated form of goosefoot (*C. berlandieri* ssp. *jonesianum*) (Smith and Funk 1985). In contrast to both sumpweed and sunflower, an increase in seed size is not a reliable indicator of domesticated status in *Chenopodium*. The relative thickness of the seed coat or testa, however, can be employed as a morphological indicator of domestication in *Chenopodium* (Smith 1985a, 1985b) in that it represents a strong selective pressure for reduced germination dormancy. While modern wild populations of *C. berlandieri* in the Eastern Woodlands have testa thickness values of 40 to 80 microns, assemblages of fruits having mean seed coat thickness values of less than 20 microns have been reported from a number of sites dating from 2300 B.P. to 1800 B.P. (Ash Cave, Edens Bluff, Russell Cave, White Bluff; Fritz and Smith 1988), and comparable thin-testa specimens have been recovered from sites dating from ca. 2500–1500 B.P. in west-central Illinois (John Roy, Asch and Asch 1985a), central Kentucky (Salts Cave, Yarnell 1974) and central Tennessee. In addition, thin-testa fruits of *C. berlandieri* ssp. *jonesianum* recovered from the Newt Kash and Cloudsplitter rock shelters in eastern Kentucky have recently been accelerator dated to 3400 ± 150 B.P. and 3450 ± 150 B.P., respectively, providing the earliest evidence for domesticated *Chenopodium* in the East (Smith and Cowan 1987). Support for these early eastern Kentucky accelerator dates on the thin-testa chenopod cultivar is provided by a direct radiocarbon age determination of 2926 ± 40 B.P. obtained on a carbonized clump of thin-testa fruits that had been placed in a woven bag before being stored in a rear wall crevice at the Marble Bluff Shelter in northwestern Arkansas (Fritz 1986a, 1986b). Pale-colored fruits of a second cultivar variety of *Chenopodium* having an extremely thin testa have also been reported from ca. 3000 B.P. contexts at the Cloudsplitter rock shelter, indicating that the early presence of two distinct crop plants of this species by the fourth millennium B.P. was likely. While the morphological similarity of seeds of these prehistoric thin-testa and pale-colored chenopod cultivars to those of the modern Mexican cultivars *C. berlandieri* ssp. *nuttalliae* cv. "chia" and "huauzontle" raises the issue of a possible prehistoric introduction of Mesoamerican domesticates into the East, it seems reasonable at present to consider them as indigenous eastern domesticates in light of the complete absence of thin-testa and pale-colored domesticated chenopods in the archaeological record of Mesoamerica.

While morphological changes in sumpweed, sunflower, and chenopod indicate that all three had been brought under domestication in the Eastern Woodlands during the period 4000–3500 B.P., there is little evidence that this process of domestication occurred within a framework of deliberate human selection, or that these domesticated plant species initially contributed substantially to the diet of fourth millennium B.P. populations. The simple, yet extremely significant step of planting seeds of these three plant species in order to expand their stand size and harvest yield, even on a very small scale, marked the beginning of cultivation and the onset of automatic selection for morphological changes in seeds within affected domestilocality plant populations.

The seeds of domesticates and indigenous cultigens alike are relatively rare in archaeobotanical assemblages dating earlier than 3000 B.P., making it difficult to establish how closely the transition to domesticated status of sumpweed, sunflower, and chenopod may have been accompanied by an increasingly important economic role. It is quite possible that indigenous cultigens and domesticates initially had significant value to fourth millennium B.P. populations as a dependable, managed, and storable late winter-early spring food supply (Cowan 1985). But based on the continuing limited occurrence of seeds of these species in archaeobotanical assemblages, it also appears likely that indigenous cultigens and domesticates did not become substantial food sources, and food production did not play a major economic role until about 2500–2000 B.P., a full 1,000 years after their initial domestication.

Development of Farming Economies: 3000–1700 B.P. (1050 B.C.–A.D. 250)

The period from 2500 to 1700 B.P. (550 B.C.–A.D. 250) witnessed the initial development, subsequent florescence, and eventual cultural transformation of the Middle Woodland Hopewellian cultural systems of the Eastern Woodlands of North America. While the large and impressive geometric earthworks, conical burial mounds, and elaborate mortuary programs of Hopewellian populations have been the focus of archaeological attention for well over a hundred years (Brose and Greber 1979), it is only within the past decade that much information has been recovered concerning Hopewellian plant husbandry systems.

While Hopewellian populations relied on wild species of plants and animals, Middle Woodland archaeobotanical seed assemblages recovered from river valley villages and upland rock shelters and caves over a broad geographical area of the Midwest and Midsouth reflect an increase in the representation (and assumed economic importance) of seeds of seven species of indigenous cultivated plants (Fig. 6.1, Table 6.2).

Four of the plant species in question—goosefoot (*Chenopodium berlandieri*) and knotweed (*Polygonum erectum*) (both fall maturing), as well as maygrass (*Phalaris caroliniana*) and little barley (*Hordeum pusillum*) (both spring maturing)—have high carbohydrate content, while the seeds of two other species, marshelder (*Iva annua*) and sunflower (*Helianthus annuus*) are high in oil or fat (Table 6.3).

Although the analysis of seed assemblages from this time period has yielded morphological evidence only for the domesticated status of four of these seven species (marshelder, goosefoot, *Cucurbita,* and sunflower), knotweed, maygrass, and little barley are classified at present as cultigens or quasi-cultigens (Yarnell 1983) for the period 3000–1700 B.P., based on geographical range extension (Cowan 1978:282; Asch and Asch 1985c:157; Yarnell 1983), archaeological abundance relative to modern occurrence, and a

Table 6.2. The Relative Abundance of Indigenous Starchy-seeded and Oily-seeded Annuals in Archaeobotanical Seed Assemblages Dating to the Period 3000–1700 B.P. in Seven Geographical Areas of the Eastern Woodlands

Category	Eastern Missouri N	%	West-Central Illinois N	%	American Bottom N	%	Central Ohio N	%	Central Kentucky N	%	Eastern Kentucky N	%	Central Tennessee N	%	Eastern Tennessee N	%
Starchy-seeded annuals																
Fall Maturing																
Chenopod	87	17.5	1176	8.6	201	26.4	392	13.6	18444	85.9	2745	52.7	2553	49.0	1431	88.0
Knotweed	29	5.8	4374	31.8	4	0.6	945	32.8	5				475	9.0		
Spring Maturing																
Maygrass	64	12.9	5155	37.5	222	29.1	598	20.7	1150	5.4	519	9.9	1868	36.0		
Little Barley	136	27.4	2040	14.9												
TOTAL	316	63.6	12745	92.8	427	56.1	1935	67.2	19599	91.3	3264	62.6	4896	94.0	1431	88.0
Oily-seeded annuals																
Fall Maturing																
Sumpweed	9	1.8	39	0.3	1	0.1	159	5.5	512	2.4	1154	22.9	3	0.1	26	1.6
Sunflower	1	0.2	35	0.3			11	.4	218	1.0	313	6.0	12	0.2	14	.9
TOTAL	10	2.0	74	0.6	1	0.1	170	5.9	730	3.4	1487	28.9	15	.3	40	2.5
Seed of Other Species	171	34.4	908	6.5	333	43.8	774	26.8	1141	5.3	468	9.0	283	5.4	156	9.5
Total Identified Seeds	497		13727		761		2879		21471		5199		5194		1627	

Sources: Seed count information for Eastern Missouri comes from the Old Monroe site (Pulliam 1986); for West-Central Illinois from the Smiling Dan site (2100–1700 B.P.) (Asch and Asch 1985a, table 15.8); for the American Bottom from Cement Hollow and Hill Lake phases (2100–700 B.P.) of the American Bottom site (Johannessen 1981, table 29; 1983, table 21); for Central Ohio from the Murphy and O.S.U. Newark sites (ca. 1750 B.P.) (Wymer 1987b); for Central Kentucky from the Salts Cave site (Vestibule JIV, Levels 4–11, 2600–2200 B.P.) (Yarnell 1974, table 16.5; Gardner 1987); for Eastern Kentucky from the Cloudsplitter site (Excavation Units I, 13–18) Cowan 1985, tables 62–63); for Central Tennessee from the McFarland and Owl Hollow sites, late McFarland and early Owl Hollow phases at Normandy Reservoir (1900–1700 B.P.) (Crites 1978, table 2.2; Kline et al. 1982, table 6); and for Eastern Tennessee from the Long Branch Phase sites (2300–1800 B.P.) at Tellico Reservoir (Chapman and Shea 1981, table 4).

Table 6.3. The Nutritive Value of Indigenous Seed Crops of the Prehistoric Eastern Woodlands of North America, Compared to Maize

| Plant Species | % Dry Basis (g/100gm) | | | | |
	Protein	Fat	Carbohydrate	Fiber	Ash
Chenopodium berlandieri[a]	19.1	1.8	47.6	28.0	3.5
Polygonum erectum[a]	16.9	2.4	65.2	13.3	2.3
Phalaris caroliniana[b]	23.7	6.4	54.3	3.0	2.1
Hordeum pusillum[c]	—	—	—	—	—
Iva annua[c]	32.3	44.5	11.0	1.5	5.8
Helianthus annuus[d]	24.0	47.3	16.1	3.8	4.0
Zea mays (maize)[c]	8.9	3.9	70.2	2.0	1.2

[a]Asch and Asch 1985a
[b]Crites and Terry 1984
[c]Asch and Asch 1978
[d]Watt and Merrill 1963

number of other "plausibility arguments" (Asch and Asch 1982:2; Smith 1985a:55).

Judging from their relative representation in seed assemblages from different areas of the Eastern Woodlands, it is likely that these seven plant species each played roles of differing importance within a developmental mosaic of regionally variable, Middle Woodland food production systems (Table 6.2). It is hoped that as additional archaeobotanical assemblages from different areas of the Eastern Woodlands are analyzed, it will be possible to determine the degree to which observed regional variability in the representation of these seven plants actually indicates different crop mixtures as opposed to simply reflecting differential preservation, recovery, or seasonality of occupation of the site being studied.

Harvesting of sunflower seeds would have been a simple matter of cutting off the small, multiple terminal discs for drying and seed removal, while squashes and gourds would have been collected and sectioned, with the flesh being dried in strips and the seeds separated for storage and consumption. Because of the diffuse distribution of seeds along stems and branches, on the other hand, entire knotweed plants were probably cut or pulled for drying and subsequent hand threshing. The terminal "seed clusters" of the other four species (marshelder, goosefoot, maygrass, little barley) could have been beaten or hand stripped into baskets (Asch and Asch 1978; Smith 1987b) or entire plants may have been harvested and flailed after drying.

Once harvested, the initial processing of seeds (other than seed stock) in all likelihood included parching to reduce insect infestation, premature germination, and mold damage (with associated accidental carbonization and loss contributing substantially to the archaeobotanical record). A number of storage contexts have been documented, including grass-lined pits, woven bags, and gourd containers.

Prior to consumption, any number of different technologies, including wooden mortars, stone slabs, and mortar holes in stone slabs were employed to crack protective seed coats, and the seeds could then have been either boiled in ceramic vessels (perhaps with a variety of other plant and animal ingredients), or perhaps ground further into a flour. Analysis of a large sample of human paleofecal material recovered from Salts Cave and dating to about 2600–2200 B.P., in fact, indicates that once cracked, the small seeds of these indigenous species were sometimes consumed without any further processing (Yarnell 1974). Providing uniquely direct and unequivocal dietary evidence of one to four meals (Watson 1974), each of the Salts Cave human paleofecal samples also indicated substantial reliance on the seeds of indigenous cultigens and cucurbits. The four domesticates contributed a full two-thirds of the total food supply of the central Kentucky cavers (chenopod, 25 percent; sunflower, 25 percent; marshelder, 14 percent; cucurbits, 3 percent) (Yarnell 1974).

Other, more indirect measures of the increasing dietary importance of cultivated crops during the period 3000–1700 B.P. include the presence of chert hoes in Hopewellian artifact assemblages (Watson 1988), an increase in pollen and macrobotanical indicators of land clearing activities (Yarnell 1974:117; Kline et al. 1982:63; Johannessen 1984:201; Delcourt et al. 1986), and changes in settlement patterns and plant storage facilities (Smith 1986).

Based on research done during the past decade, it is now evident that indigenous cultigens increased substantially in dietary importance over a broad area of the eastern United States during the period 2500–1700 B.P., and that developing food production systems helped to fuel a process of dramatic cultural change and florescence. At the same time, however, it is still not possible to track accurately, from 3000 to 1700 B.P. and in different regions, the changing relationship between Woodland period populations and indigenous cultigens in terms of increasing energy investment in land clearing, soil preparation and cultivation, and plant harvesting and processing on the one hand, and total land area under crop and annual harvest yield values on the other. Even in the absence of this kind of detailed picture of Early and Middle Woodland food production, however, it is important to emphasize that plant husbandry systems focused on indigenous seed-bearing plants having both impressive nutritional profiles and substantial, potential harvest yield values (Table 6.3; Smith 1987b, fig. 11), and had been established over a broad area of the eastern United States prior to the initial introduction of maize into the Eastern Woodlands.

Expansion of Field Agriculture: 1700–800 B.P. (A.D. 250–1150)

Although maize pollen has been recovered from earlier temporal contexts in a number of locations (Sears 1982), the earliest convincing macrobotanical evidence recovered to date for the presence of maize in the Eastern Woodlands are carbonized kernel fragments from the Icehouse Bottom site in the Little Tennessee River Valley of eastern Tennessee that have yielded a direct particle accelerator (AMS) radiocarbon date of 1775 ± 100 B.P. (A.D. 175) (Chapman and Crites 1987). In addition, Ford has recently (1987) reported direct AMS dates of 1730 ± 85 B.P. and 1720 ± 105 B.P. on maize kernels recovered from the Edwin Harness site in Ohio.

While research interest and debate will likely continue to focus in future years on the timing and route(s) of initial entry of maize into the Eastern Woodlands, and its subsequent rate and direction(s) of range extension in the East, these corn-related issues are largely peripheral to the central question of the role of food production in Woodland period cultural development. Judging from its extremely limited representation in archaeobotanical assemblages before A.D. 800, maize played a very minor role in eastern plant husbandry systems during the first 600 years after its initial introduction. But during the six-century span from A.D. 200 to 800 that maize remained a minor crop, at or below the level of archaeological visibility, the Eastern Woodlands witnessed a number of important interrelated developments in plant husbandry, technological innovation, and biocultural evolution.

Plant husbandry systems, centered on indigenous cultigens, continued to provide a significant, if difficult to quantify, percentage of the diet of populations over a broad area of the Midwest during this period (Pulliam 1987; Wymer 1987a, 1987b), and several new domesticates, in addition to maize, appear in archaeobotanical assemblages for the first time. Although not a food crop, tobacco (*Nicotiana rustica*) is known to have reached west-central Illinois by A.D. 160 (Asch and Asch 1985c) and the American Bottom east of St. Louis by A.D. 500–600 (Johannessen 1984). This cultigen was apparently introduced directly from Mexico rather than through the Southwest, and played an important ceremonial role through the remaining portion of the prehistoric sequence.

In addition, the pale-colored, extremely thin-testa variety of *Chenopodium berlandieri*, apparently brought under domestication by 1000 B.C., occurs in food processing and storage contexts in increased quantities by A.D. 400, often in association with erect knotweed. The carbonized knotweed specimens recovered from parching pit contexts exhibit an A.D. 400–1100 temporal trend of morphological change suggesting that there is at least the possibility that the plant may have been brought under domestication (size increase, smoother pericarp surface, and reduction in pericarp thickness) (Fritz 1987). Although easily recognized when recovered uncarbonized from dry cave contexts, the seeds of this pale-colored chenopod cultivar are subject to substantial morphological distortion when carbonized, making them difficult to identify and resulting in their frequent underrepresentation in archaeobotanical assemblages. Fortunately, the processing (parching?) of large quantities of seeds of this pale-colored chenopod cultivar in shallow pit features sometimes produced, as a result of over-parching, large carbonized seed masses. In addition to documenting the continued role of this pale-colored chenopod as another starchy-seed crop within eastern plant husbandry systems, parching pit seed masses such as those excavated in west-central Illinois (Newbridge, ca. A.D. 400), the American Bottom (Sampson's Bluff, Johanning, ca. A.D. 800–1000), and southeast Missouri (Gypsy Joint, ca. A.D. 1300, Smith 1978b; Fritz 1987) also indicate the mixed processing (and cultivation?) of chenopod cultigens (predominantly the pale-colored variety, but with some representation of the thin-testa type) with erect knotweed (Polygonum erectum).

The co-occurrence of these crop plants in cultural contexts 450 km and almost 1,000 years apart suggests the possible development, by A.D. 400, of an indigenous crop complex that endured, at least along the Mississippi River Valley corridor, throughout the late prehistoric period.

The growing socioeconomic importance of plant husbandry systems over a broad area of the Midwest and Midsouth during the period A.D. 200–800 is reflected by the continued representation of indigenous cultigens in archaeobotanical assemblages and increasing evidence of deliberate human selection, the tandem increase in abundance of a pale-fruit chenopod and a possibly domesticated knotweed in an apparent crop complex association, and the addition of two tropical domesticates (maize and tobacco). All this takes place against a backdrop of considerable Late Woodland cultural change.

Population levels across the Midwest and Southeast rose markedly during this six-century span, judging from the documented increase, in a number of areas, of small one- to four-structure farmstead settlements. Such household structure clusters also sometimes loosely coalesced into dispersed communities of up to perhaps 50 inhabitants and covered less than 0.5 to 1.0 ha (Smith 1986). This widespread growth and dispersal of population within river valley landscapes is consistent with a growing reliance on food production.

In at least one region (west-central Illinois) this Late Woodland population increase has been tied to earlier weaning and shorter birth intervals (and higher fertility levels) associated with the availability of soft, palatable, digestible weaning foods (Buikstra et al. 1986). It is suggested that the weaning foods in question were probably the highly nutritious, indigenous starchy-seeded annuals. Thinner, more efficient ceramic vessels (Braun 1983) facilitated their preparation through boiling.

A variety of technological innovations in field preparation and crop processing, storage, and preparation also accompanied the seemingly abrupt and widespread appearance of maize in the archaeological record of the Midwest and Southeast at A.D. 800–900 (Smith 1986). Large, well-made chert hoes begin to be exchanged over substantial distances. Technologically superior calcium carbonate (ground mussel shell or limestone) tempered ceramic assemblages diversify, with a number of new vessel forms probably functioning in the processing ("parching" pans) and cooking (large globular jars) of maize and other cultigens. Below-ground storage pits increase in size.

During the 350-year span following the abrupt movement of maize into archaeological visibility at A.D. 800, eastern North America witnessed the rapid and widespread emergence of "Mississippian" ranked agricultural societies based on maize-dominated field agriculture (Smith 1990).

Although the limited stable carbon isotope analysis done to date (Lynott et al. 1986) suggests that maize did not become a staple food source until after A.D. 1100, there is no question that continued population growth and increasing reliance on maize and other cultigens represented a central theme in the initial Mississippian emergence.

Maize-Centered Field Agriculture after 800 B.P. (A.D. 1150)

By A.D. 1150 agricultural economies dominated by corn had been established in river valleys over a broad area of the eastern United States. With the arrival of the common bean (Phaseolus vulgaris) in some regions by A.D. 1000–1200, the crop triad of maize, beans, and squash was present in the East. During the final four centuries of North American prehistory, however, the eastern United States did not witness a uniform, homogeneous pattern of maize-beans-squash agriculture. On the contrary, currently available information suggests the continuation of a developmental mosaic across the East, with different regions exhibiting variation in the types of maize being grown and its dietary importance relative to indigenous crops and wild animal and plant resources.

The post-A.D. 1000 Fort Ancient populations of the Ohio River Valley and its Ohio, Kentucky, and West Virginia tributaries represent the regional manifestation that perhaps most closely matches the common perception of prehistoric agriculture in the East.

Although well represented in preceding Middle and Late Woodland archaeobotanical assemblages from the area (Wymer 1987), indigenous seed-bearing crops are thought to have been absent from Fort Ancient food production economies (Wagner 1987).

The common bean, on the other hand, which is not documented in the Central Mississippi Valley until A.D. 1450 (Morse and Morse 1983) and has not been reported at all from the American Bottom (Johannessen 1984), is relatively abundant in Fort Ancient contexts. Wagner's measurement of almost 800 beans from Fort Ancient sites shows them to be small, ranging from 8.6 to 11.7 mm in length and 5.6 to 8.8 mm in width (Wagner 1987, table 6), and to represent several varieties.

Attempting to assess regional variation in the dietary importance of the bean is complicated, however, by its low probability of being preserved in the archaeological record. As a result, its relative abundance or absence in archaeobotanical assemblages may be misleading. In contrast, the measurement of stable carbon isotope levels in human bone allows an accurate assessment of the dietary importance of maize, and studies done to date on Fort Ancient populations suggest that they had a greater reliance on maize than populations in other areas of the East (Wagner 1987). Described by Wagner (1987) as showing little variability and being morphologically similar to the historic early-maturing northern flint corns, typical Fort Ancient corn had a tapered 8-row cob (with some 10-row and a few 12-row and 14-row cobs present) ranging from 8.0 to 12.6 cm in length and 1.3 to 1.6 cm in midpoint diameter, and having an expanded base.

In contrast to the Ohio Valley Fort Ancient pattern of heavy reliance on a single, low variability type of maize, to the apparent total exclusion of indigenous cultigens, populations in other areas of the Eastern Woodlands continued to rely, to an admittedly reduced degree, upon indigenous plants (both cultivated and wild), while at the same time growing corn that exhibited considerably greater genetic diversity (Fritz 1990). Post-A.D. 1000 archaeobotanical assemblages from west-central Alabama (Scarry 1986), eastern Tennessee (Chapman and Shea 1981), southeast Missouri (Fritz 1987), northwest Arkansas (Fritz 1986a), the American Bottom (Johannessen 1984), and west-central Illinois (Asch and Asch 1985b) all indicate the continued, if diminished, cultivation of various starchy- and oily-seeded indigenous crop plants. In addition, the early eighteenth-century description by LePage duPratz of the cultivation of "La belle dame sauvage," a small "grain," by the Natchez, both in corn fields and along the sand banks of the Mississippi, would appear to document the continued husbandry of a domesticated variety of chenopod into the historic period (Smith 1987c).

While a low variability, 8-row maize was being grown in the Ohio Valley and into the Northeast and upper Midwest after A.D. 1100, Mississippian populations in other areas of the Eastern Woodlands were growing varieties of maize exhibiting considerably greater variation in size and shape. Caches of well-preserved corn cobs recovered from northwestern Arkansas rock shelters and dating to between A.D. 1200 and 1400 contained a high frequency of robust, wide-base and wide-cupule, 10- and 12-row cobs, along with a few 14-row cobs (Fritz 1986a:184–208). Although 8-row maize was also found to be abundant in Ozark bluff shelter collections, and cobs similar to Fort Ancient maize were occasionally present, most of the 8-row cobs recovered were small with contracting bases and narrow shanks, and were likely "representatives of unfavorable growing seasons, tillers, or possibly at times, of an early season 'little corn' variety" (Fritz 1986a:207). A similar variety of maize to that which dominated the Ozark assemblages was apparently grown throughout much of the Lower and Central Mississippi Valley, as far north as the American Bottom east of St. Louis, Missouri.

In contrast to both the Ozarks and the Ohio Valley, Mississippian period populations in Alabama were apparently growing at least two varieties of maize during the period A.D. 1000–1500 (Scarry 1986), one of which exhibited morphological change through time. A predominantly 12-row variety (with some 14-row and occasional 16-row cobs) having both tapered and straight cobs persisted as a minority type, with a more abundant 8-row variety displaying a number of changes over time (decreasing mean row number, increasing cob diameter and cupule width).

These detailed analyses of large, relatively well-preserved and well-provenienced cob collections from the Ohio Valley, Ozarks, and Alabama, along with other regional studies, would seem at the present time to suggest that a latitudinal dichotomy in the type(s) of maize grown in the East was established prior to A.D. 1000, well before maize came to dominate fields and food production systems.

An 8-row variety of maize, which was probably adapted to short growing seasons and ancestral to the historic period northern flints, had been developed by A.D. 1000 within existing northern-latitude plant husbandry systems, and had been widely adopted across the Northeast, Ohio Valley, and upper Midwest. Most likely because of a climatic adaptation to northern growing seasons, this variety of maize dominated, to the apparently almost total exclusion of other varieties, the post-A.D. 1000 agricultural systems of the northern latitudes.

In contrast to the straightforward scenario of unchanging single-variety, northern-latitude, 8-row maize agriculture, the extensive and environmentally diverse, longer growing season areas to the south of the Ohio Valley (and seemingly including a Central Mississippi Valley extension as far north as the American Bottom) witnessed a far more complex and regionally variable mosaic of maize husbandry. It is likely that one or more long growing season varieties of maize had been introduced into the Southeast prior to A.D. 500, and that subsequent, perhaps relatively frequent, arrivals of new varieties, either overland from the Southwest or northern Mexico, or across the Caribbean, contributed to an expanding gene pool and the development of a greater diversity of indigenous cultivar forms (Fritz 1990).

While corn kernels, cupules, and cobs recovered from a large number of prehistoric contexts south of the Ohio Valley have been described and general developmental trends have been proposed (Blake 1987; Watson 1988), there is still much that is not known concerning the identity, developmental history, relative importance, and geographical and temporal distributions of different varieties of prehistoric eastern maize. These complex issues of prehistoric maize selection in the central and southeastern United States have been masked somewhat by the assignment of general taxonomic labels to corn collections, based largely on cob row number (often projected from kernel or cupule edge angles). Although it may never be possible to identify in archaeological contexts the full range of variation in maize cultivars documented at historic contact (Swanton 1946), the careful and selective analysis of intact cobs from storage contexts (Fritz 1986a; Scarry 1986), as well as kernels and cupules from general refuse deposits, should result in less reliance on general taxonomic gloss, and the recognition of different regional varieties, their range of morphological variation, and their developmental history.

Even though maize came to dominate fields after A.D. 1100, and perhaps contributed more than half of the population's annual caloric intake, judging from the stable carbon isotope studies done to date, the river valley farmers of the Eastern Woodlands continued to cultivate, to varying degrees, a wide range of other plants for a variety of purposes. A number of varieties of the common bean, differentially adopted into the plant husbandry systems of various areas after being introduced into the East around A.D. 1100, were cultivated as food plants, along with fleshy varieties of squash. In addition to fleshy forms of *Cucurbita pepo,* which may represent both indigenous domesticates and tropical cultivars, the green-striped cushaw (*C. mixta*) was present as far east as northwestern Arkansas bluff shelters by ca. A.D. 1430 (Fritz 1986a:153). A domesticated variety of pale-seeded amaranth (*Amaranthus hypochondriacus*) was also a late (ca. A.D. 1100) introduction into the Ozarks, but at the present time it does not appear to have been diffused any further east (Fritz 1984).

Along with these introduced tropical cultigens, a wide variety of indigenous food crops continued under cultivation in many areas (sumpweed, sunflower, knotweed, thin-testa and pale-colored varieties of chenopod, maygrass, and little barley). In addition, a number of other quasi-cultigens and weedy campfollowers were components, to varying degrees, of river valley plant husbandry systems after A.D. 1100, including Jerusalem artichoke (*Helianthus tuberosus*), maypops (*Passiflora incarnata*), amaranth, purslane, pokeweed, ragweed, chenopod, and carpetweed (Yarnell 1987).

Within this large group of husbanded plants, the emergence of maize as the primary field crop after A.D. 1100, and the associated increased importance of food production, did not represent an abrupt change in course for late prehistoric populations, but rather was an accelerated expansion of a long-established annual cycle of plant husbandry. For at least the preceding millennium, prehistoric farmers had been clearing land, preparing the soil, planting and cultivating crops, and at the end of the growing season, harvesting, processing, and storing them. Underlying this seasonal round of plant husbandry is an even older pattern of dependence upon the rich and diverse wild plant and animal resources of the eastern forests. And this basic, if regionally variable, framework of utilization of wild plants and animals also exhibits considerable continuity throughout the final 2,000 years of prehistory in the eastern United States. The maize-centered food production systems of late prehistoric groups in the East can thus best be viewed as largely compatible extensions of pre-existing, long-evolving subsistence systems. Because of the steep ecological gradient separating the river valleys of the East from intervening upland areas, the resource-rich linear floodplain corridors had attracted human hunter-gatherer groups long before plant husbandry played even a minor supporting role in subsistence systems. One of the most important attractions of these floodplain ecosystems was the localized and dependable aquatic protein sources of both main channel shoal areas and slackwater channel remnant oxbow lakes and backswamps. The need for animal protein certainly did not diminish with the post-A.D. 1100 development of maize-centered field agriculture, and as much as half of the protein of at least some river valley maize agriculturalists came from fish and waterfowl (Smith 1975, 1978a). These main channel and backswamp aquatic resource zones were separated and paralleled by linear bands of natural levee soils, coalescing to form broad meander belts in the larger river valleys. Annually replenished by floodwaters, and easily tilled, these sandy, well-drained natural levee soils were highly prized by prehistoric farmers of the Eastern Woodlands, and the fortified mound centers, villages, and small single farmsteads of post-A.D. 1100 maize agriculturalists were, with few exceptions, situated on or adjacent to these natural levee soils (Smith 1978a). Providing both easy access to protein-rich aquatic resource zones and excellent soils for simple hoe technology, these river valley natural levee ridges were extensively settled and brought under cultivation during the late prehistoric period. At European contact, early travelers provided accounts of riding for days along such natural levee ridge systems, through a landscape of farmsteads with adjacent "infield" gardens dispersed widely within larger and more extensive "outfield" field systems.

While such seventeenth- and eighteenth-century descriptions of Native American agricultural practices (as summarized by Swanton 1946 and Hudson 1976) provide valuable and tantalizing glimpses of important topics such as field preparation, planting schedules for different varieties of maize, intercropping and multiple cropping, and crop storage and preparation, there is limited evidence, either historical or archaeological, for many of the most basic aspects of late prehistoric, maize-centered agriculture.

It is tempting to assume that the small household "infield" gardens of the contact period represent the continuation of an earlier horticultural system based on indigenous plants, with the larger "outfield" cultivation plots being a late development associated with the emergence of maize agriculture. There is no reason, however, to restrict any of the indigenous crops to garden status, since the modern cultivated (sunflower) and weedy descendants of a number of them (sumpweed, knotweed, chenopod) have produced impressive harvest yield values (Asch and Asch 1978; Smith 1987b, table 5), and were more than likely grown in "outfields" prior to A.D. 1000.

Evidence of prehistoric hilling or furrowing of fields is relatively rare in the East and with a few notable exceptions (Riley 1987) is restricted to the northern margins of maize agriculture (Gallagher et al. 1985). While the large chert hoes (Brown et al. 1990) employed (along with elk scapula hoes in the Northeast and shell hoes along the Atlantic and Gulf coasts) to prepare and maintain fields through the growing season are frequently recovered, wooden artifacts associated with the planting and preparation of crops, such as digging sticks and wooden mortars, are only rarely preserved in the archaeological record (Smith 1978b). Similarly, while below-ground pits document the storage of agricultural products in some areas, evidence of above-ground storage in gourds, bags, or elevated granaries (the *barbacoas* of the Spanish) is almost nonexistent.

While artifacts and site features, along with ethnohistorical accounts, will provide occasional glimpses of prehistoric plant husbandry practices, it is the plants themselves, preserved in the ground and in museum collections, which, with careful recovery, curation, and analysis, will continue to yield further understanding of the long and complex coevolutionary relationship between human and plant populations in the Eastern Woodlands of North America.

Acknowledgments

I would like to thank C. Wesley Cowan, Gayle J. Fritz, James B. Griffin, and Patty Jo Watson for taking the time to read early drafts of this chapter. The numerous corrections and suggestions for revision that they offered substantially improved the text.

References Cited

Anderson, E.
1956   Man as a Maker of New Plants and Plant Communities. *In* Man's Role in Changing the Face of the Earth, vol. 2, edited by W. L. Thomas, pp. 763–77. University of Chicago Press, Chicago.

Asch, D., and N. Asch

1978   The Economic Potential of *Iva annua* and its Prehistoric Importance in the Lower Illinois Valley. *In* The Nature and Status of Ethnobotany, edited by R. Ford, pp. 300–341. Anthropological Papers No. 67. Museum of Anthropology, University of Michigan, Ann Arbor.

1982   A Chronology for the Development of Prehistoric Agriculture in West-Central Illinois. Paper presented at the 47th Annual Meeting, Society for American Archaeology, Minneapolis. Center for American Archaeology, Archaeobotanical Laboratory Report 46.

1985a   Archeobotany. *In* Smiling Dan, edited by B. D. Stafford and M. B. Sant, pp. 327–99. Center for American Archeology, Research Series 2. Kampsville, Illinois.

1985b   Archeobotany. *In* The Hill Creek Homestead, edited by Michael D. Conner, pp. 115–70. Kampsville Archeology Center, Research Series No. 1. Kampsville, Illinois.

1985c   Prehistoric Plant Cultivation in West-Central Illinois. *In* Prehistoric Food Production in North America, edited by R. I. Ford, pp. 149–203. Anthropological Papers No. 75. Museum of Anthropology, University of Michigan, Ann Arbor.

Blake, Leonard

1987   Corn and Other Plants from Prehistory into History in the Eastern United States. *In* Proceedings of the 1983 Mid-South Archaeological Conference, edited by D. Dye and R. Brister. Mississippi Department of Archives and History, Archaeological Report 18. Jackson, Mississippi.

Braun, D. P.

1983   Pots as Tools. *In* Archaeological Hammers and Theories, edited by J. A. Moore and A. S. Keene, pp. 107–34. Academic Press, New York.

Brose, David, and N'omi Greber

1979   Hopewell Archaeology. Kent State University Press, Kent, Ohio.

Brown, J., R. Kerber, and H. Winters

1990   Trade and the Evolution of Exchange Relations at the Beginning of the Mississippi Period. *In* The Mississippian Emergence, edited by B. Smith, pp. 251–80. Smithsonian Institution Press, Washington, D.C.

Buikstra, J., L. W. Koningsberg, and J. Bullington

1986   Fertility and the Development of Agriculture in the Prehistoric Midwest. American Antiquity 51:528–46.

Chapman, J.

1981   The Bacon Bend and Iddins Sites: The Late Archaic Period in the Lower Little Tennessee River Valley. Department of Anthropology, University of Tennessee, Report of Investigations 31. Knoxville, Tennessee.

Chapman, J., and G. Crites

1987   Evidence for Early Maize (*Zea mays*) from the Icehouse Bottom Site, Tennessee. American Antiquity 52:352–54.

Chapman, J., and A. B. Shea

1981   The Archaeobotanical Record: Early Archaic Period to Contact in the Lower Little Tennessee River Valley. Tennessee Anthropologist 6:64–84.

Chomko, S. A., and G. W. Crawford

1978   Plant Husbandry in Prehistoric Eastern North America: New Evidence for its Development. American Antiquity 43:405–408.

Conard, N., D. L. Asch, N. B. Asch, D. Elmore, H. Gove, M. Rubin, J. A. Brown, M. D. Wiant, K. B. Farnsworth, and T. B. Cook

1984   Accelerator Radiocarbon Dating of Evidence for Prehistoric Horticulture in Illinois. Nature 308:443–46.

Cowan, C. W.

1978   The Prehistoric Use and Distribution of Maygrass in Eastern North America: Cultural and Phyto-Geographical Implications. *In* The Nature and Status of Ethnobotany, edited by R. I. Ford, pp. 263–88. Anthropological Papers No. 67. Museum of Anthropology, University of Michigan, Ann Arbor.

1985   From Foraging to Incipient Food Production: Subsistence Change and Continuity on the Cumberland Plateau of Eastern Kentucky. Ph.D. dissertation, Department of Anthropology, University of Michigan. University Microfilms, Ann Arbor.

1990   Prehistoric Cucurbits from the Cumberland Plateau of Eastern Kentucky. Paper presented at the Southeastern Archaeological Conference, Mobile, Alabama.

Cowan, C.W., and B.D. Smith
1992   New Perspectives on a Free-Living Gourd in Eastern North America. Paper presented at the 15th annual meeting of the Society of Ethnobiology, Smithsonian Institution, Washington, D.C.

Crites, G. D.
1978   Plant Food Utilization Patterns during the Middle Woodland Owl Hollow Phase in Tennessee: A Preliminary Report. Tennessee Anthropologist 3:79–92.

1985   Middle and Late Holocene Ethnobotany of the Hayes Site (40MI139): Evidence from Unit 990N918E. Report submitted to the Tennessee Valley Authority. Knoxville.

Crites, G. D., and R. D. Terry
1984   Nutritive Value of Maygrass, *Phalaris caroliniana*. Economic Botany 38:114–20.

Decker, D.
1986   A Biosystematic Study of *Cucurbita pepo*. Ph.D. dissertation, Biology Department, Texas A & M University, College Station.

1988   Origin(s), Evolution, and Systematics of *Cucurbita pepo* (Cucurbitaceae). Economic Botany 42:4–15.

Decker, D., and H. D. Wilson
1986   Numerical Analysis of Seed Morphology in *Cucurbita pepo*. Systematic Botany 11:595–607.

Decker-Walters, D., T. Walters, C. W. Cowan, and B. D. Smith
1992   Isozymic Characterization of Wild Populations of *Cucurbita pepo*. Paper presented at the 15th annual conference of the Society of Ethnobiology, Smithsonian Institution, Washington, D.C.

Delcourt, P. A., and H. R. Delcourt
1981   Vegetation Maps for Eastern North America. *In* Geobotany II, edited by R. Romans, pp. 123–65. Plenum Publishing, New York.

Delcourt, P. A., H. Delcourt, P. Cridlebaugh, and J. Chapman.
1986   Holocene Ethnobotanical and Paleoecological Record of Human Impact on Vegetation in the Little Tennessee River Valley, Tennessee. Quaternary Research 25:330–49.

Doran, G., D. Dickel, and L. Newsom
1990   A 7,290-Year-Old Bottle Gourd from the Windover Site, Florida. American Antiquity 55:354–60.

Ford, R.
1985   Patterns of Prehistoric Food Production in North America. *In* Prehistoric Food Production in North America, edited by R. Ford, pp. 341–64. Anthropological Papers No. 75. Museum of Anthropology, University of Michigan, Ann Arbor.

1987   Dating Early Maize in the Eastern United States. Paper presented at the 10th Annual Conference, Society of Ethnobiology, Gainesville, Florida.

Fowler, M.
1957   The Origin of Plant Cultivation in the Central Mississippi Valley: A Hypothesis. Paper presented at the Annual Meeting of the American Anthropological Association, Chicago.

1971   The Origin of Plant Cultivation in the Central Mississippi Valley: A Hypothesis. *In* Prehistoric Agriculture, edited by S. Struever, pp. 122–28. Natural History Press, Garden City, New York.

Fritz, G.
1984   Identification of Cultigen Amaranth and Chenopod from Rock Shelter Sites in Northwest Arkansas. American Antiquity 49:558–72.

1986a   Prehistoric Ozark Agriculture. The University of Arkansas Rockshelter Collections. Ph.D. dissertation, Department of Anthropology, University of North Carolina, Chapel Hill.

1986b   Starchy Grain Crops in the Eastern U.S.: Evidence from the Desiccated Ozark Plant Remains. Paper presented at the 51st Annual Meeting of the Society for American Archaeology, New Orleans.

1987   The Trajectory of Knotweed Domestication in Prehistoric Eastern North America. Paper presented at the 10th Annual Conference, Society of Ethnology, Gainesville, Florida.

1990   Multiple Pathways to Farming in Precontact Eastern North America. Journal of World Prehistory, vol. 4, edited by F. Wendorf and Angela Close, pp. 387–435. Plenum Publishing, New York.

Fritz, G., and B. Smith
1988   Old Collections and New Technology: Documenting the Domestication of *Chenopodium* in Eastern North America. Midcontinental Journal of Archaeology 13:3–27.

Gallagher, J. P., R. Boszhardt, R. Sasso, and K. Stevenson
1985   Oneota Ridged Field Agriculture in Southwestern Wisconsin. American Antiquity 50:605–12.

Gardner, P.
1987   Plant Food Subsistence at Salts Cave, Kentucky: New Evidence. American Antiquity 52:358–64.

Harlan, J. R., J. M. J. de Wet, and E. G. Price
1973   Comparative Evolution of Cereals. Evolution 27:311–25.

Heiser, Charles B.
1988   Aspects of Unconscious Selection and the Evolution of Domesticated Plants. Euphytica 37:77–85.

1989   Domestication of Cucurbitaceae: Cucurbita and Lagenaria. In Foraging and Farming, edited by D. Harris and G. Hillman, pp. 471–81. Unwin Hyman, London.

Hudson, C.
1976   The Southeastern Indians. University of Tennessee Press, Knoxville, Tennessee.

Johannessen, S.
1981   Plant Remains from the Truck #7 Site. In Archaeological Investigations of the Middle Woodland Occupation at the Truck #7 and Go-Kart South Sites, by A. C. Fortier, pp. 116–30. University of Illinois at Urbana-Champaign, Department of Anthropology, FAI-270 Archaeological Mitigation Project Report 30.

1983   Plant Remains from the Cement Hollow Phase. In The Mund Site, by A. C. Fortier, F. A. Finney, and R. B. LaCampagne, pp. 94–103. American Bottom Archaeology FAI-270 Site Reports 5. University of Illinois Press, Urbana.

1984   Paleoethnobotany. In American Bottom Archaeology, edited by C. Bareis and J. Porter, pp. 197–214. University of Illinois Press, Urbana.

King, F. B.
1985   Early Cultivated Cucurbits in Eastern North America. In Prehistoric Food Production in North America, edited by R. Ford, pp. 73–98. Anthropological Papers No. 75. Museum of Anthropology, University of Michigan, Ann Arbor.

Kline, G. W., G. D. Crites, and C. H. Faulkner
1982   The McFarland Project: Early Middle Woodland Settlement and Subsistence in the Upper Duck River Valley in Tennessee. Miscellaneous Paper 8. Tennessee Anthropological Association.

Knox, J. C.
1983   Responses of River Systems to Holocene Climates. In Late-Quaternary Environments of the United States, vol. 2, The Holocene, edited by H. E. Wright, Jr., pp. 26–41. University of Minnesota Press, Minneapolis.

Lynott, M., T. Boutton, J. Price, and D. Nelson
1986   Stable Carbon Isotope Evidence for Maize Agriculture in Southeast Missouri and Northeast Arkansas. American Antiquity 51:51–65.

Meltzer, D., and B. D. Smith
1986   Paleoindian and Early Archaic Subsistence Strategies in Eastern North America. In Foraging, Collection, and Harvesting: Archaic Period Subsistence and Settlement in the Eastern Woodlands, edited by S. W. Neusius, pp. 3–31. Center for Archaeological Investigations, Occasional Paper No. 6. Southern Illinois University, Carbondale.

Morse, D. F., and P. A. Morse
1983   Archaeology of the Central Mississippi Valley. Academic Press, Orlando.

Munson, P. J.
1986   Hickory Silviculture: A Subsistence Revolution in the Prehistory of Eastern North America. Paper presented at a conference on Emergent Horticultural Economies of the Eastern Woodlands, Southern Illinois University, Carbondale.

Pulliam, Christopher
1987   Middle and Late Woodland Horticultural Practices in the Western Margin of the Mississippi River Valley. In Emergent Horticultural Economies of the Eastern Woodlands, edited by W. Keegan, pp. 185–200. Center for Archaeological Investigations, Occasional Paper No. 7. Southern Illinois University, Carbondale.

Riley, T.
1987   Ridged-Field Agriculture and the Mississippian Economic Pattern. In Emergent Horticultural Economies of the Eastern Woodlands, edited by W. Keegan, pp. 295–305. Center for Archaeological Investigations, Occasional Paper No. 7. Southern Illinois University, Carbondale.

Scarry, M.

1986   Change in Plant Procurement and Production during the Emergence of the Moundville Chiefdom. Ph.D. dissertation, Department of Anthropology, University of Michigan, Ann Arbor.

Sears, W. H.

1982   Fort Center. Ripley P. Bullen Monographs in Anthropology and History No. 4. University of Florida Press, Gainesville.

Smith, B. D.

1975   Middle Mississippi Exploitation of Animal Populations. Anthropological Papers No. 57. Museum of Anthropology, University of Michigan, Ann Arbor.

1978a   Variation in Mississippian Settlement Patterns. *In* Mississippian Settlement Patterns, edited by B. D. Smith, pp. 479–505. Academic Press, Orlando.

1978b   Prehistoric Patterns of Human Behavior: A Case Study in the Mississippi Valley. Academic Press, New York.

1985a   The Role of *Chenopodium* as a Domesticate in Pre-Maize Garden Systems of the Eastern United States. Southeastern Archaeology 4:51–72.

1985b   *Chenopodium berlandieri* ssp. *jonesianum:* Evidence for a Hopewellian Domesticate from Ash Cave, Ohio. Southeastern Archaeology 4:107–33.

1986   The Archaeology of the Southeastern United States: From Dalton to de Soto, 10,500 B.P.–500 B.P. *In* Advances in World Archaeology, vol. 5, edited by F. Wendorf and A. Close, pp. 1–92. Academic Press, Orlando.

1987a   The Independent Domestication of Indigenous Seed-Bearing Plants in Eastern North America. *In* Emergent Horticultural Economies of the Eastern Woodlands, edited by W. Keegan, pp. 1–47. Center for Archaeological Investigations, Occasional Paper No. 7. Southern Illinois University, Carbondale.

1987b   The Economic Potential of *Chenopodium berlandieri* in Prehistoric Eastern North America. Ethnobiology 7:29–54.

1987c   In Search of *Choupichoul,* the Mystical Grain of the Natchez. Keynote Banquet Address, Tenth Annual Conference, Society of Ethnobiology. Florida State Museum, University of Florida, Gainesville.

Smith, B. D. (editor)

1990   The Mississippian Emergence. Smithsonian Institution Press, Washington, D.C.

Smith, B., and C. W. Cowan

1987   The Age of Domesticated *Chenopodium* in Prehistoric Eastern North America: New Accelerator Dates from Eastern Kentucky. American Antiquity 52:355–57.

Smith, B. D., C. W. Cowan, and M. P. Hoffman

1992   Is It an Indigene or a Foreigner? *In* Rivers of Change: Essays on Early Agriculture, edited by B. D. Smith, pp. 67–100. Smithsonian Institution Press, Washington, D.C.

Smith, B., and V. Funk

1985   A Newly Described Subfossil Cultivar of *Chenopodium* (Chenopodiaceae). Phytologia 57:445–49.

Swanton, J. R.

1946   The Indians of the Southeastern United States. Bureau of American Ethnology, Bulletin 137. Smithsonian Institution, Washington, D.C.

Wagner, G.

1987   The Corn and Cultivated Beans of the Fort Ancient Indians. Missouri Archaeologist 47:107–36.

Watson, P. J.

1974   Theoretical and Methodological Difficulties Encountered in Dealing with Paleofecal Material. *In* Archaeology of the Mammoth Cave Area, edited by P. J. Watson, pp. 239–41. Academic Press, Orlando.

1976   In Pursuit of Prehistoric Subsistence: A Comparative Account of Some Contemporary Flotation Techniques. Midcontinental Journal of Archaeology 1:77–100.

1985   The Impact of Early Horticulture in the Upland Drainages of the Midwest and Midsouth. *In* Prehistoric Food Production in North America, edited by Richard Ford, pp. 99–148. Anthropological Papers 75. Museum of Anthropology, University of Michigan, Ann Arbor.

1988   Prehistoric Gardening and Agriculture in the Midwest and Midsouth. *In* Interpretations of Culture Change in the Eastern Woodlands during the Late Woodland Period, edited by R. Yerkes, pp. 39–67. Occasional Papers in Anthropology. Department of Anthropology, Ohio State University, Columbus.

Watson, P., and M. Kennedy
1991   The Development of Horticulture in the East-
        ern Woodlands of North America: Women's
        Role. *In* Engendering Archaeology, edited by
        J. Gero and M. Conkey, pp. 255–75. Basil
        Blackwell, Oxford.

Watt, B., and A. Merrill
1963   Composition of Foods. U.S. Department of
        Agriculture, Agricultural Handbook No. 8.

Wymer, D. A.
1987a   The Archaeobotanical Assemblage of the
        Childers Site: The Late Woodland in Perspec-
        tive. West Virginia Archaeologist 38:24–33.

1987b   The Middle Woodland-Late Woodland Inter-
        face in Central Ohio: Subsistence Continuity
        Amid Cultural Change. *In* Emergent Horti-
        cultural Economies of the Eastern Woodlands,
        edited by W. Keegan, pp. 201–16. Center for
        Archaeological Investigations, Occasional Pa-
        per No. 7. Southern Illinois University, Car-
        bondale.

Yarnell, R.
1969   Contents of Human Paleofeces. *In* The Prehis-
        tory of Salts Cave, Kentucky, edited by P. J.
        Watson. Illinois State Museum, Reports of In-
        vestigations No. 16. Springfield.

1974   Plant Food and Cultivation of the Salts Cavers.
        In Archaeology of the Mammoth Cave Area,
        edited by P. J. Watson, pp. 113–22. Academic
        Press, Orlando.

1976   Early Plant Husbandry in Eastern North
        America. *In* Cultural Change and Continuity,
        edited by Charles E. Cleland, pp. 265–73. Ac-
        ademic Press, New York.

1978   Domestication of Sunflower and Sumpweed in
        Eastern North America. *In* The Nature and
        Status of Ethnobotany, edited by R. Ford, pp.
        285–99. Anthropological Papers No. 67. Mu-
        seum of Anthropology, University of Michi-
        gan, Ann Arbor.

1983   Prehistoric Plant Foods and Husbandry in
        Eastern North America. Paper presented at the
        48th Annual Meeting of the Society for Amer-
        ican Archaeology, Pittsburgh.

1987   A Survey of the Prehistoric Crop Plants in
        Eastern North America. Missouri Archaeolo-
        gist 47:47–60

Yarnell, R., and J. Black
1985   Temporal Trends Indicated by a Survey of Ar-
        chaic and Woodland Plant Food Remains from
        Southeastern North America. Southeastern
        Archaeology 4:93–106.

7

Earliest Plant Cultivation in the Desert Borderlands of North America

PAUL E. MINNIS

Western North America stretches from the Great Plains to the intermountain west and finally to parts of the Pacific coast. This region encompasses a great diversity of natural environments and native peoples. Yet the earliest use of domesticated plants and reliance on agriculture is largely restricted to the Desert Borderlands, an area that includes the southwestern United States and much of northern Mexico (commonly considered the "Greater Southwest" by North Americans and the "Northwest" by Mexicans, both politico-centric terms).

This chapter focuses exclusively on this region. Although there is evidence for a few domesticated plants, some agriculture, and significant humanly caused environmental manipulation in aboriginal California (Treganza 1947; Lewis 1973; Wilke and Fain 1974; Bean and Lawton 1976; Lawton et al. 1976; Wright and Bailey 1983; Moratto 1984), the often dense prehistoric human populations and prehistoric cultural complexity were based largely on the collection of naturally available resources (e.g., acorns, sea mammals, and fish), not on the cultivation of cultigens. Cultivated plants were also important in the Great Plains, but their introduction appears to have been later than in the Desert Borderlands. Furthermore, prehistoric Great Plains agriculture has a closer affinity to Eastern Woodlands agriculture than to the west (see Smith, this volume).

What is the theoretical significance of the study of agricultural origins in the Desert Borderlands? Most research on the origins of cultivated plants and agriculture focuses on areas of pristine domestication—the first domestication of plants in places where no other crops have been previously domesticated. The Near East and Mesoamerica are the two most thoroughly documented cases of pristine domestication (e.g., Flannery 1973; Harlan 1975; McClung, this volume; Miller, this volume). Because there are little significant archaeological data on the early domestication of indigenous crops, the Desert Borderlands cannot readily be used to discuss the origin of domesticated plants.

As Harris (1977), Flannery (1973), and others have pointed out, agriculture is the outcome of processes that can involve different factors. That is, the principle of equifinality applies to the origins of plant cultivation: different processes can lead to the same result. In addition to pristine domestication, processes such as secondary domestication or transdomestication (Hymowitz 1972) have occurred more commonly than pristine domestication. One very common process is *primary crop acquisition,* the first use of crops that were domesticated elsewhere. The prehistory of the Desert Borderlands is a reasonably well-documented example of this process (Minnis 1985a).

I briefly summarize our present knowledge about the plants cultivated by prehistoric inhabitants in the study area. Although this discussion includes a consideration of the dating of the first prehistoric cultivated plants in the region, it must also cover other issues in order to make sense of the factors involved in early cultivation. Such factors include: (1) the congruence between a hunting-and-gathering diet and a diet utilizing cultigens, (2) environmental manipulation by hunters and gatherers, (3) the relationship between human population mobility and small-scale cultivation, (4) casual agriculture and the labor cycle, and (5) agricultural yields and risks of small-scale plant cultivation. A consideration of all these factors should allow for a more detailed discussion of the transformation from hunter-gatherer to agriculturalist in arid to semiarid regions.

In short, I argue that the first cultivation of plants by Archaic period peoples in the American Southwest and northern Mexico was not as major a change in ecological, economic, or social relationships as has been assumed. Due to the level of cultivation, the patterns of seasonal movement, and the nature of the crops used, cultivation fit well within the existing environmental, economic, and social context. Furthermore, casual agriculture was quickly adopted throughout the Desert Borderlands because it offered a low cost, low effort way to increase the economic security of Archaic peoples. Not until at least one millennium (and perhaps as many as three) after the first use of cultivated plants in the Desert Borderlands did agriculture become the economic mainstay for many prehistoric populations (sometime between A.D. 200 and A.D. 1000).

In contrast, Wills (1988, 1990) argues that this position ignores documented changes in Late Archaic prehistory that correlate with the first use of crops, especially in the Mogollon region. Much of this disagreement seems simply a question of the scale of change; changes did occur in the human condition with the use of crops, but these changes did not lead immediately to an agriculturally dependent economy.

Primary Crop Acquisition

Pristine domestication is the initial evolution of mutualistic relationships between people and plants. The outcome of these interactions results in changes in both plant and human populations. Factors involved in primary crop acquisition can also include those important in pristine domestication, such as changes in human population size and in natural environmental fluctuation. In addition, primary crop acquisition can involve unique factors, such as the availability of established farming techniques and domesticated species (LeBlanc 1982; Dennell, this volume). Therefore, primary crop acquisition can occur differently from pristine domestication. For example, Berry (1982, 1985) argues that the transition from a solely hunting-and-gathering to an agriculturally based subsistence economy on the Colorado Plateau was a rather abrupt occurrence. Plant cultivation dramatically increased the food supply in an area otherwise impoverished in edible biomass. By studying a range of circumstances leading to the early cultivation of crops, we can begin to understand the variety of factors responsible for the widespread use of cultivated plants.

Archaic period cultivation, particularly of maize, seems to have fit well within the already established subsistence strategies, settlement patterns, and social organization. This fit, however, does not constitute an adequate explanation for why these people began to alter their activities to include cultivation. Two explanations for the origin and spread of agriculture in the Desert Borderlands are considered briefly.

The first type of explanation is the "model of necessity." Variants of this explanation have been used to account for pristine domestication and other processes of agricultural origins, including those in the study area (e.g., Binford 1968; Flannery 1973; Cohen 1977; Glassow 1972, 1980; Rindos 1984; Barker 1985). Advocates of these models argue that ecological interrelationships with hunters and gatherers that led to the cultivation of crops were responses to problems in securing an adequate food base. Such stresses can result from a variety of factors, such as increases in population and/or decreases in naturally available food supplies.

The second type of explanation is the "model of opportunity." This explanation posits that plant cultivation, which led to domestication and agriculture, was a result of the advantages offered by cultivation. Such advantages include greater control of the timing of food collection and increased yields. Models of opportunity take many forms, but the one of most interest is a modification of LeBlanc's (1982) discussion of the origins of intensive agriculture in the Desert Borderlands. He suggests that the introduction of already domesticated plants and their associated technology and the knowledge of planting, tending, harvesting, storing, and consuming provided a greater return and a more secure food supply than hunting and gathering alone. LeBlanc's argument concerns the origin of intensive agriculture, not the first use of plant domesticates, so that it is not directly applicable to the study of first cultigen use in the Desert Borderlands. Furthermore, domesticated plants occur in the study area well before the associated agricultural technology. Nevertheless, LeBlanc's model can be transformed (although he may not agree with this modification) to emphasize that the presence of domesticates themselves and of casual agriculture offered a more secure and abundant economy than hunting and gathering did. Once these plants became available, they were quickly added to the Archaic repertoire of foodstuffs. As indicated in a description of Western Apache cultivation, low intensity farming in the Desert Borderlands can be a very dependable and secure economic strategy (Buskirk 1986); therefore, models of opportunity deserve consideration.

Although not mutually exclusive, the two types of models have different implications for Archaic prehistory. Table 7.1 presents a comparison of expectations for each model (these expectations are discussed in Minnis 1985a).

The Plants

Until recently, plants cultivated in the Desert Borderlands fell into two categories (Table 7.2). The first, and the one that receives the greatest attention, consists of plants domesticated first in Mesoamerica and later adopted by prehistoric peoples to the north. These plants include maize (*Zea mays* var. *mays*), various beans (*Phaseolus vulgaris, P. lunatus, Canavalia ensiformis*), squashes (various species of *Cucurbita*), gourd (*Lagenaria siceraria*), amaranth (*Amaranthus hypochondriacus*), and cotton (*Gossypium hirsutum* var. *punctatum*). Some of these plants were and continue to be the most widely used and important crops of the Desert Borderlands.

The second group of domesticates consists of a few poorly known plants that may have originally been domesticated within the Desert Borderlands. In-

Table 7.1. Two Models of Primary Crop Acquisition

Characteristic	Model of Necessity	Model of Opportunity
Speed of the spread of cultivation	slower	higher
First use of crops correlates with population density	higher	lower
First use of crops correlates with resource abundance	higher	lower
First use of crops correlates with resource instability	higher	lower
Areal extent of initial cultigen use	smaller	larger
Probability of plant manipulation before cultivation	higher	lower
Crops introduced as an assemblage	lower	higher

Source: Adapted from Minnis (1985a).

cluded in this category are devil's claw (*Proboscidea parviflora* var. *parviflora*) (Nabhan et al. 1981; Bretting 1982); the tepary bean (*Phaseolus acutifolius* var. *latifolius*) (Nabhan and Felger 1984); and panic grass (*Panicum sonorum*) (Nabhan and de Wet 1984).

Recently accumulated evidence indicates that other native plants of the Desert Borderlands may have been cultivated, if not domesticated. These plants include the century plant or agave (*Agave murpheyi* or *A. parryi*) (Fish et al. 1985); little barley (*Hordeum pusillum*) (Adams 1986); Rocky Mountain beeweed (*Cleome serrulata*); and many other plants for which we do not have adequate identifications (Winter 1974; Ford 1981; Winter and Hogan 1986). In addition, there is evidence of a wide range of agricultural techniques used by prehistoric groups in the region (e.g., Fish and Fish 1986). Because of the diversity of plants that were manipulated and the variety of cultivation techniques that were developed, research programs must shift from an emphasis on finding the earliest cultigen remains to a focus on addressing a broad range of questions about the environmental, cultural, and historical contexts of cultivation and the causes for the adoption of cultigens in the Desert Borderlands. Moreover, we must begin to deal with intraregional diversity within the Desert Borderlands. The Sonoran Desert, for example, is quite different from the Colorado Plateau, and at some level, these differences must have been significant.

Table 7.2. Domesticated Plants in the Desert Borderlands

Latin Name	Common Names	Comments
Zea mays	corn, maize, Indian corn	This cultigen was the most important crop of the prehistoric peoples of the Desert Borderlands. Nearly 300 Precolumbian varieties were developed in the New World. Uses: Maize is an extremely efficient producer of carbohydrates. Its vegetative tissue has many uses in construction and as a tinder, and maize pollen and meal have religious significance for many Native American groups.
Phaseolus vulgaris	common bean	This bean was the earliest and most widely grown cultivated (Leguminosae) legume in the Desert Borderlands, and its remains have been found in many archaeological sites. Uses: Beans are an especially important protein source.
P. acutifolius var. *latifolius*	tepary bean	The bean is drought resistant and was grown in the more arid areas of the Desert Borderlands, particularly the Sonoran Desert. Uses: see common beans.
P. lunatus	lima bean, sieve bean	This bean was not widely cultivated. Current archaeological evidence suggests that it was a relatively late crop in the Desert Borderlands. Uses: see common bean.
P. coccineus	scarlet runner bean, Aztec bean	This legume was not widely cultivated and may not have been a prehistoric crop in the study area. Uses: see common bean.
Canavalia ensiformis	jack bean	This cultigen was not grown widely and is commonly found in sites within the Sonoran Desert. Uses: see common bean.
Cucurbita pepo	pumpkin, squash	This was the earliest and most widely grown cucurbit. Uses: The fruit flesh, seeds, and flowers are edible. Hard-shelled varieties can be fabricated into containers, much like the gourd.
C. moschata	warty squash	Warty squash was grown prehistorically, but its distribution was restricted to the warm deserts. Uses: see pumpkin.
C. mixta	green-striped	Remains of this cucurbit have been recovered from archaeological sites, especially those in the warm deserts in the study area. Uses: see pumpkin.
Lagenaria siceraria	bottle gourd	The hard-shelled fruit of this plant has been an important part of the material cultural inventory of cultures in the Desert Borderlands. Due to poor archaeological preservation and curation of prehistoric gourd containers, relatively few remains are recovered from archaeological sites. Uses: The mature fruits can be made into many types of containers. The immature fruits, seeds, and flowers are edible.
Amaranthus hypochondriacus	amaranth	Although a very important Mesoamerican cultigen, few examples of this plant have been found in the Desert Borderlands but it is reported to have been grown by ethnographically documented groups. Due to the difficulty in documenting the presence of cultivated amaranths in the archaeological record, this plant may have been more common than is suggested by the archaeological record. Uses: Amaranth provides protein-rich seeds and edible greens. Some varieties have been used as a source of dye.
Gossypium hirsutum var. *punctatum*	cotton	Before A.D. 900–1000, prehistoric cotton production focused on the Sonoran Desert, but later cotton was more widely grown elsewhere in the study area. Uses: In addition to the seed hairs used in weaving, the seeds themselves are edible.
Hordeum pusillum	little barley	Recent recovery of naked seeds of this grass, both in the Sonoran Desert and Eastern Woodlands of North America, indicate that this plant may have been a cultigen. Uses: Presumably, the grains were consumed.
Panicum sonorum	panic grass	This grass is cultivated by several groups in the Desert Borderlands. As of yet, however, no archaeological examples have been discovered. Uses: The grains are a good source of carbohydrates.
Proboscidea parviflora	devil's claw	The history of this domesticate is not well known. The presence of white seeds and unusually long fruit awns indicate human manipulation. No clearly prehistoric examples of this cultigen have been documented. Uses: Long strips from the fruits are used in basket weaving and the seeds are edible.
Agave parryi	century plant, agave, Mescal, maguey	Recent evidence of agave fields in southern Arizona strongly indicates the cultivation of agave, perhaps *A. parryi*. Uses: The flowering stalks and "hearts" are consumed after roasting. Juice can be fermented. Fibers are used in weaving and for cordage.
A. murpheyi	century plant, agave	See: *A. parryi*.

Notes: In addition to the above-listed cultigens, various authors (e.g., Yarnell 1977; Winter 1974; Winter and Hogan 1986; Fowler 1986) have suggested that many other plants may have been cultivated or significantly manipulated by native population in the Desert Borderlands and adjacent Great Basin. Some of the plants listed below may also have been cultivated prehistorically: *Cleome serrulata, Nicotiana attentuata, N. trigonophylla, Solanum jamesii, S. trifolium, Lycium pallidum, Physalis longifolia, P. hederaefolia, P. foetnens* var. *neomexicana, Rumex hymenosepalus, Asclepias* spp., *Cirsium neomexicana, Sphaeralcea coccinea, Lepidium* spp., *Allium* spp., *Datura* spp., *Amaranthus powellii, A. lecocarpus, Mirabilis multiflora, Helianthus* sp., *Chenopodium berlandieri, Atriplex argentea, Dichelostemma pulchella, Cyperus* sp., *Eragrostis orcuttiana, Agropyron trachycaulum, Elymus* sp., *Rorippa curvisiliqua, Echnochloa* sp., *Sophia* sp., *Oryzopsis hymenoides*.

Natural Environment and Prehistory

The American Southwest and northern Mexico encompass a large number of plant communities (Brown 1982), and an understanding of this variation is critical for modeling differences in the role of early cultivation among the peoples of this region. In general, two broad topographic zones can be defined: deserts and major continental mountain ranges.

Deserts include the arid Sonoran Desert in southern Arizona and northern Sonora and the hot Chihuahuan Desert that centers on southern New Mexico, much of Chihuahua, and parts of extreme western Texas. The "cold" Great Basin deserts cover much of northern Arizona, northern New Mexico, southern Utah, and southwestern Colorado. To the east the deserts grade into the Great Plains, to the north the Desert Borderlands abut the Great Basin and Rocky Mountains, and to the south they border on the humid zones (e.g., in Tamaulipas). Major mountain ranges include the Rocky Mountains and associated ranges in the American Southwest and the Sierra Madre Occidental of northern Mexico. Numerous small mountain ranges are scattered throughout the Desert Borderlands.

As a semiarid to arid region, the Desert Borderland is generally characterized as marginal for plant cultivation. Water is often present in low quantities or with poor quality, precipitation can fluctuate greatly from year to year, growing season temperature is often very high, growing season length can be short in the mountains, surface temperature may be lethally high, and deep fertile soils are infrequent.

This broad characterization of the Desert Borderlands as a marginal area for agriculture may be correct when viewed from a Midwest North American agribusiness perspective. Indeed, many prehistoric agriculturalists on occasion seemed to have faced major problems sustaining an adequate food supply in the face of changing environmental and social conditions (Cordell 1984; Minnis 1985b; Dean et al. 1985). For small groups with limited field requirements, however, areas within the Desert Borderlands would have had excellent farming potential. By excellent, I mean locations that would have provided abundant and dependable harvests. Potential fields with adequate water (either subsurface or runoff) and fertile soil may be overlooked by agronomists and Euro-American farmers because they are perceived as being too small or out of the way (e.g., in small mountain valleys). Consequently, the agricultural

environment of the Desert Borderlands is much better than normally perceived; ethnohistorically documented groups have successfully farmed throughout the entire region.

The prehistoric climate and environment of the Desert Borderland has changed. Largely because of tree-ring data, the post-300 B.C. record is rather well documented (e.g., Euler et al. 1979). This is not the case for the earlier time period (2000–1000 B.C.) when cultigens were first used in the area. Analysis of packrat middens, geomorphology, and palynology are the three major approaches for paleoenvironmental reconstruction during this time period, and their results do not necessarily agree (e.g., Van Devender and Spaulding 1979; Betancourt and Van Devender 1981; Hall 1985). In general, it appears that by 2000 B.C. the major postglacial vegetational changes had occurred and the basic natural environments present just before European contact had been established. The one exception may be that the Chihuahuan Desert shrub communities had not yet expanded to their present distributions. This suggests that the climate at this time was generally similar to that at present. While the general environment at 2000–1000 B.C. was similar to that at A.D. 1500, we do not have an adequate understanding of smaller scale, yet important, environmental changes when domesticated plants were first used in the study area. Clearly, we cannot now effectively evaluate the precise natural environmental contexts of the first use of crops.

The earliest cultigens appear during the Late Archaic period (2000–1000 B.C.). Human groups in the Desert Borderlands during this period conform to a general pattern of low population density, hunter-gatherer bands with flexible social organization and mobility patterns (for a summary of late Archaic archaeology, see Woodbury and Zubrow 1979). The earliest sedentary agricultural strategies occur no earlier than A.D. 200 and, perhaps, as late as A.D. 700. There are, however, recent interpretations suggesting that sedentary populations inhabited the Tucson Basin of southern Arizona during the Late Archaic (e.g., Fish et al. 1990).

Time and Location of Early Cultivation

After five decades of research, archaeologists have identified a large number of sites with possible evidence of early (Archaic period) cultigens (Table 7.3 and Fig. 7.1). The most important of these sites is Bat

Table 7.3. Archaeological Sites with Possible Early Maize from the Desert Borderlands

Site	Location	Date[a]	Comments
Armijo Rockshelter	Arroyo Cuervo region, New Mexico	1050–550 B.C.	Maize pollen was recovered from one of four profiles. This is unreliable for three reasons: (1) no site report has been published; (2) few pollen grains were recovered; and (3) dating is indirect through stratigraphy. References: Schoenwetter, personal communication, 1979; Irwin-Williams, personal communication, 1979
Atlatl Cave	Chaco Canyon, New Mexico	2900 B.C.	The shallow deposit of this cave yielded maize macroplant remains as well as the remains of beans and squash. The shallow deposits, woodrat disturbance, and the lack of a full report limits the reliability of this example. References: Mathews and Neller 1979.
Bat Cave	San Augustin Plains, New Mexico	800–460 B.C.	Much controversy surrounds the maize remains from this site. See Berry (1982, 1985) and Wills (1985, 1988) for discussion of this site. References: Smith 1950; Dick 1954, 1965; Mangelsdorf 1954, 1974; Mangelsdorf and Smith 1949; Mangelsdorf et al. 1967; Wills 1985, 1988, 1990; Long et al. 1986.
Black Mesa sites	Black Mesa, Arizona	see comments	Berry (1982, 1985) assigns pre-A.D. sites with maize sites for sites from this project. However, Smiley (1985) believes that these sites are post-A.D. 1. References: Berry 1982, 1985; Smiley 1985.
Chaco alluvial site	Chaco Canyon, New Mexico	5050–3850 B.C.	One pollen grain was identified from this site and as such is an unreliable example of early maize. References: Hall 1977.
Chaco sites	Chaco Canyon, New Mexico	at least 770 B.C.	Sites with macroplant remains of maize and squash, and maize pollen were recovered. The maize pollen was much older than the macroplant remains. References: Simmons 1984, 1986.
Cienega Creek	east-central Arizona	unclear	See Berry and Wills for discussions of maize pollen from this site. References: Berry 1982, 1985; Haury 1957; Martin and Schoenwetter 1960; Wills 1985, 1988.
Cienega Creek, Empire Valley	southeastern Arizona	older than 50 B.C.	Seven pollen grains (out of 64,000 identifications) were maize. As most records of pollen, this case is an ambiguous record of early maize. References: Martin 1963
Cordova Cave	Pine Lawn Valley, New Mexico	300 B.C.	The radiocarbon dating of this early example is unreliable. References: Cutler 1952; Kaplan 1963; Wills 1988.
County Road	Hay Hollow Valley, Arizona	1000 B.C.	This site has not been published so it cannot be used as a reliable example of early maize. References: Berry 1982; Plog 1974; Real 1965; Smiley 1985.
Cowboy Cave	Canyonlands National Park, Utah	A.D. 395	A cache of maize from this site was at first dated to 2000 B.C. based on cultural similarities with other sites. However, the earliest directly dated maize is A.D. 396. References: Jennings, 1975, 1980.
Double Adobe site	southeastern Arizona	older than 1900 B.C.	As Martin suggests, the single maize pollen from this site cannot be considered a trustworthy example of early maize. References: Martin 1963.
En Medio Rockshelter	Arroyo Cuervo region, New Mexico	1550 B.C.	The site has yet to be published, so the integrity of the maize pollen cannot be assessed. References: Irwin-Williams, pers. comm., 1979; Irwin-Williams and Tompkins 1968; Schoenwetter, personal communication, 1979.
Fresnal Rockshelter	Tularosa Basin, New Mexico	800–400 B.C.	Maize macroplant remains have been recovered from this site. Recent direct dating of maize provides the earliest date of 800–400 B.C. Study of this collection continues. References: Bohrer 1982; Carmichael 1982; Human Systems Research 1973; Long et al. 1986.
Hay Hollow site	Snowflake, Arizona	unclear	Both maize pollen and macroplant remains were recovered. Radiocarbon dates range from 470 B.C. to A.D. 305. References: Berry 1985; Bohrer 1972; Fritz 1974; Martin 1967.
Jemez Cave	Jemez Mountains, New Mexico	1000–400 B.C.	Maize macroplant remains were recovered during the 1930s excavation and have been directly dated to 1000–400 B.C. References: Alexander and Reiter 1935; Ford 1968, 1975; Long et al. 1986.
Kenton Caves	Cimarron County, Oklahoma	unknown	These caves yielded maize macroplant remains that Mangelsdorf characterized as being like those from Bat Cave. No stratigraphic excavations were made, so the age of these remains cannot now be assessed. References: Lintz and Zabawa 1984; Renaud 1929, 1930.

Continued next page

Table 7.3.—*Continued*

Site	Location	Date[a]	Comments
LA 17337	Chaco Canyon region, New Mexico	unknown	See "Chaco sites."
LA 181093	Chaco Canyon region, New Mexico	unknown	See "Chaco sites."
LA 18091	Chaco Canyon region, New Mexico	unknown	See "Chaco sites."
La Palma site	Tucson Basin, Arizona	Late Archaic	Maize pollen from a mano was recovered. References: Dart 1985.
LoDaisKa Rockshelter	near Denver, Colorado	unclear	Berry discusses why this is an unreliable example of early maize. References: Berry 1982, 1985; Irwin and Irwin 1959, 1961.
Matty Canyon site	southeastern Arizona	270 B.C.	Three maize pollen grains (out of 8,000 identifications) were noted. References: Martin 1963.
Milagro site	Tucson, Arizona	850 B.C.	Maize macroplant remains have been directly dated, but the dates have yet to be released. Wood charcoal from the same provenience as the maize dated to 850 B.C. References: Huckell and Huckell 1984; Huckell, personal communication, 1986.
NA 14646	St. Johns, Arizona	185 B.C.	Maize macroplant remains were recovered from a structure with radiocarbon dates clustering around 185 B.C. References: Berry 1985.
O Block Cave	Pine Lawn Valley, New Mexico	unclear	Two radiocarbon dates (830 B.C. and 650 B.C.) have been obtained from this site. The remains have not been published so their relationship to the dates cannot be assessed. References: Martin et al. 1952.
O'Haco Cave	Chevelon Creek, Arizona	3000–1000 B.C.	The earliest stratum with maize cannot be dated more precisely, and the shallowness of the deposits makes stratigraphic control difficult. References: Bruier 1977.
Ojala Cave	near Los Alamos, New Mexico	670–590 B.C.	This small cave yielded two maize kernels, both of which came from the 240-250 cm level. The two radiocarbon dates from this stratum were 670 B.C. and 590 B.C. No final report has been issued, so its value cannot be judged effectively. References: Traylor et al. 1977.
Placitas Arroyo site 5	near Hatch, New Mexico	80 B.C.	Pollen and macroplant remains of maize were recovered, but no detailed site report has been published. References: Morenon and Hayes 1978.
Sheep Camp Shelter	Chaco Canyon region, New Mexico	250 B.C.	See "Chaco sites."
Swallow Cave	near Casas Grandes, Chihuahua	unknown	Maize remains were recovered from deposits that may predate the ceramic Mogollon levels (before A.D. 200). Dating is uncertain, and a comprehensive site report has not been published. Other caves in the region have deposits dating to post-A.D. 1000 times. References: Minnis, personal observation, 1989; Mangelsdorf 1974; Mangelsdorf and Lister 1956.
Tabeguache Cave	Montrose County, Colorado	older than 10 B.C.–A.D. 10	The maize remains have not been dated, but recent dendrochronological dates from the upper levels suggest a possible pre-A.D. date for the maize-bearing levels. References: Dean, personal communication, 1980; Hurst 1940, 1941, 1942; Hurst and Anderson 1949.
Tucson Basin sites	Tucson Basin, Arizona	unknown	Numerous, recently excavated sites in and around Tucson have yielded early maize, and it has been suggested that Late Archaic populations were more sedentary than previously thought. References: for a general discussion, see Fish et al. 1990.
Tumamoc Hill site	Tucson, Arizona	520 B.C.	The earliest maize macroplant remains date to 520 B.C. References: Fish et al. 1986.
Tularosa Cave	Pine Lawn Valley, New Mexico	1000–400 B.C.	Maize macroplant remains were recovered from the lowest level and have been directly dated to 1000–400 B.C. References: Cutler 1952; Long et al. 1986; Wills 1988.
Valencia site	Tucson Basin, Arizona	810 B.C. and 250 B.C.	Maize pollen was recovered from a grinding stone in a pithouse. Morphologically similar pithouses date to 810 B.C. and 250 B.C. References: Doelle 1985.

[a]Radiocarbon dates are provided without standard deviations. Consult original sources for more precise information on the temporal placement of these maize remains.

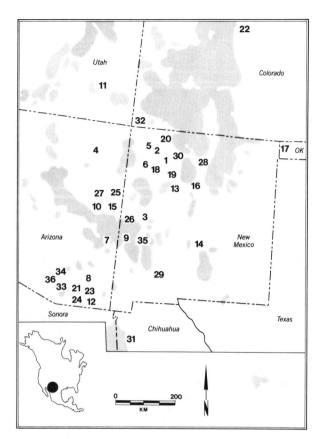

Fig. 7.1 Location of archaeological sites reported to have early maize remains in the Desert Borderlands. 1, Armijo rockshelter; 2, Atlatl Cave; 3, Bat Cave; 4, Black Mesa sites; 5, Chaco alluvial sites; 6, Chaco Archaic sites; 7, Cienega Creek site; 8, Cienega Creek site, Empire Valley; 9, Cordova Cave; 10, County Road site; 11, Cowboy Cave; 12, Double Adobe site; 13, En Medio rockshelter; 14, Fresnel rockshelter; 15, Hay Hollow site; 16, Jemez Cave; 17, Kenton caves; 18, La 17337; 19, La 1803; 20, La 18091; 21, La Paloma site; 22, LoDaisKa rockshelter; 23, Matty Wash site; 24, Milagro site; 25, Na 14646; 26, O Block Cave; 27, O'Haco rockshelter; 28, Ojala Cave; 29, Placitas Arroyo site; 30, Sheep Camp rockshelter; 31, Swallow Cave; 32, Tabeguache Cave; 33, Tucson Basin sites (other); 34, Tumamoc Hill site; 35, Tularosa Cave; 36, Valencia site.

Cave, where the first significant assemblage of prehistoric maize was recovered (Dick 1954, 1965; Mangelsdorf 1954, 1974; Mangelsdorf and Smith 1949; Mangelsdorf et al. 1967; Wills 1985, 1988, 1990). Largely on the basis of Bat Cave, archaeologists came to regard the introduction of maize in the American Southwest as having occurred sometime around 2000 B.C. (e.g., Woodbury and Zubrow 1979). This date was then used to evaluate other early maize remains. Recent reconsideration of previously published examples of early maize, particularly Berry's critique (1982,

1985), has raised serious questions about the dating of the earliest maize in the Desert Borderlands.

Very recent and often unpublished radiocarbon dates of actual maize specimens and new excavations are beginning to provide a more secure base for dating the first cultigens. The best date for the earliest evidence of maize in the Desert Borderlands is around 1000 B.C. (see Table 7.3). Numerous examples of directly dated maize fall near this time period and include Sheep Rock Shelter (macroplant remains), Tumamoc, Milagro, Bat Cave, and Jemez Cave. The best example of maize remains earlier than 1000 B.C. is the pollen from the Chaco Canyon region that may date to 2000 B.C. (Simmons 1986). Unless other secure cases of 2000 B.C. maize are forthcoming, the 1000 B.C. date is the best approximation for the first evidence of cultigens because of the relatively large number of directly dated specimens from around this time.

The dating of the earliest cultigens other than maize is less certain. Before the recent program of directly dating cultigen remains, it was thought that maize was domesticated much earlier than beans and squash. The sequence of the first evidence of crops is now unclear. Long et al. (1986) present directly dated radiocarbon dates for squash and beans. Several examples of squash may be as old as maize and perhaps are older. The oldest date for Bat Cave *Cucurbita pepo* is approximately 1500–1100 B.C., and the oldest squash from the Chaco area is 1400–700 B.C. (Simmons 1986). Long et al. (1986) report that the oldest dates for beans are those from Tularosa Cave (1000–400 B.C.) and Bat Cave (400–300 B.C.). Therefore, as Long et al. (1986:1) point out, "present data cannot resolve the differences between the timing of first appearance of *Zea mays, Cucurbita pepo,* and *Phaseolus vulgaris.*"

Early maize has been found in many areas of the Desert Borderlands. Most earlier examples of Archaic age crops came from sites at relatively high elevations (1800–2000 m). This led to the commonly held belief that early maize was not well adapted to low moisture conditions, and that the availability of a drought resistant maize was a major factor in the spread of agriculture in the Desert Borderlands (e.g., Martin and Plog 1974; Whalen 1973). However, the recent recovery of Archaic maize from the deserts around Tucson indicates that these conclusions are unwarranted (see Table 7.3). The modern maize variety morphologically closest to Archaic maize, Chapalote, is "adapted to a wide range of altitudes" and, in fact, "it produces best at low elevation"

(Wellhausen et al. 1952:56), suggesting that early Chapalote-like maize could be grown under many conditions.

As in many areas of the world, too much research focuses solely on the date of the earliest cultigens present in the Desert Borderlands. Whether the first evidence of cultigens occurs at 2000 B.C. or at 1000 B.C. is not as important a question as it appears. Whatever date is accurate, there was a long period of time during which plants were cultivated by peoples who practiced a basic hunting-and-gathering economy with few, if any, major changes in their lifestyle. If 2000 B.C. is the correct date—and assuming that relatively sedentary farming did not develop until A.D. 200—then the period of casual cultivation lasted over two millennia. If cultigens were first grown around 1000 B.C., then this period is only 1,200 years. The long period of casual cultivation by hunting-and-gathering populations deserves significant attention. Rather than simply dating early remains, investigators should focus more attention on the context and causes of the economic transformation to plant cultivation.

Context of Casual Cultivation

In order to indicate how limited cultivation conforms to hunting and gathering, I consider five topics: (1) types of cultivation and environmental manipulation, (2) changes in diet with use of cultigens, (3) casual agriculture and mobility, (4) labor requirements of casual agriculture, and (5) effects of cultivation scale on the risk of food stress. The Western Apache are discussed as an example of how a basically hunting-and-gathering group in the Desert Borderlands incorporated casual agriculture into its way of life. The Apache, however, should not be considered a direct analogue for the Archaic. The historic Apache faced a very different sociopolitical setting from the Archaic peoples; they also had horses, which can change the logistics of mobility. Instead, this ethnographic example can provide insights into the relationship between casual agriculture and hunting and gathering.

Variation in Cultivation and Environmental Manipulation

Domesticated plants are at one end of a continuum of human manipulation of plants and their ecological relationships (Ford 1985). At the other end are the less enduring manipulations of plant distribution, abundance, biochemistry, and morphology. For example, the prehistoric inhabitants of Mesa Verde pruned conifers (Nichols and Smith 1965), the Cahuilla prune mesquite trees to increase their harvest (Bean and Saubel 1972), the Tarahumara manipulate mustards (Bye 1979), and prehistoric groups in northern Arizona seem to have burned squawbush (*Rhus trilobata*) to increase the number and length of straight branches (Bohrer 1983). Thus, in considering early plant cultivation, we need to be sensitive to a full range of possible changes caused by human manipulation of plant populations, although these less drastic changes can be harder to detect in the archaeological record than are domesticates. By ignoring variation in the degree of manipulation, we too often make simple contrasts between no manipulation on one hand and full domestication on the other. The intermediate ranges are particularly important when discussing early cultivation. The considerable range of plant manipulation by numerous groups throughout the Desert Borderlands and Great Basin has been summarized (Winter 1974; Winter and Hogan 1986; Fowler 1986).

We know that Archaic peoples in the Desert Borderlands manipulated their environment (Winter and Hogan 1986; Hall 1985); the squawbush example cited above dates to Archaic times. Some archaeologists have assumed that prehistoric hunter-gatherers in the Desert Borderlands, in contrast to prehistoric agriculturalists, had no effect on the natural environment. This is highly unlikely. I suspect that we will find that hunting-and-gathering populations in the Desert Borderlands actively manipulated their environments, as has been documented for other areas of the world, including North America. The well-documented burning of vegetation by native Californians is a nearby example (e.g., Lewis 1973).

Diet and Initial Cultivation

There is a general similarity between the first cultigens and important naturally available resources in the region. The three major groups of early cultigens were a grass (maize), a legume (*Phaseolus vulgaris*), and a cucurbit (*Cucurbita pepo*). There are analogues to these cultigens in the native vegetation of the Desert Borderlands. Many edible wild grasses, such as dropseed (*Sporobolus*) and Indian ricegrass (*Oryzopsis hymenoides*), were present. We cannot accurately assess the importance of grass seeds in Archaic diets, but these seeds have been recovered from Archaic sites. They are, however, conspicuously absent from some Archaic cave sites in Utah (Winter and Wylie 1974; Winter and Hogan 1986). A variety of legumes were present in the Desert Borderlands. Chief among these are mesquite and screwbean (*Prosopis* spp.), various species in the genus *Acacia,* other tree legumes from the Sonoran Desert (e.g., *Olynea tesota, Cercidium microphyllum,* and *Cercidium floridum*), and a variety of herbaceous legumes. Similarly, several species of cucurbits were naturally present in this region (e.g., *Cucurbita foetidissma, C. digitata,* and *C. palmata*). *C. foetidissma* seeds were recovered from Cordova Cave, presumably from Archaic deposits (Kaplan 1963). Because naturally available resources similar to the first cultivated plants were present in the Desert Borderlands, the introduction of these crops may not have required novel or unusual processing techniques (Wetterstrom 1978).

Intraregional variation in the distribution of these plants may be important. For example, legumes are much more abundant in the Sonoran and Chihuahuan deserts than in the areas to the north, and cucurbits are more abundant in the southern parts. Consequently, we might expect that the addition of cucurbits and the protein-rich legumes to the Archaic food inventory would have had a greater impact on economic and social patterns in the areas north and east of the hot deserts (Berry 1982, 1985). Unfortunately, we have yet to investigate intraregional differences in terms of the consequences of cultivated plant use in the Desert Borderlands, but generally we can conclude that these first cultigens represented known resources that did not require novel techniques of collection, preparation, and storage.

While the general nature of the cultivated and wild resources may have been similar, there were important differences. Obvious among them were a probable decrease in nutrient quality with an increased use of cultigens, a decrease in the consumption of toxic biochemicals often present in natural plants and less abundant in crops (e.g., legumes and cucurbits), an increase in the variety of foodstuffs available, and an increase in the human control of the amount, location, mix, and timing of food supply availability.

Mobility and Casual Cultivation

A false dichotomy can be made between the mobility of hunting and gathering and the sedentary demands of agriculture. Hunters and gatherers, particularly those in semiarid to arid regions (excluding groups in particular environments with high carrying capacities, such as much of California and the Northwest Coast of North America), are characterized as having a great deal of mobility in order to exploit naturally available resources efficiently. By contrast, agricultural societies are viewed as relatively sedentary because of mobility constraints due to plant cultivation. Hence, plant cultivation is often seen as inconsistent with a hunting-and-gathering economy. Recent reconsideration of some prehistoric Puebloan farming communities, however, has argued for more mobility than previously assumed (e.g., Nelson and LeBlanc 1986).

Rather than viewing subsistence economies in this dichotomous fashion, we need to conceive of a continuum from solely hunting and gathering to intensive agriculture with many intermediate strategies involving degrees of residential and logistic mobility in between. Such a perspective allows us to realize that the constraints of mobility due to plant cultivation are not as great as some may think, and that there are many ethnographic examples of societies that combine elements of both strategies.

Many native societies of the Desert Borderlands cultivated plants. The best-known examples, the Pima of Arizona, the Yumans of the Colorado River, the Puebloans of Arizona and New Mexico, and many groups in the Sierra Madre Occidental of northern Mexico (e.g., Opata, Tarahumara, etc.), derived the bulk of their food from domesticated plants (Ortiz 1979, 1983; Winter 1974).

In addition, there were many groups who relied on cultigens for a small part of their food supply (Ortiz 1979, 1983; Winter 1974; Castetter and Bell 1941, 1952), among them various Apache groups. A study of these groups can provide some lessons on how small-scale plant cultivation need not interfere with the mobility commonly associated with hunting-and-gathering economies in semiarid to arid environments. I discuss only one group, the eighteenth- and nineteenth-century Western Apache of east-central and southeastern Arizona (Reagan 1929; Goodwin 1935; Castetter and Opler 1936; Gallagher 1977; and, particularly, Buskirk 1949, 1986).

The Western Apache were quite mobile during their yearly cycle, exploiting habitats ranging from 300 m to 1500 m in elevation. Hunting contributed about 35 to 40 percent of the food supply, foraged plants yielded about the same percentage, and cultivated crops provided around 25 to 30 percent. Winter camps were located in the southern and lower elevation areas where they hunted and gathered. In March or April, Apache families moved their residences to farming sites in the higher elevations to the north of their winter camps. During late spring, summer, and early fall, they farmed, hunted, and gathered resources, such as pinyon nuts, acorns, and juniper fruits. Crops were harvested in late August to October, and the bands then moved back to their winter residences.

Maize was the principal crop, although other aboriginally available cultigens, including gourds, squashes, and beans, were grown. European-introduced crops, such as wheat, watermelons, tobacco, chile, some fruit trees, sunflowers, and many garden vegetables, were also tended. Devil's claw and goosefoot (*Chenopodium* sp.) were also planted, although they may not have been true domesticates, and cotton may also have been cultivated by the Western Apache.

Fields were small. Goodwin (1935:163) estimates that each field averaged "about half an acre, often less." Each household owned four to six fields, but usually only two or three were planted each year. Several types of fields were used. Easily irrigated sites on the floodplain were the most common locations. Fields watered by springs and seeps were also planted. Finally, *ak chin,* or places where alluvial fans from small drainages widen and are occasionally watered by runoff, were used. At high elevations, dry farming was practiced. Buskirk (1949, 1986) emphasizes that adequate water was the primary constraint on the selection of field locations.

Fields were planted from late March through July, with April and early May being the preferred planting time. Lower elevation fields could be planted before higher elevation sites. Field maintenance consisted of preparing the soil, irrigating the field a few times, planting the crops, and weeding once or twice. After the crops were tall enough to compete with weeds, the fields were usually left alone until harvest. Alternatively, less mobile household members might stay to watch the fields. Goodwin summarizes the farming cycle:

Preparation of the field and planting took about a month. All members of the family were expected to help if needed. When the corn was about three feet tall, most of the people moved away for the summer to harvest the various wild plant foods. In September they returned to harvest and store the crops, this again taking about a month's time. (1935:63)

Western Apache families depended on farming to varying degrees. Households in some locations did not farm at all, whereas in other areas nearly all households farmed. Several factors contributed to this variation. According to Buskirk (1949, 1986), there were environmental constraints on farming in some locations that lacked an adequate number of field sites. Navajo raiding directed against some Apache groups precluded their substantial investment in agriculture. Sociopolitical motivations also affected agricultural productivity. As Buskirk (1986) noted, some households grew more maize to increase the surplus that could be given away, thus enhancing prestige.

This single example provides clear documentation that farming is compatible with mobile hunting and gathering. Field preparation, planting, maintenance, and harvesting were embedded in an economy that was largely based on hunting and gathering and seasonal movement. As is discussed below, small-scale farming can be very advantageous to hunting-and-gathering populations.

Labor Cycle and Casual Cultivation

The labor demands of small-scale cultivation, as illustrated by the Western Apache, can be quite compatible with hunting and gathering. Compatibility, however, depends upon both the timing of labor demands and the absolute amount of labor required.

Western Apache farming activities were concentrated during two periods: planting and harvesting. Planting and tending of the young crops took approximately three to four weeks, starting as early as late March. Few naturally available plant resources would have been ready to collect during the planting season. Therefore, there seems to have been little scheduling conflict between cultivation and hunting-gathering, especially if the same segment (such as age- or sex-based groups) of the population did most of the foraging and agricultural work. The timing of the harvest probably presented even less of a problem. If crop harvesting conflicted with the collection of naturally available foodstuffs, then the cultigens could have been left in the field until the other resources had been collected; one characteristic of domesticates is their lack of natural dispersal mechanisms.

As exemplified by the Western Apache, small-scale plant cultivation can easily coexist with the labor demands of mobile hunting and gathering. In fact, in this case, subsistence labor devoted to agriculture seems to have been invested at a time when surplus labor would otherwise have been available. If a similar situation occurred during the Late Archaic, as I assume it did, then this advantage probably was a major reason for the fast spread of plant cultivation.

Risk and Casual Cultivation

Traditional agricultural strategies are often organized to reduce the risk of crop failure (Barlett 1980). Growing many different cultigens and cultivars, utilizing different fields located in different microenvironments, and growing crops that are well adapted to a given location are all agricultural strategies to reduce risk.

Although there are areas of excellent agricultural potential for small-scale farmers in the Desert Borderlands, the well-documented fluctuations in precipitation during prehistory caused periodic crop failure (Cordell 1984; Dean et al. 1985; Minnis 1985b). Historic records for eastern New Mexico, for example, show crop failures for maize grown by dry farming occurring one out of every three years (Stanten et al. 1939).

If traditional economies minimize risk and if the Desert Borderlands have a high probability of crop failure, why would prehistoric groups begin to cultivate crops? I suggest that the answer is in the scale of plant cultivation. Much of the Desert Borderlands may have presented a significant risk of crop failure for agricultural groups. Precipitation is too little or too variable, soils can be infertile, and some field locations are susceptible to short growing seasons. The original cultivators of the region, with their small populations and field requirements, would have selected only the best field locations, which are present throughout the Desert Borderlands. Small-scale cultivators would have had less difficulty with plant disease than more intensive agriculturalists would have had. Small scattered fields would have deprived pests of a dense host population. These farmers, therefore, would have minimized the risk of crop failure.

Thus, casual cultivation could have been practiced without an initial disruption to the existing economic system, would not have required much effort, and could have utilized only field locations that had the greatest probability of dependable harvests. In short, low intensity agriculture would have provided excellent economic insurance. With only minimal effort, the subsistence base would have been diversified by the addition of one or more resources. Furthermore, the hunter-gatherer-farmer would have had more control over domesticates than over naturally available resources in ensuring the availability of food.

Crop Dispersal

Although many factors critical to pristine domestication can also have been important in primary crop acquisition, the latter process involves two unique factors necessary to understanding the transition from hunting and gathering to cultivation. The first is the interregional spread of crops. Because the most important crops in the Desert Borderlands were initially domesticated in Mesoamerica, the Borderlands offer an excellent case study. Crops did not disperse as an assemblage but as individual cultigens, a process that may differ from other examples of primary crop acquisition (see Dennell, this volume). The second is the length of time involved in the introduction of new crops. A consideration of this factor sheds light on the reasons for the initial use of cultigens. There is a long history of dispersal of cultivated plants between Mesoamerica and the Desert Borderlands.

Any discussion of the transfer of crops, agricultural technology, and knowledge about plant cultivation between Mesoamerica and the Desert Borderlands is limited by the fact that the prehistory of the area between central Mexico and the American Southwest is largely unknown, particularly for the Archaic period. The best-documented early crop assemblages in Mexico are located 1500 km (Tamaulipas) and 2000 km (Tehuacan) from Bat Cave. Thus, there is evidence for early cultigens in their area of initial domestication and for their peripheral distribution in desert North America but little in the intermediate regions.

Mesoamerican Crops in the Desert Borderlands

Many plants were domesticated in Mesoamerica; Harlan (1975) lists sixty-seven. Only a few of them, however, were important and widely grown: maize, various cucurbits, gourd, several genera of beans, amaranths, chenopods, chiles, cotton, and avocado. Only ten of the Mesoamerican crops on Harlan's list were cultivated prehistorically in the Desert Borderlands. Thus, the transfer of crop assemblages between the two areas appears to have been limited. Eliminating those tropical cultigens that would not grow in the Desert Borderlands, however, one finds that many of the most important Mesoamerican crops (e.g., maize, beans, and cucurbits) were grown in the prehistoric Desert Borderlands.

There are several possible exceptions (amaranth and scarlet runner bean) and two known exceptions (chiles and chenopods). Very few prehistoric domesticated amaranth remains have been identified from the Desert Borderlands (e.g., Bohrer 1962; Sauer 1967), although the seeds of wild amaranth are commonly found in archaeobotanical assemblages. This limited distribution may be due to the fact that the obvious difference between domesticated and non-domesticated amaranth—the color of the seed coat, which is light in cultivated amaranth and dark in native amaranths—is lost during carbonization. Few prehistoric scarlet runner beans (Ford 1981) and no prehistoric domesticated chenopods have been identified in archaeological collections from the Desert Borderlands, although, like amaranth, wild chenopods are very common. Since prehistoric domesticated chenopods have been found in eastern North America (Smith, this volume), the Desert Borderlands appears to have been an island in the New World where chenopods were not cultivated, a pattern that may not survive further scientific inquiry. The absence of prehistoric chiles from the study area has always been puzzling. All chile remains from archaeological sites postdate Spanish contact, yet native chiles are present in parts of the Desert Borderlands (Nabhan 1985) and, after contact, domesticated chiles became an integral part of southwestern cuisines. No satisfactory explanation has been offered for this pattern.

The fact that many Mesoamerican crops dispersed into the Desert Borderlands does not mean that prehistoric peoples simply incorporated cultivars developed elsewhere; crop evolution continued after crops arrived in the American Southwest and northern Mexico. Many unique cultivars, such as a Hopi maize that can be planted very deeply (Collins 1914), were developed in the region.

By contrast, crop dispersal between the Desert Borderlands and areas to the east did not occur. Jerusalem artichoke (Helianthus tuberosus), the cultivated sunflower (Helianthus anuus), sumpweed (Iva annua var. macrocarpa), and other plants domesticated in the Eastern Woodlands were not grown by prehistoric groups in the Desert Borderlands, although the distances between the southern Plains and the study area

are no greater than those between Mesoamerica and the Desert Borderlands. However, crop dispersal routes from east to west traverse substantial topographic and climatic diversity.

It is clear that there was a sustained movement of crops between Mesoamerica and the Desert Borderlands. The absence of many Mesoamerican crops, particularly chiles and chenopods, in the study area, however, suggests that there were also biological and cultural barriers between the two areas. This is particularly striking when one considers the transfer of crops following Spanish contact. A whole range of domesticates was added, from tree crops (e.g., peaches) to staples (e.g., wheat) to condiments (e.g., garlic). Some of these cultigens were originally Mesoamerican crops (e.g., chile). The number of crops added within a short period of time after contact was much greater than the number of prehistoric Mesoamerican cultigens grown during the entire prehistoric period.

Order and Timing of Crop Dispersal

The rate at which crops developed in one area and were cultivated in another is a critical aspect in the study of primary crop acquisition. Was the dispersal of crops and agricultural technology a slow or rapid process? To answer this question, I briefly examine the first occurrences of crops in Mesoamerica and the Desert Borderlands.

As mentioned earlier, squash (Cucurbita pepo) and beans (Phaseolus vulgaris) may have been introduced as early as maize (ca. 1000 B.C.). Cotton and the relatively rare amaranth were later introductions. Cotton was not grown before A.D. 300–500 and the earliest amaranth appeared much later. Thus within a thousand years, probably starting at 1000 B.C., the most important food plants had been incorporated into the subsistence assemblages in the Desert Borderlands.

The order of the first occurrence of crops in the Desert Borderlands appears to have been different from that in Mesoamerica. The crop sequences from the two major areas studied archaeologically in Mesoamerica—the Tehuacan Valley and Tamaulipas—however, are not the same either. These differences may be indicative of regional variation and/or sampling biases. Furthermore, it is often difficult to determine the level of domestication in these early plants. Maize, squash, and beans are not the earliest cultivated plants in the sequences. Lagenaria is the

Table 7.4. Earliest Evidence for Selected Mesoamerican Cultigens

Plant	Date	Location
maize	5000–3000 B.C.	Tehuacan
cotton	5000–3000 B.C.	Tehuacan
beans	4000–2000 B.C.	Tehuacan and Tamaulipas
squash (C. pepo)	7000–5000 B.C.	Tehuacan and Guila Naquitz
gourd	7000–5000 B.C.	Tehuacan, Tamaulipas and Guila Naquitz

Source: Flannery 1973. See also McClung de Tapia, this volume.

earliest domesticate from Tamaulipas, and chile and avocado are the earliest from Tehuacan. Similarly, the sequence of first occurrence is not the same for Tamaulipas, Tehuacan, or the Desert Borderlands, suggesting that crops dispersed individually, not as complexes.

As one would expect, there is a lag between the domestication of a plant and its cultivation in remote areas. This lag appears to be in the range of 2,000 to 6,000 years. This can be seen by comparing the dates in Table 7.4 with those for the earliest cultigens in the Desert Borderlands. It is difficult to identify the possible reasons for this lag. Some factors may be biological. Teosinte, the probable ancestor of maize, will not reproduce sexually under the long day-length of northern latitudes, and the most northerly teosinte is now found in southern Chihuahua (Wilkes 1967). Therefore, the selection of varieties of maize that could reproduce under the long day-light conditions must have taken many generations. We simply do not know how quickly this could have been accomplished, particularly if it were done through casual agriculture.

Biological factors alone, however, cannot explain the lag in crop dispersal. One would not expect human populations in regions distant from the area of domestication to cultivate crops until they were productive. Although obvious morphological changes in teosinte showing selection toward a domesticated plant are documented as early as 7000 B.C., an assemblage of productive maize types (Nal-Tel/Chapalote) is not documented in the Tehuacan sequence until around 2000–1000 B.C. (Mangelsdorf et al. 1967). Therefore, the earliest maize types (proto-Chapalote and Chapalote) in the Desert Borderlands occur shortly after their appearance in Mesoamerica. The

lag between the domestication of maize and its appearance in outlying regions is relatively short, much less than the initial estimate of 2,000 to 4,000 years.

Other cultigens did not disperse as rapidly as maize. The lag for cotton is at least 3,000 years and the lag for squash is also surprisingly long. If the oldest squash remains from the Desert Borderlands date to 2000 B.C., then the lag is 3,000 to 5,000 years. I would expect that much earlier squash remains will be recovered from sites in the study area. Of course, the possibility that a cucurbit was domesticated in Eastern North America complicates this analysis (see Smith, this volume).

Thus, the rate of geographical transfer of crops appears to vary for different cultigens. Maize and beans, for example, seem to have been grown in the study area within one or two millennia of the appearance of productive cultivars in Mesoamerica. Others, such as squash, gourd, and cotton, appear to have transferred more slowly. These patterns further suggest that individual crops and cultivars diffused independently of each other, not as an assemblage. Consequently, we see very different patterns of crop movement, and study of the factors responsible for these differences will illuminate the processes of agricultural origins.

Based on current data, the crops that seem to have dispersed initially from Mesoamerica to the Desert Borderlands are food plants (maize, squash, and beans). Cotton and gourd arrived relatively late. This pattern suggests that the reasons for the first use of cultigens in the study area dealt with the need for food. This is particularly interesting in light of the fact that in many areas with pristine domestication, non-food plants, such as gourds, were very early domesticates.

Summary and Conclusions

Study of the earliest agriculture in the Desert Borderlands is valuable for understanding: (1) the prehistory of the region, (2) the evolution of particular cultigens, and (3) the process of primary crop acquisition, one of many processes that can be involved in transition from hunting and gathering to an agricultural economy.

The earliest reliable evidence for cultigens in the Desert Borderlands is around 1000 B.C. Maize, beans, and a squash were the first cultigens. On the basis of current evidence, one cannot ascertain the sequence of the introduction of these crops, but I suspect that cucurbits may be as early or earlier than maize and beans, as they are in other areas of North America. During the following 3,000 years, many other domesticates were added to the inventory of cultigens grown by indigenous populations in the region. A few of these plants were native and were probably domesticated regionally (*Panicum sonorum, Proboscidea parviflora,* and *Phaseolus acutifolius* var. *latifolius*). The most important prehistoric crops, however, came from Mesoamerica.

After its arrival in the Desert Borderlands, the cultivation of maize, the most important crop, spread rapidly. More and more examples of directly dated cultigen remains show that this crop expanded quickly into many regions—from the Sonoran Desert to the Colorado Plateau. This dispersal pattern suggests that the maize population was sufficiently variable to be grown in very different regions and that it was a useful crop. Because there are very few dates for early squash and beans, we cannot determine whether a similar pattern of rapid dispersion occurred with them.

I have argued that the introduction of domesticated plants into the Desert Borderlands fits relatively easily within the existing environmental, economic, and social context. The example of the Western Apache illustrates how casual cultivation can articulate with a mobile hunting-and-gathering lifestyle. Their crops were similar to naturally available foods so that food preparation techniques and cuisine need not have been radically altered. For the Western Apache, agriculture was of low intensity and easily integrated into established patterns of seasonal mobility. The most sustained agricultural activity took place at planting, a time of otherwise low labor demands; harvesting would not have presented major scheduling problems. Because their plants lacked natural dispersal mechanisms, they were left in the field to be harvested after the naturally available wild plant foods were collected.

The fact that casual cultivation articulated well with Western Apache subsistence economy does not mean that it was unimportant. As with any economy, the Western Apache had to contend with occasional periods of low food availability. Agriculture allowed them greater control over their food supply. For example, it increased the stability of their provision base. Thus, casual cultivation functioned as a form of insurance when important foodstuffs were unavailable or available only in reduced amounts (Minnis 1985a).

There was a long period (1,000 to 3,000 years) of casual agriculture when plant cultivation was incorporated into a basically hunting-and-gathering lifestyle, perhaps similar to that of the Western Apache. Thus, the introduction of crops themselves cannot be the sole catalyst for the transition to intensive agriculture, which seems to have occurred between A.D. 200–1000. Elsewhere I have argued that constraints on economic options due to increasing human populations were a primary factor in the use of intensive agriculture in the Desert Borderlands (Minnis 1985a).

Many of the most important Mesoamerican crops were ultimately grown in the Desert Borderlands, but the rate of dispersal from Mesoamerica to the Borderlands varied for each cultigen. The most quickly dispersed species (maize and beans) were food plants; non-food cultigens (e.g., gourd and cotton) were rather late introductions. This pattern suggests that the need for food was a primary reason that crops were first grown in the Desert Borderlands. Such a case is different from many examples of pristine domestication.

As shown in Table 7.1, there are different expectations for the two prevailing explanations for primary crop acquisition. Most of these expectations cannot be evaluated adequately, because there are no sufficiently precise studies of resource stability, resource abundance, Archaic population density, and environmental manipulation by prehistoric hunters and gatherers in the study area. The study of the first use of domesticated plants by Archaic peoples in the Desert Borderlands, however, can provide an understanding of one of the many processes characterizing the transition from hunting and gathering to agriculture.

Acknowledgments

Numerous individuals provided information critical for this paper at various stages of preparation: Vorsila Bohrer, Jeffrey Dean, William Doelle, Marcia Donaldson, Richard Ford, Bruce Huckell, Cynthia Irwin-Williams, Christopher Lintz, Charles Miksicek, Thomas O'Laughlin, James Schoenwetter, Mark Stiger, Chip Wills, Mark Wimberly, and Joseph Winter. Patricia Gilman, Patty Jo Watson, and C. Wesley Cowan read various drafts.

References Cited

Adams, Karen R.
1986   A Model to Assess New World *Hordeum pusillum* (Barley) Characteristics in Natural and Human Habitats. Paper presented at the 9th Annual Ethnobiology Conference, Albuquerque.

Alexander, Hubert G., and Paul Reiter
1935   Report on the Excavation of Jemez Cave, New Mexico. University of New Mexico Bulletin No. 278.

Barker, Graeme
1985   Prehistoric Farming in Europe. Cambridge University Press, Cambridge.

Barlett, Peggy F.
1980   Agricultural Decision Making: Anthropological Contributions to Rural Development. Academic Press, New York.

Bean, Lowell B., and Harry Lawton
1976   Some Explanations for the Rise of Cultural Complexity in Native California with Comments on Proto-Agriculture and Agriculture. *In* Native Californians: Theoretical Retrospective, edited by L. Bean and T. Blackburn, pp. 19–48. Ballena Press, Ramona, California.

Bean, Lowell B., and Katherine S. Saubel
1972   Temalpakh. Malki Museum Press, Morongo Indian Reservation, California.

Berry, Michael S.
1982   Time, Space, and Transition in Anasazi Prehistory. University of Utah Press, Salt Lake City.

1985   The Age of Maize in the Greater Southwest: A Critical Review. *In* Prehistoric Food Production in North America, edited by R. Ford, pp. 279–308. Anthropological Papers No. 75. Museum of Anthropology, University of Michigan, Ann Arbor.

Betancourt, Julio L., and Thomas R. Van Devender
1981   Holocene Vegetation in Chaco Canyon, New Mexico. Science 214:656–58.

Binford, Lewis R.
1968   Post-Pleistocene Adaptations. *In* New Perspectives in Archaeology, edited by S. Binford and L. Binford, pp. 313–41. Academic Press, New York.

Bohrer, Vorsila L.
1962   Ethnobotanical Material from Tonto National Monument. *In* Archaeological Studies at Tonto National Monument, edited by L. Caywood, pp. 75–114. Southwest Monuments Association Technical Series No. 2.

1972   Paleoecology of the Hay Hollow Site. Fieldiana 63:1–30.

1982   Former Dietary Patterns as Determined from Archaic Age Plant Remains from Fresnel Shelter, Southcentral New Mexico. The Artifact 19(304):41–50.

1983   New Life from Ashes: The Tale of the Burnt Bush (*Rhus trilobata*). Desert Plants 5:122–24.

Bretting, Peter
1982   Morphological Differentiation of *Proboscidea parviflora* subsp. *parviflora* (Martyniaceae) under Domestication. American Journal of Botany 69:1531–37.

Brown, David E.
1982   Biotic Communities of the American Southwest-United States and Mexico. Desert Plains 4 (Nos. 1–4).

Bruier, Frederick L.
1977   Plant and Animal Remains from Caves and Rock Shelters of the Chevelon Canyon, Arizona: Methods for Isolating Cultural Depositional Processes. Ph.D. dissertation, Department of Anthropology, University of California, Los Angeles.

Buskirk, Winfred
1949   Western Apache Subsistence Economy. Ph.D. dissertation, Department of Anthropology, University of New Mexico, Albuquerque.

1986   The Western Apache: Living on the Land before 1950. University of Oklahoma Press, Norman.

Bye, Robert A., Jr.
1979   Incipient Domestication of Mustards in Northwest Mexico. The Kiva 44:237–56.

Carmichael, David L.
1982   Fresnal Shelter, New Mexico: Preliminary Dating and Evidence for Early Cultigens. Paper presented at the 47th Annual Meeting of the Society for American Archaeology, Minneapolis.

Castetter, Edward F., and Willis H. Bell
1941   Pima and Papago Indian Agriculture. University of New Mexico Press, Albuquerque.

1952   Yuman Indian Agriculture. University of New Mexico Press, Albuquerque.

Castetter, Edward F., and Morris E. Opler
1936   The Ethnobiology of the Chiracahua and Mescalero Apache. University of New Mexico Bulletin No. 297.

Cohen, Mark Nathan
1977   The Food Crisis in Prehistory. Yale University Press, New Haven.

Collins, G. N.
1914   A Drought-Resistant Adaptation in Seedlings of Hopi Corn. Journal of Agricultural Research 1:293–302.

Cordell, Linda S.
1984   Prehistory of the Southwest. Academic Press, Orlando.

Cutler, Hugh C.
1952   A Preliminary Survey of Plant Remains of Tularosa Cave. In Mogollon Cultural Continuity and Change, the Stratigraphic Analysis of Tularosa and Cordova Caves, by Paul Martin, J. Rinaldo, E. Bluhm, Hugh Cutler, and R. Grange, pp. 461–79. Fieldiana: Anthropology 40.

Dart, Allen
1985   Archaeological Investigations at La Paloma: Archaic and Hohokam Occupations at Three Sites in the Northeastern Tucson Basin. Institute for American Research Anthropological Papers No. 4.

Dean, Jeffrey S., Robert C. Euler, George J. Gumerman, Fred Plog, Richard Hevly, and Thor N. V. Karlstrom
1985   Human Behavior, Demography, and Paleoenvironment on the Colorado Plateaus. American Antiquity 50:537–54.

Dick, Herbert W.
1952   Evidence for Early Man in Bat Cave and the Plains of San Augustin, New Mexico. Proceedings of the 29th Congress of Americanists 3:158–63.

1954   The Bat Cave Pod Corn Complex: A Note on its Distribution and Archaeological Significance. El Palacio 61:138–44.

1965   Bat Cave. School of American Research Monograph 27.

Doelle, William H.
1985   Excavations at the Valencia Site, a Prehistoric Hohokam Village in the Southern Tucson Basin. Institute for American Research Anthropological Papers No. 3.

Euler, Robert C., George J. Gumerman, Thor N. V. Karlstrom, Jeffrey S. Dean, and Richard H. Hevly
1979   The Colorado Plateaus: Cultural Dynamics and Paleoenvironment. Science 205:1089–1101.

Fish, Suzanne K., Paul R. Fish,, and John Madsen
1990   Sedentism and Mobility in the Tucson Basin prior to A.D. 1000. In Perspectives on Southwestern Prehistory, edited by P. Minnis and C. Redman, pp. 76–91. Westview Press, Boulder, Colorado.

Fish, Suzanne K., Paul R. Fish, Charles Miksicek, and John Madsen
1985   Prehistoric Agave Cultivation in Southern Arizona. Desert Plants 7:107–13.

Fish, Paul R., Suzanne K. Fish, Austin Long, and Charles Miksicek
1986   Early Corn Remains from Tumamoc Hill, Southern Arizona. American Antiquity 51:563–572.

Flannery, Kent V.
1973   The Origins of Agriculture. Annual Reviews of Anthropology 2:271–310.

Ford, Richard I.
1968   Jemez Cave and its Place in an Early Horticultural Settlement Pattern. Paper presented at the 33rd Annual Meeting of the Society for American Archaeology, Santa Fe.

1975   Re-excavation of Jemez Cave, New Mexico. Awanyu 3:13–27.

1981   Gardening and Farming before A.D. 1000: Patterns of Prehistoric Plant Cultivation North of Mexico. Journal of Ethnobiology 1:6–27.

Ford, Richard I. (editor)
1985   Prehistoric Food Production in North America. Anthropological Papers No. 75. Museum of Anthropology, University of Michigan, Ann Arbor.

Fowler, Catherine S.
1986   Subsistence. In Handbook of North American Indians, vol. 11, Great Basin, edited by W. D'Azevedo, pp. 64–97. Smithsonian Institution Press, Washington, D.C.

Fritz, John M.
1974   The Hay Hollow Site Subsistence System. Ph.D. dissertation, Department of Anthropology, University of Chicago.

Gallagher, Marsha V.

1977   Contemporary Ethnobotany among the Apache of the Clarkdale, Arizona, Area, Coconino and Prescott National Forest. U.S. Forest Service, Southwestern Region, Archaeological Report No. 14.

Glassow, Michael A.

1972   Changes in the Adaptation of Southwestern Basketmakers: A Systems Perspective. *In* Contemporary Archaeology, edited by M. Leone, pp. 289–302. Southern Illinois University Press, Carbondale.

1980   Prehistoric Agricultural Development in the Northern Southwest: A Study in Changing Patterns of Land Use. Ballena Press Anthropological Papers No. 16. Ballena Press, Socorro, New Mexico.

Goodwin, Grenville

1935   The Social Divisions and Economic Life of the Western Apache. American Anthropologist 37:55–65.

Hall, Stephen A.

1977   Late Quaternary Sedimentation Paleoecologic History of Chaco Canyon, New Mexico. Geological Society of America Bulletin 88:1593–1618.

1985   Quaternary Pollen Analysis and Vegetational History of the Southwest. *In* Pollen Records of Late-Quaternary North American Sediments, edited by V. Bryant and R. Holloway, pp. 95–123. American Association for Stratigraphic Palynologists Foundation, Dallas.

Harlan, Jack R.

1975   Crops and Man. American Society of Agronomy, Madison, Wisconsin.

Harris, David R.

1977   Alternative Pathways toward Agriculture. *In* Origins of Agriculture, edited by C. Reed, pp. 179–244. Mouton Publishers, The Hague.

Haury, Emil W.

1957   An Alluvial Site on the San Carlos Indian Reservation. American Antiquity 23:2–27.

Huckell, Bruce, and Lisa Huckell

1984   Excavations at Milagro, an Archaic Site in the East Tucson Basin. Unpublished report on file with the City of Tucson.

Human Systems Research, Inc.

1973   Technical Manual, 1973 Survey of the Tularosa Basin. Human Systems Research, Inc., Tularosa, New Mexico.

Hurst, C. T.

1940   Preliminary Work at Tabeguache Cave-1939. Southwestern Lore 6:48–62.

1941   The Second Season in Tabeguache Cave. Southwestern Lore 7:48–62.

1942   Completion of Work in Tabeguache Cave. Southwestern Lore 8:7–16.

Hurst, C. T., and Edgar Anderson

1949   A Corn Cache from Western Colorado. American Antiquity 14:161–67.

Hymowitz, T.

1972   The Trans-Domestication Concept Applied to Guar. Economic Botany 26:49–60.

Irwin, Henry J., and Cynthia C. Irwin

1959   Excavations at the LoDaisKa Site. Proceedings of the Denver Museum of Natural History No. 8.

1961   Radiocarbon Dates from the LoDaisKa Site. American Antiquity 27:114–17.

Irwin-Williams, Cynthia C., and S. Tompkins

1968   Excavations at En Medio Rock Shelter, New Mexico. Eastern New Mexico University Contributions in Anthropology Vol. 1, No. 2.

Jennings, Jesse D.

1975   Preliminary Report: Excavations of Cowboy Cave, June 3–July 26, 1975. Manuscript on file at the Department of Anthropology, University of Utah, Salt Lake City.

1980   Cowboy Cave. University of Utah Anthropological Papers No. 104.

Kaplan, Lawrence

1963   Archeoethnobotany of Cordova Cave. Economic Botany 17:350–59.

Lawton, Harry W., Phillip W. Wilke, Mary De Decker, and William M. Mason

1976   Agriculture among the Paiute of Owens Valley. Journal of California Archaeology 3:13–50.

LeBlanc, Steven A.

1982   The Advent of Pottery in the Southwest. *In* Southwestern Ceramics: A Comparative Review, edited by A. Schroeder, pp. 27–52. Arizona Archaeologist 15.

Lewis, Harry T.

1973   Patterns of Indian Burning in California: Ecology and Ethnohistory. Ballena Press, Ramona, California.

Lintz, Christopher, and Leon George Zabawa
1984   The Kenton Caves of Western Oklahoma. *In* Prehistory of Oklahoma, edited by R. Bell, pp. 161–74. Academic Press, New York.

Long, Austin, R. I. Ford, D. J. Donahue, A. J. T. Jull, T. W. Linick, and Ted Zabel
1986   Tandem Accelerator Dating of Archaeological Cultigens. Paper presented at the 51st Annual Meeting of the Society for American Archaeology, New Orleans.

Mangelsdorf, Paul C.
1954   New Evidence on the Origin and Ancestry of Maize. American Antiquity 19:409–10.

1974   Corn: Its Origin, Evolution, and Improvement. Harvard University Press, Cambridge.

Mangelsdorf, Paul C., Herbert W. Dick, and Julian Camara-Hernandez
1967   Bat Cave Revisited. Botanical Museum Leaflets, Harvard University 17(6).

Mangelsdorf, Paul C., and Robert H. Lister
1956   Archaeological Evidence on the Evolution of Maize in Northwestern Mexico. Botanical Museum Leaflets, Harvard University 17(6).

Mangelsdorf, Paul C., Richard S. MacNeish, and Walton C. Galinat
1967   Prehistoric Wild and Cultivated Maize. *In* The Prehistory of the Tehuacan Valley, vol. 1, edited by D. Byers, pp. 178–200. University of Texas Press, Austin.

Mangelsdorf, Paul C., and C. Earle Smith, Jr.
1949   New Archaeological Evidence of the Evolution of Maize. Botanical Museum Leaflets, Harvard University 13:213–47.

Martin, Paul S.
1963   The Last 10,000 Years. University of Arizona Press, Tucson.

1967   The Hay Hollow Site (200 B.C.–A.D. 200). Field Museum of Natural History Bulletin Vol. 38, No. 5.

Martin, Paul S., and Fred Plog
1974   The Archaeology of Arizona. Natural History Press, Garden City, New York.

Martin, Paul S., John R. Rinaldo, Elaine Bluhm, Hugh C. Cutler, and R. Grange, Jr.
1952   Mogollon Cultural Continuity and Change: The Stratigraphic Analysis of Tularosa and Cordova Caves. Fieldiana: Anthropology No. 40.

Martin, Paul S., and James Schoenwetter
1960   Arizona's Oldest Cornfield. Science 133:33–34.

Mathews, T. W., and E. H. Neller
1979   Atlatl Cave: Archaic-Basketmaker II Investigations in Chaco Canyon National Monument. *In* Proceedings of the First Conference on Scientific Research in the National Parks, edited by R. Linn, p. 873. National Park Service Transactions and Proceedings Series 5.

Minnis, Paul E.
1985a   Domesticating People and Plants in the Greater Southwest. *In* Prehistoric Food Production in North America, edited by R. Ford, pp. 309–39. Anthropological Papers No. 75. Museum of Anthropology, University of Michigan, Ann Arbor.

1985b   Social Adaptation to Food Stress: A Prehistoric Example. University of Chicago Press, Chicago.

Moratto, Michael J.
1984   California Archaeology. Academic Press, New York.

Morenon, E. Pierre, and Thomas Hays
1978   New Evidence from the Jornada Branch: Excavations in the Placitas Arroyo. Paper presented at the 43rd Annual Meeting of the Society for American Archaeology, Tucson.

Nabhan, Gary P.
1985   Gathering the Desert. University of Arizona Press, Tucson.

Nabhan, Gary P., and Richard S. Felger
1984   Teparies in Southwestern North America: A Biogeographical and Ethnohistorical Study of *Phaseolus acutifolius*. Economic Botany 32:2–19.

Nabhan, Gary P., and J. M. J. de Wet
1984   *Panicum sonorum* in Sonoran Desert Agriculture. Economic Botany 38:65–68.

Nabhan, Gary P., Alfred Whiting, Henry Dobyns, Richard Hevly, and Robert Euler
1981   Devil's Claw Domestication: Evidence from Southwestern Indian Fields. Journal of Ethnobiology 1:135–64.

Nichols, Robert F., and David G. Smith
1965   Evidence of Prehistoric Cultivation of Douglas-fir at Mesa Verde. *In* Contributions of the Wetherill Mesa Project, edited by D. Osborne, pp. 57–64. Society for American Archaeology Memoir No. 19.

Nelson, Ben A., and Steven A. LeBlanc
1986   Short-Term Sedentism in the American South-
        west: The Mimbres Valley Salado. University
        of New Mexico Press, Albuquerque.

Ortiz, Alfonso
1979   Southwest. Handbook of North American In-
        dians, vol. 9. U.S. Government Printing Of-
        fice, Washington, D.C.

1983   Southwest. Handbook of North American In-
        dians, vol. 10. U.S. Government Printing Of-
        fice, Washington, D.C.

Plog, Fred
1974   The Study of Prehistoric Change. Academic
        Press, New York.

Reagan, A. B.
1929   Plants Used by the White Mountain Apache
        Indians of Arizona. Wisconsin Archaeologist
        8:143–61.

Real, L.
1965   A Preliminary Report on the County Road Site
        Artifact Distribution Analysis. Manuscript on
        file with the Field Museum of Natural His-
        tory.

Renaud, Etienne B.
1929   Archaeological Research in Northeastern New
        Mexico and Western Oklhoma. Colorado Sci-
        entific Society Proceedings 12:113–50.

1930   A Summary of Prehistoric Cultures of the Ci-
        marron Valley. El Palacio 29:123–29.

Rindos, David
1984   The Origins of Agriculture. Academic Press,
        Orlando.

Sauer, Jonathan D.
1967   The Grain Amaranths and their Relatives: A
        Revised Taxonomic and Geographic Survey.
        Annals of the Missouri Botanical Gardens
        54:103–37.

Simmons, Alan H.
1984   Archaic Prehistory and Paleoenvironments in
        the San Juan Basin, New Mexico: The Chaco
        Shelters Project. University of Kansas, Mu-
        seum of Anthropology, Project Report Series
        No. 53.

1986   New Evidence for the Early Use of Cultigens
        in the American Southwest. American Antiq-
        uity 51:73–89.

Smiley, Francis Edward IV
1985   The Chronometrics of Early Agricultural Sites
        in Northeastern Arizona: Approaches to the
        Interpretation of Radiocarbon Dates. Ph.D.
        dissertation, Department of Anthropology,
        University of Michigan, Ann Arbor.

Smith, C. Earle, Jr.
1950   Prehistoric Plant Remains from Bat Cave. Bo-
        tanical Museum Leaflets, Harvard University
        14:157–80.

Staten, Glen, D. R. Burnham, and John Carter, Jr.
1939   Corn Investigations in New Mexico. New
        Mexico State University, Agricultural Exper-
        iment Station Bulletin No. 260.

Traylor, Diane, Nancy Wood, Lyndi Hubbel, Robert
Scaife, and Sue Weber
1977   Bandelier: Excavations in the Flood Pool of
        Cochiti Lake, New Mexico. Manuscript on file
        with the National Park Service, Southwestern
        Cultural Resource Center, Santa Fe.

Treganza, A. E.
1947   Possibilities of an Aboriginal Practice of Agri-
        culture among the Southern Diegueno. Amer-
        ican Antiquity 12:169–73.

Van Devender, Thomas R., and W. Geoffrey Spaulding
1979   Development of Vegetation and Climate in the
        Southwestern United States. Science 204:701–
        10.

Wellhausen, Edward J., L. M. Roberts, E. Hernandez X.,
and Paul C. Mangelsdorf
1952   Races of Maize in Mexico. Harvard Univer-
        sity, Bussey Institute, Cambridge.

Wetterstrom, Wilma
1978   Cognitive Systems, Food Patterns, and Pal-
        eoethnobotany. In The Nature and Status of
        Ethnobotany, edited by R. Ford, pp. 81–95.
        Anthropological Papers No. 67. Museum of
        Anthropology, University of Michigan, Ann
        Arbor.

Whalen, Norman M.
1973   Cochise Site Distribution in the San Pedro
        River Valley. The Kiva 40:203–11.

Wilke, Phillip J., and D. N. Fain
1974   An Archaeological Cucurbit from Coachella
        Valley. The Journal of California Anthropol-
        ogy 1:110–14.

Wilkes, H. Garrison

1967   Teosinte: The Closest Relative of Maize. Bussey Institute, Harvard University, Cambridge.

Wills, Wirt H. III

1985   Early Agriculture in the Mogollon Highlands of New Mexico. Ph.D. dissertation, Department of Anthropology, University of Michigan, Ann Arbor.

1988   Early Prehistoric Agriculture. School of American Research Press, Santa Fe.

1990   Cultivating Ideas: The Changing Intellectual History of the Introduction of Agriculture in the American Southwest. *In* Perspectives on Southwestern Prehistory, edited by P. Minnis and C. Redman, pp. 319–22. Westview Press, Boulder.

Winter, Joseph C.

1974   Aboriginal Agriculture in the Southwest and Great Basin. Ph.D. dissertation, Department of Anthropology, University of Utah. Salt Lake City.

Winter, Joseph C., and Paul F. Hogan

1986   Plant Husbandry in the Great Basin and Adjacent Northern Colorado Plateau. *In* Anthropology of the Desert West: Essays in Honor of Jesse D. Jennings, edited by C. J. Condie and Don D. Fowler, pp. 117–44. University of Utah Anthropological Papers No. 10.

Winter, Joseph C., and Henry G. Wylie

1974   Paleoecology and Diet at Clyde's Cave. American Antiquity 39:303–15.

Woodbury, Richard B., and Ezra B. W. Zubrow

1979   Agricultural Beginnings, 2000 B.C.–A.D. 500. *In* Handbook of North American Indians, vol. 9: Southwest, edited by A. Ortiz, pp. 43–60. Smithsonian Institution Press, Washington D.C.

Wright, Henry A., and Arthur W. Bailey

1983   Fire Ecology. Wiley-Interscience, New York.

Yarnell, Richard A.

1977   Native Plant Husbandry North of Mexico. *In* Origins of Agriculture, edited by C. Reed, pp. 861–75. Mouton Publishers, The Hague.

8

The Origins of Agriculture in Mesoamerica and Central America

The origin of agriculture in Mesoamerica and Central America has received somewhat less attention in recent years than in the late 1960s and early 1970s, when there was a great surge in research, data interpretation, and theoretical discussion. One reason for this apparent decline in interest was a decrease in the scale of archaeological research during the 1980s, which was due to economic considerations, a reduction of foreign participation in Mexican field research during part of the decade, and an increasingly complex political climate in much of Central America. In addition, research on the development of a sedentary lifestyle was somewhat diminished in favor of greater attention to the nature of social and economic organization in already-established sedentary communities. This latter situation is especially true in the case of research in the central highlands of Mexico.

At present, a number of publications summarize the most important current theoretical positions on the process of domestication of major plants that supported sedentary populations in Mesoamerica and a large part of Central America outside Mesoamerica proper. The most important contributions to our understanding of these processes and their relationship to prehistoric sociocultural and economic development have been made by MacNeish (e.g., 1958, 1964b, 1967b, 1971, 1972) and more recently by Flannery (1968, 1973, 1976, 1985, 1986). As for the theoretical aspects, these authors have been directly or indirectly responsible for many of the models, as well as discussions of the causes underlying the evolutionary processes of plant domestication.

Archaeological and archaeobotanical data from sites in Mesoamerica and Central America strongly support the position that the complexities of such processes as the domestication of certain plants, the development of agriculture as a dominant economic activity, and the institution of associated sociocultural changes cannot be appreciated if these processes are viewed simply as local versions of a worldwide phenomenon. Some authors would argue that, although the plants in question and the details of their relationships with human beings and other species might differ, similar evolutionary processes occurred in many parts of the world at approximately the same time (e.g., Cohen 1981). Such a broad level of generalization, however, ignores the specific relationships that permit us to hypothesize how, in fact, such processes got underway and continued. As Flannery (1972, 1973, 1985, 1986) has pointed out, models that most accurately describe these processes in terms of differ-

EMILY MCCLUNG DE TAPIA

143

ent geographical situations are likely to be inadequate outside the local context to which they apply. In other words, the processes themselves are different. While comparison may be productive, there is no need to insist that all regions must somehow conform to the structure of a single general model.

Geographic Area

The term "Mesoamerica" refers specifically to the cultural "superarea" defined by Paul Kirchoff (1943). Its extent was based on the geographical distribution—at the time of the Spanish Conquest in Mexico—of a series of 45 disparate cultural traits, ranging from such elements as cultivation and the use of specific crops to cannibalism, confession, and matrilineal clans.[1] According to Sanders and Price (1968:6) and Palerm and Wolf (1972:150,157), the area lies between 10° and 22° north latitude. It includes the modern political entities of Mexico south of the Panuco-Lerma drainage (from the mouth of the Panuco River on the Gulf Coast to the Rio Grande de Santiago on the Pacific side), Guatemala, El Salvador, Belize, and the western part of Honduras, Nicaragua, and Costa Rica to an approximate boundary extending from the mouth of the Ulua River and Lake Yojoa in Hondu-

ras to the Gulf of Nicoya in Costa Rica (Fig. 8.1). Kirchoff originally placed the rest of Central America in the "circum-Caribbean" superarea to the southeast of Mesoamerica (Jiménez Moreno 1977:471–72). Archaeological evidence suggests that this area remained essentially marginal to the mainstream of domestication of plants that became the principal seed crops in most of Mesoamerica. It is important to point out, however, that very little archaeobotanical evidence that could further illuminate the development and spread of cultivation has been reported from sites in this marginal area (see Smith 1987).

Mesoamerica and Central America are areas of great environmental and ecological diversity. Although Mesoamerica lies within the tropics, altitude tends to offset latitude insofar as climatic characteristics are concerned. Most of Mesoamerica fails to manifest the characteristics typically associated with tropical areas, such as constantly high temperatures, substantial precipitation, luxurious vegetation, etc., although these elements are not entirely absent, especially in the coastal areas of the Gulf of Mexico and the Caribbean. Palerm and Wolf (1972:150) have described this area not as tropical but as a climatic "mosaic" with predominantly cold, temperate, or warm *subcálido* climatic types. The unifying element from an ecological point of view is the suitability of the

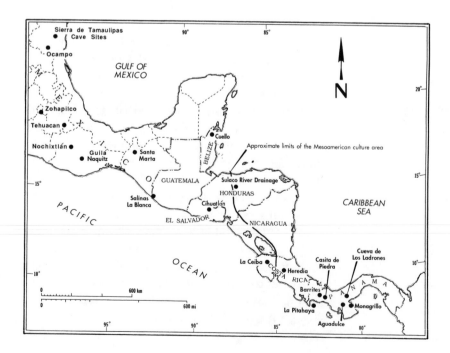

Fig. 8.1. Location of Mesoamerican and Central American sites mentioned in the text.

area for rainfall agriculture (except for some particularly arid zones).

In very general terms, three natural regions, which include several subdivisions exhibiting particular characteristics that were significant for specific prehispanic cultural and socioeconomic developments (West 1964:364–65,368), can be distinguished (Fig. 8.2):

1. Extra-tropical dry lands and adjacent subhumid areas of northern Mexico. These areas are characterized by definite winter and summer temperature seasons, xerophytic plant assemblages, and predominantly arid-type land forms.

2. Cool tropical highlands, including the Sierra Madre Occidental and Oriental of Mexico. The area from Mexico to Nicaragua is characterized by oak-conifer forests of North American affinity and further south in Costa Rica and Panama by a highland oak-laurel-myrtle rain forest related to South American plant assemblages.

3. Warm tropical lowlands of Mexico and Central America. This area comprises many individual units with varied tropical plant assemblages, which range from true lowland rain forest to tropical deciduous woodland and savannah and tropical scrub, depending on local topographic, climatic, and edaphic factors. The feature that unifies this area is climate, which is characterized by relatively high temperatures and year-round, frost-free conditions.

The northern frontier of Mesoamerica has never been clearly defined archaeologically or ethnographically and, undoubtedly, its exact position varied at different points in the prehispanic past. The northern limit of the culture area as defined by Kirchoff for the sixteenth century, however, formed an important ecological division based on the potential for rainfall agriculture to the south of it, with the exception of specific desert zones, and the predominance of hunter-gatherer groups to the north (Palerm and Wolf 1972:150). The ecological boundary did not always coincide with the cultural frontier, whose position depended in large part on cultural and political developments in the heart of central Mexico.

Rather than distinguishing agriculturalists from hunter-gatherers as in the north, the southern frontier separated agricultural groups with different cultural traditions and diverse levels of sociopolitical and economic development. To the northeast of the Nicoya-Ulua line (the Caribbean side) are extensive coastal plains characterized by high temperatures and precipitation and dominated by tropical rainforest vegetation. To the southwest (the Pacific side) is a broken landscape with high valleys, greater climatic variability in terms of temperature and rainfall, and a prolonged dry season in certain areas.

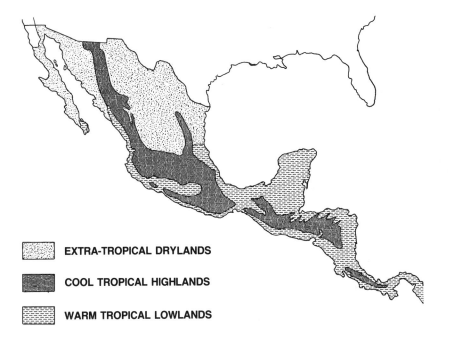

Fig. 8.2. Natural regions of Mesoamerica and Central America.

EXTRA-TROPICAL DRYLANDS

COOL TROPICAL HIGHLANDS

WARM TROPICAL LOWLANDS

History of Archaeological and Botanical Investigations

Early investigations tended to utilize botanical data, particularly evidence from pollen cores, to reconstruct past environmental conditions relevant to human settlement and exploitation. For example, Sears (1951) used pollen data from the bed of Lake Texcoco as evidence for climatic changes in the Basin of Mexico. On the basis of shifts in the relative frequencies of oak, pine, and herbaceous pollen, combined with stratigraphic information, he suggested that the transition from the Archaic to Teotihuacan I (Early Formative to Terminal Formative/Early Classic) (see Table 8.1) in the Basin took place during a prolonged dry period characterized by low oak and high pine pollen. He believed this transition was preceded by a moister period favoring the formation of a rich black soil and was followed by another moister period corresponding to the Aztec occupation. Sears (1951)

suggested that periods of increased moisture implied higher lake levels and greater precipitation.

Lorenzo (1956:41) later argued that the decline in arboreal pollen was, in fact, related to human activities and population increase rather than to significant shifts in precipitation. Sanders (1965:29–30) interpreted Kovar's analysis of pollen cores from sites in the Teotihuacan Valley in the northeastern Basin of Mexico, particularly the spring of El Tular near Atlatongo, to support both Sears's and Lorenzo's arguments. On the basis of relative frequencies of herbaceous and arboreal pollen, Sanders suggests that precipitation and lake level were at a maximum during the Early and Middle Formative, and that a reduction in rainfall and lake level followed during the Late and Terminal Formative. The drier conditions presumably favored changes in subsistence adaptations. Sanders interpreted the Teotihuacan period (Classic) as being favorable for agriculture, especially when intensive techniques such as irrigation were in-

Table 8.1. Mesoamerican Chronology Showing Terminological Variations

Modified Period Nomenclature		Date	Traditional Period Nomenclature	Date	Central Mexico (variant)
Late Horizon		A.D. 1350–1520	Late Postclassic	A.D. 1200–1520	Postclassic
			Early Postclassic	750–1200	
Second Intermediate		750–1350			
Middle Horizon	2	500–750	Late Classic	500–750	
	1	300–500			Classic
First Intermediate	5	150–300	Early Classic	150–500	
	4	100 B.C.–A.D. 150			
	3	300–100 B.C.	Terminal Formative (Terminal Preclassic)	300 B.C.–A.D. 150	
	2	650–300	Late Formative (Late Preclassic)	650–300	
	1	1150–650	Middle Formative (Middle Preclassic)	1050–650	
Early Horizon	2	1300–1150			
	1	1500–1300	Early Formative (Early Preclassic)	1500–1050	
Initial Ceramic		2500–1500	Incipient Cultivators and Food Collectors (Archaic)	7000–2000	
Upper		8000–2500			
Lithic Middle		10,000–8000	Early Hunters	10,000–7000 B.C.	
Lower		30,000–10,000 B.C.			

Source: Willey 1962; Sanders and Price 1968; Tolstoy 1978; Price 1976; Sanders, Parsons, and Santley 1979.

corporated into the agricultural system, but suggested that the climate and lake level returned to Terminal Formative conditions during the Early Toltec phase (Early Postclassic). Sanders (1965:30) believed that the Late Toltec phase (Early Postclassic), however, was characterized by a return to more favorable conditions. Recent research at Zohapilco (Tlapacoya) has provided some additional evidence on the nature of environmental conditions prevalent in the prehistoric Basin of Mexico (Niederberger 1976, 1979, 1987).

Pollen data were also considered in early studies directed toward locating the progenitor of cultivated maize and its homeland. Five pollen grains identified as maize at a depth of some 70 m (from prehuman levels) in a drill core from Mexico City were considered to represent wild maize; the pollen was interpreted at that time as evidence that the ancestor of maize was neither teosinte, nor *Tripsacum,* nor a cross between the two (Barghoorn et al. 1954). More recently, Beadle (1982:26–27) has argued that the specimens are, in fact, too large to represent a hypothetical wild maize. Using the size of known archaeological maize cobs together with a correlation between cob length and pollen diameter, he suggests that the pollen grains are probably diploid grains produced by a tetraploid teosinte rather than diploid grains produced by wild maize; moreover, they could represent modern contamination of the drill core.

Rejecting the possibility of contamination, Sears (1982) instead concludes that the sediments that contained these grains were sunken remnants of an Archaic period context rather than Pleistocene deposits (see also Galinat 1985:273). Ludlow-Weichers et al. (1983:239–40) call attention to the possibility that the variability in maize pollen size could be a response to such factors as day-length and temperature, as well as laboratory preparation techniques (see Flannery 1973:294), all of which suggest that large samples are necessary to support interpretations based on pollen size.

In his search for the homeland of maize, MacNeish undertook excavations at Santa Marta Rockshelter in Chiapas (MacNeish and Peterson 1962). Charcoal from cave floors 2 and 3 date the Santa Marta complex at 6780 B.C. ± 400 (M980) and 5360 B.C. ± 300 (M979) (MacNeish 1964a:414). Studies of maize pollen from the site indicate that it was not sufficiently old enough to confirm that the region was an area where cultivated maize may have developed. MacNeish (1967b:307) reported that the Santa Marta in-

habitants derived as much as 70 percent of their diet from wild plants and only about 30 percent from animals, and they left no evidence of agriculture. (He provides, however, no basis for calculating these percentages and says little about the chronological distribution of maize pollen.) MacNeish (1967b:308) suggests that the absence of maize pollen in the Santa Marta pollen profile before ca. 2300 B.C., as well as the absence of manos and metates at certain other sites, indicates that some lowland groups between ca. 5000 and 2300 B.C. did not practice maize agriculture.

At present, the paleoethnobotanical evidence for early plant cultivation and eventual domestication in Mesoamerica is limited. Recently summarized by Flannery (1985:243;1986:6), these data are all derived from Mexican contexts: (1) dry caves in the Sierra de Tamaulipas (MacNeish 1958), (2) dry caves in the Sierra Madre near Ocampo, Tamaulipas (Mangelsdorf et al. 1964), (3) dry caves in the Valley of Tehuacan, Puebla (Byers 1967), (4) the open-air site of Zohapilco (Tlapacoya) where carbonized or waterlogged material was found (Niederberger 1976, 1979, 1987), and (5) cave sites, including Guilá Naquitz, in the Oaxaca Valley (Flannery 1976, 1986). It goes without saying that the excavations that have provided meaningful plant evidence cover a limited geographic area. Furthermore, the samples obtained are frequently minute, and their interpretation is often hindered by the absence of sufficient corroborating data.

Archaeological data from Tamaulipas and the Tehuacan Valley are significant not because they point to the origin of agriculture in Mesoamerica but because they document the transition from a subsistence pattern based on food procurement to one dependent largely on food production in specific regions. Therefore, they provide an important basis for continued investigation, such as that recently reported from Guilá Naquitz (Flannery 1986). A period of incipient agriculture was first recognized from archaeobotanical evidence recovered from caves in Tamaulipas (MacNeish 1958; Whitaker 1957). Although it is now evident that the transition to agriculture documented in Tamaulipas is chronologically later than that in the Tehuacan Valley, the evidence points to the importance of localized developments for understanding the broader picture. It is probable that regional specializations existed from an early time. In Tamaulipas, for example, cucurbits (*Cucurbita*) and beans (*Phaseolus vulgaris*) were apparently the earliest cultivars, before maize (*Zea mays*) became important. Maize seems to have appeared earlier in

Tehuacan than elsewhere,[2] although teosinte (*Z. mexicana*), an important relative of maize, is reported from Zohapilco in the southern Basin of Mexico at approximately the same time (ca. 5000 B.C.) (Lorenzo and González 1970). On the other hand, recent evidence from Guilá Naquitz suggests that squash (*Cucurbita pepo*) was cultivated in Oaxaca from a very early date (ca. 8000 B.C.). By around 2000 B.C., the complex of major cultivated plants now recognized to comprise the fundamental Mesoamerican diet had been established. At the same time, the transition from preceramic to ceramic-using populations in Mesoamerica had taken place, suggesting an increase in sedentary living. Table 8.2 summarizes the earliest occurrences of important plant remains in the archaeological record, and Tables 8.3 and 8.4 describe the principal archaeobotanical remains recovered from prehistoric sites in Mesoamerica and Central America.

Another area that MacNeish (1967a:4) originally considered to be a potential source of domesticated maize was the Rio Balsas drainage in western Mexico. Both Chalco-type teosinte (*Zea mexicana* ssp. *mexicana*) and Balsas-type teosinte (*Z. mexicana* ssp. *parviglumis*) are present in this area (Guzmán 1981). The former apparently clusters in the north at somewhat higher altitudes, whereas the latter occurs more frequently in the south at lower altitudes (Rovner 1984:2). The most recent attempt to document the possible association of human subsistence activity with the apparent natural distribution of teosinte in Jalisco was unfortunately inconclusive (Rovner 1984). A limited survey of the region covering the western margin of the Central Mexican Plateau failed to detect traces of preceramic sites that, once excavated, might provide evidence for plant utilization, specifically the exploitation of teosinte and the evolution of maize in this area (I. Rovner, personal communication n.d.). The only archaeobotanical study reported from Jalisco—an analysis of phytoliths from the site of Guachimonton, Teuchitlan, in the central part of the state—indicates the presence of maize in a much later context (400 B.C.–A.D. 700) (Zurita 1987:70).[3]

Major Mesoamerican Domesticated Plants

The most important plants domesticated in various parts of Mesoamerica have been discussed by Flannery (1973, 1985) and Smith (1987), among others. The genera that have been subject to the most intensive study are briefly discussed below.

Maize (*Zea mays*)

The origin of maize (*Zea mays*) has distinguished itself as one of the most controversial topics among botanists around the world. This topic has generated more than a century of debate, based originally on sparse evidence and confusing descriptive reports (see Iltis and Doebley 1984) and later on the results of detailed field and laboratory observations, cytological testing, and the study of dated archaeological specimens. The debate continues to draw attention because no definitive answer has as yet been found.

A vast literature is available on the subject of the so-called maize problem. Flannery (1973, 1985, 1986) provides a comprehensive summary of the major positions. According to the prevailing view, maize is the result of natural hybridization and frequent backcrosses with wild teosinte (*Z. mexicana*), its nearest wild relative (Galinat 1971, 1985; Beadle 1972, 1977; Iltis 1983). Most of the current research is focused on determining which subspecies of teosinte is the most probable ancestor (Iltis and Doebley 1984). According to the other position, developed by Mangelsdorf (1947, 1974; Mangelsdorf and Reeves 1938), cultivated maize is the descendent of an extinct, wild pod popcorn whose small hard kernels were individually enclosed in glumes. Annual teosinte was believed to have resulted from the hybridization of maize with a wild grass (*Tripsacum*). Mangelsdorf later modified this position in light of new evidence (see Flannery 1973:292; Mangelsdorf 1986). Wilkes (1972) pointed out that although teosinte and *Tripsacum* may occur in the same habitat, they do not hybridize naturally or under laboratory conditions. A comparison of pollen structures also fails to support the position that annual teosinte is a hybrid derivative of maize and *Tripsacum* (Beadle 1982; Ludlow-Weichers et al. 1983; Mangelsdorf 1986:72).

Several factors have been cited to support the more widely accepted position that teosinte is the ancestor of maize: (1) free and frequent hybridization between maize and teosinte under natural conditions, (2) identical chromosome number (n = 10) and, apparently, structure, (3) important anatomical similarities, and (4) similar morphological characteristics and overlapping size ranges for the pollen of both plants (Galinat 1971; Beadle 1982; Iltis 1983; Iltis and Doebley

Table 8.2. First Appearance of Plants in Four Regions of Mesoamerica

Plant Genera	Region			
	Tehuacan Valley	Valley of Oaxaca (Guila Naquitz)	Tamaulipas	Basin of Mexico
Setaria	ca. 7000 B.C. (poss. domestic by 6000 B.C.)		ca. 3500 B.C.	
Zea mexicana (teosinte)		pollen ca. 7400–6700 B.C.		kernels, ca. 5000 B.C.
Z. mays	ca. 5050 B.C. (cobs)		ca. 2500 B.C.	pollen, ca. 5200–2000 B.C.
Cucurbita pepo (squash)	ca. 5200 B.C.	ca. 8000 B.C.	ca. 7000 B.C.	
C. mixta ("cushaw")	ca. 5000 B.C.			
C. moschata ("butternut")	ca. 4500 B.C. (?)			
Cucurbita sp.				ca. 5200–2000 B.C.
Phaseolus sp. (wild runner bean)		ca. 8700–6700 B.C.	ca. 7000–5500 B.C.	
P. coccineus (scarlet runner bean)	ca. 200 B.C.			
P. vulgaris (common bean)	ca. 5000–3500 B.C.		ca. 4000–2300 B.C.	
P. acutifolius (tepary bean)	ca. 3010 B.C.			
Persea americana (avocado)	ca. 7200 B.C.			
Capsicum annuum (chile pepper)	ca. 6500 B.C. (wild) domestic by ca. 4121 B.C.			
Amaranthus sp. (amaranth)	ca. 4500 B.C.			ca. 5200–2000 B.C. (also Cheno-ams pollen)
Lagenaria (bottle gourd)	ca. 5050 B.C.	ca. 7000 B.C.	ca. 7000 B.C.	
Chenopodium sp.				ca. 5200–2000 B.C. (also Cheno-ams pollen)
Sechium sp. (chayote)				ca. 5200–2000 B.C.

Source: Flannery 1973: 286–87; Kaplan 1981; Mangelsdorf, MacNeish, and Galinat 1967; Niederberger 1976, 1979; Smith 1987.

1984). Positions diverge, however, on the way in which the characteristic maize cob evolved (e.g., Iltis 1983; Galinat 1985).

Another major research question is where the process of domestication began. The principal types of teosinte are found in semiarid highland regions that, in the past, presumably extended at least from Chi-huahua to southern Guatemala. The Chalco type, which is native to the Basin of Mexico, seems to be among those most closely related to maize, whereas the types found in Guerrero and southern Guatemala are considered much less similar (Wilkes 1972:1073). Piperno's study of phytoliths from subspecies of maize and teosinte also support the view that the

Table 8.3. Archaeobotanical Remains from Archaeological Sites in Mesoamerica

Species

Region or Site	Period of Occupation	Zea mays (Z. mexicana)	Leguminosae	Phaseolus spp.	Capsicum sp.	Physalis sp.	Solanum sp.	Cucurbita sp.	Lagenaria sp.	Opuntia spp.	Agave spp.	Amaranthus sp.	Chenopodium sp.	Persea americana	Portulaca sp.	Sechium sp.	Gossypium sp.	Salvia sp.	Argemone sp.	Cyperus sp.	Scirpus sp.	Oxalis sp.	Prunus capuli	Prosopis sp.	Byrsonima sp.	Nicotiana sp.	Manihot sp.
S.W. Tamaulipas Phases																											
Infiernillo	7000–5000 B.C.				*			*	*	*	*																
Ocampo	5000–2200			*	*			*	*	*																	
Flacco	2200–1800	*		*				*	*			*															
Guerra	1800–1400			*	*			*	*			*															
Mesa de Guaje	1400–500	*(*)?		*	*			*	*			*					*									*	
Palmillas–San Lorenzo	ca. A.D. 100–1300	*						*	*								*									*	
San Antonio	1300–1750	*		*	*			*	*								*										
S. Central Tamaulipas Phases																											
La Perra	3000–2200 B.C.	*		*				*	*		*																
Laguna	500 B.C.–A.D. 0	*		*				*	*																		
Los Angeles	A.D. 1200–1780	*						*		* *																	
Basin of Mexico (North)																											
Loma	ca. 650–300 B.C.	*	*	*	*	*	*	*	*	*	*	*	*		*			*		*	*	*	*				
Torremonte El Arbolillo (Tolstoy)	EH-FI[b]	*	*	*						*																	
Cuanalan	400 B.C.–A.D. 100	*a	*	*	*	*	*	*	*	*	*	*										*					*
Tula	Early Postclassic	*a								*	*	*	*a	*	*		*	?	*				*	*		*	
Teotihuacan	100–750	*	*	*						*	*	*	*	*	*					*	*		*				
Otumba	Early Postclassic											*	*														
Basin of Mexico (South)																											
Tlapacoya	ca. 5000 B.C.	(*)										*	*a														
Zohapilco	ca. 5500–2000	**a						*		*		*	*			*		?									
Santa Catarina	EH-FI[b]	*		*															*								
Coapexco	EH-FI[b]	*		*																							
Terremote	EH-FI[b]	*	*	*			*																				
Terremote	400–200	*		*			*																				
Itzapalapa	Late Aztec A.D. 1350–1520	*	*		*	*	*	*	*	* *	*	*	*	*	* *					*	*	*	*				
Ch-Az-195	Early Aztec	*				*		*		*	*	*	*	*							*		*				
Valley of Temascalzingo																											
San Jose Ixtapa	Early Postclassic	*	*			*							*														
Tehuacan Valley Phases																											
Ajuerdo	10,000–7000 B.C.	*								*		*		*										*			

Continued next page

Table 8.3.—Continued

Region or Site	Period of Occupation	Zea mays (Z. mexicana)	Leguminosae	Phaseolus spp.	Capsicum sp.	Physalis sp.	Solanum sp.	Cucurbita sp.	Lagenaria sp.	Opuntia spp.	Agave spp.	Amaranthus sp.	Chenopodium spp.	Persea americana	Portulaca sp.	Sechium sp.	Gossypium sp.	Salvia sp.	Argemone sp.	Cyperus sp.	Scirpus sp.	Oxalis sp.	Prunus capuli	Prosopis sp.	By. rosorima sp.	Nicotiana sp.	Manihot sp.
Tehuacan Valley (cont.)																											
El Riego	7000–5000	*			*			*		*	*	*		*			*							*			
Coxcatlan	5000–3400	*		*	*			*	*	*	*	*		*										*			
Abejas-Purron	3400–1500	*		*	*			*	*	*	*	*		*			*										
Ajalpan	1500–850	*				*		*		*	*						*							*			
Santa Maria	850–200	*		*	*			*	*	*	*	*		*			*							*			
Palo Blanco	200 B.C.–A.D. 700	*		*	*			*	*	*	*	*		*			*							*			
Venta Salada	700–1540	*		*	*			*	*	*	*	*		*			*							*			
Valley of Oaxaca																											
Guilá Naquitz	8900–6700 B.C.			*	*			*	*	*	*			*?													
Cueva Blanca	3295	*		*																							
San José (phase)	1150–850	(*)						*						*													
Fábrica San José	850–550	*		*	*			*		*	*	*	*	*	*												
Nochixtlan Valley																											
Phases																											
Cruz	700–200 B.C.	*	*																								
Ramos	200 B.C.–A.D. 250/300	*																									
Las Flores	300–900/1100																										
Natividad-Convento	1000-Post-Conquest	*		*	*									*													
Chiapas																											
Media Luna	200 B.C.–A.D. 1	*		*		*?			*?																		
Cuatro Haches	200 B.C.–A.D. 1	*[a]			*?	*?		*																			
Santa Marta	ca. 2500 B.C.	*[a]									*																
Coastal Guerrero	ca. 300–2300 B.P.	*[a]																									
Central Guerrero																											
Xochipala	ca. 1000 B.C.–A.D. 300	*																									
Guatemala																											
Salinas La Blanca	1000–850 B.C.	*																									
Lake Petenxil	ca. 2000 B.C.	*[a]																									
El Salvador																											
Chalchuapa	ca. 400 B.C.–A.D. 300	*												*													
Joya de Cerén	ca. 300–900	*																									
Cihuatan	after 900																										

[a] pollen
[b] overall period of occupation (Early Horizon–First Intermediate) for four sites dating to ca. 1700–875 B.C. excavated by Tolstoy (1978).

Table 8.4. Archaeobotanical Remains from Sites in Honduras, Costa Rica, and Panama

Sites	Approximate Occupation	Acrocomia sp.	Scheelia sp.	Byrsonima sp.	Hymenaea courbaril	Zea mays	Phaseolus vulgaris	Ipomaea batatas	Pouteria sp.	Simarouba sp.	Spondias sp.	Perea americana
Honduras												
Sulaca River Valley (Salitron Viejo, Guarabuqui, Manacal, etc.)	Late Classic	*		*		*	*		*	*	*	
Costa Rica[a]												
Barrial de Heredia	200 B.C.–A.D. 400					*						
Severo Ledesma	100 B.C.–A.D. 500					*						
La Fábrica	A.D. 400–700					*						
Aguacaliente	A.D. 800–1300					*						
La Ceiba	A.D. 800–1300					*	*					*
Panama												
Casita de Piedra	4600–300 B.C.	*	*	*	*							
El Trapiche	4600–300 B.C.	*	*	*	*							
Cerro Punta	A.D. 200–400	*			*	*	*					
Barriles	A.D. 600	*				*	*	*				
La Pitahaya	A.D. 800–1100	*	*			*						

Sources: Sulaco River Valley, Honduras (Lentz 1983), Costa Rica (Smith 1987), and the Rio Chiríqui Drainage, Panama (Smith 1980).
[a]Smith (1987:29) cites the presence of coyol (Acrocomia sp.) from Costa Rica but specifies no particular site.

Chalco type (*Z. mexicana* ssp. *mexicana*) is more similar to maize than the Balsas type (*Z. mexicana* ssp. *parviglumis*), suggesting that the Chalco type is the ancestor of maize (1984:370). The notable similarity between Chalco-type teosinte and maize, however, may also be due in large part to their continual hybridization and backcrossing.

Recent field research has resulted in the discovery (and, in one case, the rediscovery) of a series of teosinte varieties native to the Rio Balsas region in the modern state of Jalisco. Among the varieties reported were *Zea perennis* and *Z. diploperennis*, as well as *Z. mexicana* ssp. *mexicana* and *Z. mexicana* ssp. *parviglumis* (Guzmán 1981; Iltis 1983; Iltis and Doebley 1984). Mangelsdorf (1986:77–78) has recently postulated that *Z. diploperennis* is, in fact, the species of teosinte (instead of *Tripsacum*) that hybridized with his hypothesized extinct wild maize. He apparently made the suggestion to account for the development of traits that enhanced the success of maize: a good root system, strong stalks, and resistance to disease. According to Mangelsdorf's scheme, the better known annual teosintes, such as Chalco and Balsas, presumably arose as offspring from this hybridization process.

Archaeological evidence for teosinte continues to be too sparse to permit more than a brief description of its occurrences and frequencies. Although Mangelsdorf (1986:75) has interpreted its relative absence in the archaeological record as a clear indication of its later development, other explanations are at least as plausible, if not more so. These range from poor preservation conditions to the use of teosinte in as yet undiscovered sites where domestication may have been in process. Finally, few remains would be expected if the kernels were being ground and consumed, as was likely in at least some areas (see Beadle 1972:10).

Beans (*Phaseolus* spp.)

The major changes that affected wild populations of *Phaseolus* include: (1) increased seed permeability, which reduced the amount of soaking required for cooking, (2) the development of limp, straight, non-shattering pods, rather than twisted pods, which reduced the seed loss during harvesting, and (3) changes from perennial to annual growth in certain varieties. Archaeological remains, however, do not document these kinds of changes.

One of the best indicators of bean domestication is

an increase in seed size. Unfortunately, archaeological specimens of the common bean (*Phaseolus vulgaris*) do not demonstrate such a transition. Kaplan (1981:251) suggests that size increases must have occurred quite early in Mexico (by 7,000 years ago). These changes probably took place during a period of food gathering or early cultivation in which predation by animals was reduced by careful storage, thereby permitting natural selection to operate on the existing polygenic system. The scarlet runner bean (*P. coccineus*) was apparently domesticated much later (ca. 2,000 years ago) than the common bean (Kaplan 1965).

Kaplan (1981:246) comments that beans are probably underrepresented in the archaeological record because of harvesting methods. Wild *P. vulgaris* is not known from Mesoamerican sites, in spite of the presence of wild types in the same areas, although other wild bean species were gathered (Kaplan 1981:251). Specimens of wild runner beans (ancestors of the domesticated variety) were recovered from contexts dating to 7000–5500 B.C. in the Ocampo region in Tamaulipas (Kaplan 1965, 1981; Flannery 1986:6,7). Remains of runner beans were more abundant at Guilá Naquitz, but they belong to a wild variety that has no known domesticated descendants (Flannery 1986:6–7).

Squash and Pumpkins (*Cucurbita* spp.)

Cucurbits are usually poorly preserved among archaeological remains. The features considered most indicative of domestication are characteristics of the peduncle (stem), which is rarely preserved. The earliest remains are seed specimens, which often occur in extremely low frequencies because they are likely to have been consumed. The flesh of wild varieties is practically inedible or nonexistent.

Clear archaeological evidence that documents the development of cultivated forms from wild species of squashes and gourds (Smith 1987:86) is presently unavailable. Seeds of wild cucurbits were reported from caves in Tamaulipas and Oaxaca dating to between 8000 and 7000 B.C. Seed specimens resembling those of squash (*C. pepo*) were recovered from Tamaulipas and Tehuacan (ca. 5200 B.C.). Seeds of the green striped cushaw (*C. mixta*) were found as early as 5000 B.C. in Tehuacan, but peduncles of this species that suggest domestication do not appear until 3000 B.C. (Flannery 1973:301). In this context, the specimens recovered from preceramic levels at Guilá Naquitz are particularly significant because cultivated *C. pepo* has been identified from seeds and peduncles dating to ca. 8750–7840 B.C. Originally, this species was believed to have been domesticated in northern Mexico and the southern United States on the basis of its presence in the Ocampo deposits at Tamaulipas and at sites in the American Southwest (Whitaker and Cutler 1986:275; see also Ford 1981:7;1985:346). As Smith (1986:274) points out, while the early date for the cultivation of *C. pepo* in Mesoamerica (and the New World) is in itself interesting, its real significance lies in the amount of time and human selection activity that preceded its appearance.

Phytoliths from *Cucurbita* rinds have been identified from the site of Sitio Sierra in central Panama (340 B.C.–A.D. 500). They have not, however, been clearly established as domesticated squash. The apparent absence of wild species of *Cucurbita* in Panama suggests that the development of a productive maize, bean, and squash complex had occurred by 300 B.C. (Piperno 1985b:250).

The cultivated bottle gourd (*Lagenaria siceraria*) has been reported from Tamaulipas (ca. 7000 B.C.), Tehuacan (ca. 5050 B.C.), and Guilá Naquitz (ca. 7000 B.C.). Flannery (1986:6) discusses the possibility that the bottle gourd was in reality the first New World domesticate.

Chile Peppers (*Capsicum* spp.)

The cultivated forms of chile peppers (*C. annuum*) are derived from wild forms in central Mexico (Smith 1987:86). The earliest archaeological remains date to the El Riego phase (7000–5000 B.C.) in Tehuacan (Smith 1987:86) and the Infiernillo phase (7000–5000 B.C.) in southwestern Tamaulipas (Mangelsdorf et al. 1964:512). Species identification requires intact stems, a rare occurrence among archaeological specimens. Chile seeds are usually found carbonized. Smith (1987:87) observes that no *Capsicum* remains recovered from Guilá Naquitz could be assigned to the preceramic levels of the site; furthermore, no *Capsicum* seed remains have been identified at the Central American sites that he has studied.

Avocado (*Persea americana*)

The earliest archaeological occurrence of avocado is documented from the Tehuacan Valley during the Ajuereado phase (ca. 8000 B.C.) (Smith 1987:88). Archaeobotanical specimens increase in frequency and size during the Tehuacan sequence, suggesting eventual cultivation (see Smith 1966, 1969 for another view). Specimens were recovered from preceramic levels at Guilá Naquitz but are regarded as possible Postclassic intrusions (Smith 1986:267). Numerous avocado pits were recovered from somewhat later contexts (1400–450 B.C.) at the site of Fábrica San José in Oaxaca (Ford 1976).

Other Cultivated Plants

Other genera that were undoubtedly significant include foxtail grass (*Setaria* sp.), *alegría* or amaranth (*Amaranthus* spp.), and *huauhtzontli* or goosefoot (*Chenopodium* spp.). *Setaria* was recovered from numerous contexts in the Tehuacan Valley as early as the Ajuereado phase (Callen 1967; Smith 1967) and in Tamaulipas during the Ocampo phase (ca. 4000 B.C.) (Callen 1963). Both Callen (1967:287) and Smith (1967:232) interpret its abundance as evidence that it may have been cultivated by the El Riego phase (ca. 7000–5000 B.C.). *Setaria*, however, was not destined to become an important domesticate in Mesoamerica. Flannery (1968:94–95; 1973:196–97) suggests that it may have been replaced by maize once maize was sufficiently productive to be an acceptable alternative. To date, there are few archaeobotanical data to document such a transition. Nevertheless, there is evidence for a gradual reduction in *Setaria* as maize acquires more importance during the Coxcatlan phase at the Tehuacan sites (Smith 1967:236).

Species of *Amaranthus* and *Chenopodium* are exceptionally productive from the standpoint of human nutrition because of their high protein content, particularly because so much of the plant is edible: leaves, inflorescences, and seeds. There is now considerable evidence for these genera, although they have been recovered mainly from archaeological sites in Central Mexico and most of the remains are associated with later time periods—Middle Formative through Classic (see Table 8.1).

The present evidence for their cultivation and eventual domestication is ambiguous at best. On the basis of the characteristics of exceptionally well-preserved inflorescences, Sauer (1969:81) reported domesticated *Amaranthus* in Tehuacan deposits dated to approximately 4000 B.C. Unfortunately, preservation conditions are not as favorable at most other sites where archaeological plant remains have been studied. Normally, the only remains of these genera that have been recognized are seeds, and those are usually carbonized. This eliminates the possibility of establishing comparative morphological criteria, such as seed coat color and certain details of seed coat ornamentation, that are useful for identification. Until recently, no attempt was made to study the effects of cultivation and eventual domestication on the morphology of seeds from the Mexican species of *Amaranthus*. Relative seed size and seed frequency in samples were the only indicators of their possible role as agricultural components. A similar situation holds true for *Chenopodium*. Seeds pertaining to this genus occur in the same preservation state as *Amaranthus*, often at the same sites (although its apparent absence in the Tehuacan Valley is an exception).

Research currently underway at the National University of Mexico is directed toward evaluating the traits discussed by Asch and Asch (1977, 1985), Wilson (1980), and Smith (1984) for recognizing domesticated *Chenopodium* seeds and for developing parallel techniques for identifying *Amaranthus*. The research involves detailed morphological and anatomical studies of seeds of the indigenous Mexican species of both genera.

The eventual domestication of species of both genera in Mesoamerica occurred later than the main components of the traditional subsistence base already mentioned. Early evidence comes from the Tehuacan Valley (amaranth only, ca. 10,000–7000 B.C.), Zohapilco (ca. 5200 B.C.), and southwestern Tamaulipas (ca. 2200–1800 B.C.). They are clearly present in a number of established sedentary communities from Early, Middle, and Late Formative contexts (Reyna Robles and González Quintero 1978; McClung de Tapia et al. 1986), as well as in Classic period urban centers, such as Teotihuacan (McClung de Tapia 1987). Their presence in paleoethnobotanical remains from numerous sites over a long chronological sequence suggests that they were undoubtedly important components of the Mesoamerican subsistence base from an early time, although considerable quantitative analysis remains to be carried out.

Numerous other plants, especially fruits, were probably cultivated from an early date in different parts of Mesoamerica and Central America. Some examples include sapote blanco (*Casimiroa edulis*), sapote negro (*Diospyros digyna*), and ciruela (*Spondias mombin*), all of which were recovered from sites in the Tehuacan Valley as early as ca. 5000 B.C. and from sites in Oaxaca. Ciruela also occurs at sites in Guatemala and Honduras (Smith 1987:88–89). Smith suggests that these fruit trees were introduced into the Tehuacan Valley, where they were probably cultivated because they would have required irrigation during the dry season in order to survive.

Many xerophytic plants, such as maguey or century plant (*Agave* spp.), nopal or prickly pear (*Opuntia* spp.), and pitahaya (*Lemaireocereus* spp.), were very likely cultivated from an early time, although it is not clear what an appropriate archaeological indicator of cultivation would be, other than seed size, which is extremely variable within a single fruit. Smith (1987:89) suggests that these plants may have been among the earliest cultivated because they are propagated easily.

Palm fruits are relatively frequent in the archaeological record. Coyol (*Acrocomia mexicana*) is reported from Tehuacan (ca. 5000 B.C.), where it was undoubtedly cultivated because it must be irrigated during the dry season. Coyol is also reported among archaeological plant remains from Belize, Honduras, Costa Rica,[4] and Panama (Smith 1980; Lentz 1990). Seed remains from Panama are identified as *Scheelia* sp., and fragments of pejibaye palm (*Bactris gassipaes*) are reported from Costa Rica and Panama. Palm fruits apparently were important components of prehispanic Central American diets prior to the introduction of maize (Smith 1987:91–92).

Little has been said about root crops for the obvious reason that archaeobotanical remains are virtually absent in Mesoamerica and Central America. It is important, however, to consider the potential impact of root crop agriculture, particularly in the humid tropics of this region. Root crops are consistently underemphasized largely because of the scarcity of direct evidence in the form of tangible plant remains. Fragments of stem tissue identified as *Manihot* by Callen (1967:285–86) in coprolites from Tehuacan suggest that its earliest appearance occurred during the Santa Maria phase (900–200 B.C.). A seed of wild *Manihot* has been reported by Smith from a Lagunas phase context (ca. 500 B.C.–A.D. 1) in Tamaulipas (in MacNeish 1958:146).

Pollen analysis may be a more appropriate way of demonstrating the presence of some vegetatively reproduced crops. In western Panama, for example, the Gatun Lake pollen sequence suggests that manioc and sweet potatoes were cultivated substantially later (ca. A.D. 150) in this area than in northern South America (Linares de Sapir and Ranere 1971:351).[5]

Finally, the distribution of plant remains in the archaeological record may be more useful as an indicator of where agriculture did not begin than where it did begin. As Smith (1987:95) points out, "for the greatest number of cultivated American species, we can positively say that crops did not originate where the archaeological remains have been found."

This observation has several implications. Little archaeological research and fewer archaeobotanical studies have been conducted in the areas of Mesoamerica where plant domestication is thought to have taken place. In many cases, we have only a general idea of how and when the domestication of certain plants occurred. In some cases, we have not definitely recognized the wild ancestors of modern domesticates. If we add to this the changes in wild plant distribution patterns after several millennia of environmental modification, our picture becomes still more complicated. Furthermore, current archaeological investigations in Mesoamerica are not guided by such criteria, and archaeobotanical studies are frequently the byproducts of research that has been designed and carried out with other interests in mind. Finally, chance and preservation are relevant as well. Excellent preservation conditions have favored the discovery of most early cultivated and domesticated plants. The absence of evidence for the transition from wild to cultivated and domesticated forms may be largely the result of such factors as inadequate preservation, selection of the wrong sites or the wrong part of the right site, and prehistoric consumption of the remains we would hope to find.

Culture History and the History of Cultivation

A substantial part of the data on which the Mesoamerican model is based comes from extensive archaeological investigations carried out in the Tehuacan Valley. Consequently, it seems appropriate to summarize the general outline of the gradual transition from nomadic hunting to a greater emphasis on plant food collection and, ultimately, to plant cultivation in that region. The salient features of the Tehuacan cultural phases and related subsistence activities have been described by MacNeish (e.g,, 1964b, 1967b, 1971, 1972, 1981).

It is useful at this point to distinguish between plant cultivation and domestication as biological and cultural processes on one hand and agriculture as a subsistence activity on the other because they are not necessarily the same. Implicit in one of MacNeish's

articles (1971) is a general developmental scheme for the Tehuacan Valley, which, although it may not be the most adequate descriptive framework, does emphasize the kinds of distinctions I think can be useful. In his graphic descriptions of microband/macroband activities during the early phases of the Tehuacan sequence, he refers to: plant collecting (with seasonal variants), barranca horticulture, hydro-horticulture, barranca agriculture, irrigation agriculture, and orchard culture (MacNeish 1971:310–14). In his view, the earliest step toward cultivation may have simply been a matter of returning consistently to the same area from year to year to collect specific plants. This would have led to some clearing (weeding) and a general improvement of the area, which would have provided a new artificial environment for seed or fruit crops. This, in turn, would have led to various genetic changes in the seed or fruit population. Unfortunately, MacNeish fails to explain why primitive populations would begin weeding stands of wild plants and to outline the processes by which environmental modification might lead to genetic changes in particular plants. He suggests (1971:312) that these processes may have led first to actual cultivation, which perhaps began as seeds were taken from one area and simply dropped in another, and eventually proceeded to intentional planting, seed selection, and horticulture. Once horticulture was well underway with a great variety of cultivated plants, primitive populations may have shifted to the intensive cultivation of a few plants (such as maize, chile, beans, and squash), although they may have continued the seasonal exploitation of other food resources. This process was continually elaborated upon until full-time agriculturalists were settled in sedentary communities in the valley. One may focus on the subsistence system, as MacNeish does, or on the biological aspects of the transition from plant collecting (which may just involve tolerance of and, occasionally, care for stands) to cultivation (which may involve care and, possibly, selection for preferred characteristics) and finally to domestication (which may involve the careful selection and manipulation that results in the plant's dependence upon human intervention in the reproductive cycle).

Such sequences are not necessarily evolutionary schemes. Rather, they depend on such interrelated factors as the environmental conditions that predominate in a particular region, the nature of plants that are or can be adapted to such circumstances, the human perception of those plants, and the organizational and technological potential for their exploitation.

Thus, the relationship between plant cultivation and agriculture as a subsistence system is not a necessary one. Certain plants may be cultivated without providing the basic component of the human diet and, more importantly, without necessarily provoking genetic changes that occasionally result from selection and that ultimately render plants dependent on humans for their propagation. On the other hand, genetic changes may take place without the influence of cultivation. I suspect that a careful examination of the ethnographic and historical literature, together with archaeological data from different parts of the world, would provide us with a large number of cases of mixed subsistence economies in which different plants produced under varied conditions of cultivation (and not necessarily domestication) comprise varying proportions of diets, reflecting so many combinations that terms like agriculture become so general as to be of little use. Thus, especially in specific cases, it is wise to be cautious in using terms like cultivation and domestication on one hand and agriculture on the other because they are not always synonymous.

Tehuacan Valley

During the earliest cultural phase in the Tehuacan Valley sequence, the Ajuereado phase (10,000–7000 B.C.), the inhabitants of the region appear to have lived in small, nomadic family groups (microbands) comprising approximately four to eight people, who changed their campsites seasonally according to the availability of food resources. Hunting was the major subsistence activity and, possibly, the only one during the winter (MacNeish 1971:307), although some wild plant foods were also collected. As they pursued these subsistence activities, MacNeish argues, the inhabitants acquired a close familiarity with the local "microenvironments" and their food potential at different times of the year. This gradual accumulation of knowledge coincided with Late Pleistocene climatic variation that substantially altered the regional environment. As grasslands and water sources in the valley's center were reduced, many animals presumably became extinct (e.g., horse and antelope) and others that congregated in the decreasing grasslands were easily killed. The lack of sufficient meat resources presumably led to an increased exploitation of plant foods, although hunting never completely

ceased. During the latter part of the phase, the substitution of smaller game, such as deer and cottontail rabbits, supplanted the previous emphasis on horse, antelope, and jack rabbits. Other small game, such as gophers, rats, turtles, and birds, continued to be "collected" throughout the remainder of the phase.

During the succeeding El Riego phase (7000–5000 B.C.), the settlement pattern was characterized by nomadic families who lived in dry season camps much like the previous phase, but tended to congregate in larger groups (macrobands) during the spring and summer wet seasons. MacNeish at one time described these groups as possibly being "patrilineal with some sort of weak temporary leadership in the hands of a male and perhaps some concept of territoriality" (1964:41). Plant collection increased during this period and there is some evidence for the use of plants that were later domesticated (e.g., squash, chile, avocado, etc.).

During the Coxcatlan phase (5000–3400 B.C.), the nomadic microband/macroband pattern continued, but the macrobands were much larger. MacNeish (1971, 1972) calls this pattern "seasonal macro-microbands." More plants were utilized, including cultivated and domesticated species. Maize appears in the archaeological record during this phase (refer to Note 2) and the relative quantity of meat in the diet declines.

"Central-based bands" developed during the Abejas phase (3400–2300 B.C.), the final preceramic phase of the sequence. This settlement pattern is characterized by the occupation of microband hunting camps during the dry season and semipermanent macroband settlements on river terraces in the valley. The proportion of agricultural plants exploited was significant and, according to MacNeish, yielded a stable subsistence, thereby allowing cultural development and technological improvements to take place (1971:313). These processes presumably continued during the following Purron phase (2300–1500 B.C.), about which little is known except that pottery makes its initial appearance.

During the Ajalpan phase (1500–850 B.C.), full-time agriculture was established and irrigation was introduced. The population of the valley settled in small semipermanent and permanent villages. During later phases in the valley sequence, covering a period from approximately 850 B.C. to A.D. 1540, population increased and a more complex settlement pattern developed (MacNeish 1971, 1972). The Tehuacan Valley phases described above are the most relevant here, because it is during this time span that the major ecological and sociocultural changes related to the introduction and development of agriculture took place. Figure 8.3 summarizes the trends in subsistence components proposed by Mac-Neish for this region (1967b:301).

Guilá Naquitz, Valley of Oaxaca

The excavation of Guilá Naquitz in the Mitla area of the Oaxaca Valley provides a context within which to study in detail the behavior of prehistoric hunter-gatherers. The preceramic occupation of the cave site corresponds to ca. 8900–6700 B.C. and is contemporary with the Late Ajuereado and part of the El Riego phases described for the Tehuacan Valley (Flannery 1986:38).

Flannery provides the following description of the cultural context within which the occupants of the cave can be considered:

During the Early Ajuereado phase, under the cooler and drier conditions of the Late Pleistocene period, the occupants of the southern Mexican highlands hunted native wild horse, pronghorn antelope, and jackrabbit, sometimes by communal drives. . . . With warming tempera-tures at the onset of the Holocene, the thorn-scrub-cactus forest reconstructed by Smith . . . spread over Oaxaca and Tehuacan, and many Pleistocene animals disappeared; so, apparently, did communal game drives. The early Archaic witnessed the establishment of the microband-macroband foraging pattern originally defined by MacNeish. . . . The Late Ajuereado foragers in Tehuacan and the Naquitz phase foragers in Oaxaca may have been among the first to display this early Archaic settlement-subsistence pattern. What followed in both areas was incipient agriculture. (Flannery 1986:39)

Flannery (1986:39) further suggests that the preceramic hunter-gatherer population of the Oaxaca Valley was characterized by family units comprising two to five members, which in turn were a part of a local group averaging 25 members that might coalesce at larger campsites during periods of more abundant resources. Local groups were related through marriage with others to form an "effective breeding population" of 175 or more people distributed throughout neighboring valleys. Finally, a so-called "dialect tribe" of at least 500 people sharing a mutually intelligible dialect extended over a somewhat broader area.

The Guilá Naquitz inhabitants, therefore, represent family microbands that occupied the site from

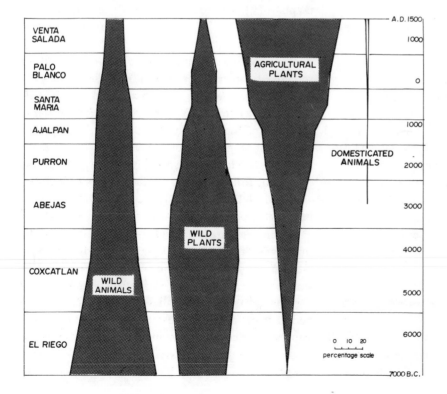

Fig. 8.3. Hypothesized proportions of subsistence components in the Tehuacan Valley inhabitant's diet. The width of the vertical lines is roughly proportional to the amount of cultivated plants in diet. (From MacNeish 1967b:301)

late August or early September until late December. It is unclear whether their presence at the site followed the breakup of a macroband or whether it was part of a foraging strategy pursued during years when resources were insufficient to allow macroband formation. Hunting was apparently of little significance to the cave inhabitants. Other sites in the Mitla region, such as Cueva Blanca (microband) and Gheo-Shih (macroband), provide a broader picture. Although the chronological differences in their occupations do not permit the formulation of a definite subsistence-settlement pattern involving these or similar sites, settlement types that are similar to the Tehuacan Valley sequence can be recognized (Flannery 1986:41–42).

Zohapilco and the Southern Basin of Mexico

Excavation of the site of Zohapilco near the modern community of Tlapacoya on the edge of the ancient shore of Lake Chalco included the recovery of pollen samples and macrobotanical remains. The site was apparently occupied by macrobands of hunter-gatherers who appeared to have been semisedentary and, possibly, sedentary because of the abundance of food resources available from the lake, lakeshore, and surrounding piedmont zones (Niederberger 1976, 1979). Pollen studies suggest an early humid period with abundant vegetation at ca. 5500–3500 B.C. (Playa phase). Several kernels of Chalco-type teosinte were found associated with materials of this phase, suggesting the possibility of an incipient agricultural stage (Lorenzo and González 1970; Niederberger 1986:112). The site was abandoned following a devastating volcanic eruption and was reoccupied around 2500 B.C. (Zohapilco phase, ca. 2500–2000 B.C.) by a horticultural population that exploited a broad range of plants, including *Amaranthus*, *Physalis*, *Capsicum*, and *Cucurbita*, represented by seed remains, and maize, represented by pollen.

Other Regions of Mesoamerica and Central America

Archaeobotanical evidence for plant cultivation and agriculture south of the modern border of Mexico is sparse. Some botanical remains in the form of plant impressions were recovered from levels dated to the Cuadros phase (1000–850 B.C.) at the site of Salinas la

Blanca near Ocos on the Pacific coast of Guatemala (Coe and Flannery 1967:70–72). Maize was recognized, as were pollen grains of *Trema* sp., a shrub that typically invades abandoned cornfields. Other fruits, including avocado, were also identified from impressions. The abundance of coastal, estuary, and riverine products (shellfish, fish, reptiles, etc.), tropical fruits, and maize apparently permitted a sedentary existence; maize agriculture need not have been a dominant part of the subsistence economy.

Phytolith data from Middle-Late Preclassic period deposits (900–300 B.C.) indicate the presence of maize cultivation at the site of El Balsamo on the Pacific coast of Guatemala (Pearsall 1982:866).

Maize pollen was identified in cores taken from the sediments of Lake Petenxil and other sites in the Peten, Guatemala, suggesting the presence of agricultural populations in that area as early as ca. 2000 B.C. (Cowgill and Hutchinson 1962; also compare Rice 1976; Wiseman 1978).

Analysis of phytoliths and macrobotanical remains from Swasey phase deposits (2000–1000 B.C.) at the site of Cuello in Belize also indicate the presence of maize (Pearsall 1982:866).

The earliest archaeobotanical evidence for maize in El Salvador consists of a carbonized cob from Chalchuapa in a Late Preclassic context dating to ca. 400 B.C.–A.D. 300 (Sheets 1982:102). Sheets also mentions maize impressions from the Classic period (A.D. 300–900) at Joya de Ceren (1982:107) and maize phytoliths from the Postclassic or Historic period at Cihuatlan (1982:116).

Plant remains from sites in the Sulaco River Valley of north-central Honduras (Lentz 1983) are also quite late, corresponding to Late Classic period contexts (ca. A.D. 600–900). These specimens suggest a subsistence pattern in which fruit collection is combined with maize cultivation.

The species reported by Lentz (see Table 8.4) tend to be similar to those reported by Smith (1980) from the Chiriqui Province in western Panama. Two cave sites in Panama, Casita de Piedra and El Trapiche, dating to ca. 4600–2400 B.C. provide early evidence for localized food collection. A later site in the Cerro Punta area (ca. A.D. 200–400) provides evidence of cultivated plant remains, and specimens obtained from Barriles (ca. A.D. 600) and La Pitahaya (ca. A.D. 800–1100) include mainly palm fruits and a limited number of maize kernels. These data suggest a prehistoric Central American subsistence pattern that reflects adaptation to localized resources, supplemented to some extent by maize cultivation.

Piperno (1984) studied phytolith samples from four sites in Central Panama—three rockshelters and a floodplain agricultural village—that were occupied from 6500 B.C. to A.D. 500. Carbonized palm kernels and carbonized wood were recovered from preceramic contexts in the rockshelters (Cueva de los Vampiros, Cueva de los Ladrones, and Aguadulce), but domesticated macrobotanical remains (maize and common beans) were found only in first millennium B.C. deposits at the village site of Sitio Sierra. Maize pollen was reported, however, from several preceramic and early ceramic deposits at the Cueva de los Ladrones shelter. Piperno (1984:375) argues that maize pollen together with phytoliths in preceramic contexts indicates that maize was cultivated throughout most of the sequence at this site, where the earliest occupation dates to 4910 B.C. An analysis of the differences in size and type of phytoliths associated with maize suggests that different types were grown at the sites mentioned above and that new productive types were introduced into Central Panama after 1000 B.C. (Piperno 1984:375).

Pollen identified from cores taken in the Lake Gatun area of the Canal Zone in Panama includes maize in combination with other Gramineae, Compositae, and herbaceous weeds dated to ca. 3100 B.C.–A.D. 200 (Linares de Sapir and Ranere 1971:351). A deeper core was thought to contain "wild" maize (4 grains), dated to ca. 5300–4300 B.C. (see Ludlow-Weichers et al. 1983:241). Pollen data from this area were later compared with phytoliths extracted from the same proveniences (Piperno 1985a). The presence of phytoliths from maize and other herbaceous plants dated to 2850 B.C. indicate that agriculture and forest clearance occurred about 1,000 years earlier than the pollen sequence suggests (2000 B.C.) (Piperno 1985a:18).

Cultural developments on the Pacific side of the Isthmus of Panama, where "chiefdoms" were believed to have developed by ca. 300 B.C.–A.D. 300 (Linares de Sapir and Ranere 1971:352), contrast sharply with those on the Atlantic side. Excavations carried out in the Bocas area (Linares de Sapir and Ranere 1971:353) indicate the presence of small hamlets whose inhabitants practiced hunting, fishing, and shellfish gathering in combination with long fallow forest agriculture. This pattern presumably includes the periodic movement of small segments of the population into dispersed settlements.

Models for the Origin of Agriculture

We have come a long way from the "accidental" domestication hypotheses for the origins of agriculture. Arguments for the "readiness" of human populations to begin cultivating plants based on their familiarity with their surroundings are no longer considered adequate. Even potentially testable assumptions based on climatic modifications and consequent variation in the availability of resources are clearly insufficient by themselves.

Flannery's Systems Approach

As Flannery (1968) points out, primitive people are rarely adapted to whole environments, or even to "microenvironments" within a defined region (Coe and Flannery 1964), but rather to a relatively small number of plants and animals whose habitats crosscut several physiographic zones. He developed a model in which particular plants and animals were seen as the focal point of a number of procurement systems, each of which necessitated some degree of specialized technology, such as appropriate tools, storage facilities, etc.

These systems in the central highlands of Mexico initially involved the exploitation of maguey, cactus fruits, tree legumes, white-tailed deer, and cottontail rabbits. Two principal regulatory mechanisms helped to keep the system going and to counteract deviation from the established subsistence pattern: *seasonality* (i.e., certain resources were available only at certain times of the year) and *scheduling* (i.e., certain resources were preferable to others that were available at the same time). Such mechanisms were said to maintain a "negative feedback" process that promoted equilibrium and worked to prevent deviation from a stable situation over a long period of time. Seasonality and scheduling prevented the intensification of any single procurement system to the point where the resource was threatened, and permitted the maintenance of a sufficiently high level of procurement efficiency so that there was little pressure for change. When change did occur, it was attributed to a "positive feedback" process, which amplified deviation from the stable pattern and caused the system to be restructured in the direction of a higher level of stability.

In order for such a deviation-amplifying mechanism to get underway, Flannery postulates that a relatively insignificant or accidental "kick" occurred.

He suggests that the necessary stimulus may have been provided by genetic changes in one or two of the plants useful to human groups but of minor importance in relation to other resources that were more heavily depended upon. For example, the wild ancestor of *Zea mays* underwent modification as a consequence of natural crossbreeding and subsequent evolution, resulting in an increase in cob size, cob number, and kernel row number, and in the loss of glumes that enclosed individual kernels. Beans also underwent changes that resulted in their becoming more permeable in water and thus easier to prepare, and in the development of limp pods that would not shatter when ripe, facilitating their harvesting (Flannery 1968:94). Thus, a natural increase in yield as a consequence of minor genetic changes in certain plants could have enhanced their attractiveness as potential human subsistence resources, and could be further incremented by human selection.

Such changes were thought to initiate a positive feedback process. As Flannery states (1968:95), "the more widespread maize cultivation, the more opportunity for favorable crosses and backcrosses; the more favorable genetic changes, the greater the yield; the greater the yield, the higher the population and hence the more intensive cultivation." Therefore, the procurement of wild grass, and later cultivated forms, grew at the expense of other procurement systems. Because a spring planting and a fall harvest became necessary, maize procurement competed to some degree with the collection of spring- and fall-ripening resources, and with deer hunting during the summer rainy season.

A functional association between band size and resources is demonstrated by the tendency for populations to disperse in small, scattered microbands during seasons when few resources were available and to coalesce in certain areas during seasons or years when desirable plants were abundant and required immediate harvesting. Consequently, human demography was modified by positive feedback stimulated by early maize and bean cultivation, and macrobands gradually increased in size as the duration of their stay in certain areas lengthened (Flannery 1968:96).

Rescheduling of other procurement systems became necessary, depending on the extent to which cultivation could be undertaken in different parts of Mesoamerica. In regions where maize cultivation could be carried out year round, certain seasonal resources decreased in importance or were abandoned, and attention was turned toward other resources that did not conflict with agricultural activities. In areas where maize cultivation was practiced only during the rainy season, dry season resources could still be collected as intensively as year-round resources.

MacNeish (1971, 1972) later adopted the systems terminology introduced by Flannery but used it as a descriptive framework rather than as an explanatory model. He continued to believe that major changes in human adaptation occurred when they did because human populations had accumulated enough ecological and subsistence knowledge to readapt to new environmental conditions—a position similar to that expressed by Braidwood (1960:44) for the Near East. Flannery, on the other hand, attributed the usefulness of a model based on cybernetic concepts to its ability to treat prehistoric cultures as systems and to examine the mutual causal processes that amplify initially small, seemingly insignificant deviations from a pre-existing pattern into major cultural change.

The model is still essentially supported by the evidence from the Tehuacan and Oaxaca valleys and, possibly, from the Basin of Mexico, although pertinent data from that region have never been adequately published. While specific plant and animal resources may differ, and certain environmental conditions may affect their relative abundance and distribution, a systems framework can meaningfully describe the transition from an emphasis on certain resources to a dependence on others. As Flannery (1986:22) suggests, such a scheme "forces one to be explicit about the relationships among variables, rather than just listing them." Recent efforts by Flannery (1986:19–28; 435–38; 501–507) and his colleagues (e.g., Reynolds 1986:439–500) involve the application of increasingly more sophisticated analytical techniques, such as computer simulation, to bring the model closer to reality and to experiment with the effect of different variables on the system. One of the most important aspects of Flannery's investigations in the Oaxaca Valley has been the attention to detail in data recovery in order to reduce sample bias and increase representation, drawbacks frequently attributed to data recovered from the Tehuacan excavations (Stark 1981:354–55; Flannery 1986:303). While data recovery was not entirely flawless, a serious attempt to reduce sampling error was made.

Flannery currently views the origin of Mesoamerican agriculture as a consequence of human adaptation to the unpredictable availability of subsistence resources during the gradual replacement of late Pleistocene vegetation by Holocene floral communities, which included the ancestors of many plants that were later domesticated. Growing hunter-gatherer populations, whose mobility was increasingly limited by societal constraints (e.g., territoriality), sought solutions other than emigration and thus turned to such strategies as diversification in resource exploitation, increased use of storage techniques, technological improvements in food-processing equipment, and "an increasing trend from foraging to logistically based collection" (Flannery 1986:14). Flannery (1986:14) contends that "what these alternatives seemingly shared was a goal of resiliency, risk reduction, amelioration of environmental extremes, and an increase in resource predictability." He sees plant cultivation, then, as the "ultimate collecting strategy," a way of evening out the differences between good and bad years.

With respect to other approaches, such as Binford's (1968) density equilibrium model or Cohen's (1981) overpopulation argument, Flannery (1986:11) points out that the empirical data available from Mesoamerica suggest that incipient agricultural populations were so small that these models are seemingly irrelevant. Although populations may have increased, they did not reach the limits of available resources.

Other Biological Approaches to Plant Domestication

Another important view developed in recent years argues essentially that cultivation resulted from the intensification of symbiotic relationships between people and the plants they consumed in a coevolutionary process. According to this view, agriculture is the result of domestication, which is in itself a process independent of human volition (Rindos 1984). Flannery (1986:15–16) observes, however, that there are many documented cases in both the Old World and the New World in which plant domestication follows cultivation. The concept of symbiosis as an important biological relationship, however, should not be ignored.

Flannery (1986:16) suggests that this view would be more appropriately applied to hunter-gatherer populations than to incipient agriculturalists. I would argue, however, that agricultural populations utilize plants that are at different stages on the "pathway" toward domestication. Some plant species never reach this ultimate stage, although they may be of considerable importance for subsistence. Domestication can be viewed as a process that has occurred and

continues to take place under diverse conditions at different rates.

Certain types of plants display great plasticity in their ability to adapt to variable conditions. For example, the annual habit, together with a tendency to colonize disturbed habitats, are important characteristics (see Hawkes 1969:21). In semiarid Mesoamerica there is substantial annual variability in temperature and humidity. In order to guarantee success, plants need to germinate and grow quickly when the summer rains begin and to complete a full life cycle before the fall-winter dry season gets underway. During the dry period, seeds lie dormant awaiting suitable conditions for germination. Under such conditions, there is strong selection for big seeds with ample food reserves or for very large quantities of seeds. These features allow wild plants, such as weeds, to survive difficult conditions without competition from others that seek more favorable habitats.

Furthermore, Galinat (1985:272) notes that although many of the grasses that have become important food plants have both perennial and annual species, it is usually the annuals that are brought under domestication (1985:272). In Mexico today, many plants exploited as greens constitute a significant dietary complement. Often referred to as "quelites," a generic term that normally encompasses undomesticated edible greens, they are usually derived from young, tender, annual herbs. While weedy annual species of *Chenopodium*, *Amaranthus*, and *Portulaca* (purslane) are often exploited as quelites (Bye 1981:109), these genera have also been domesticated and are important food resources.

Bye (1981:119) proposes a more general pathway to domestication, emphasizing the biological aspects of weedy plants, such as annuals, and the impact of human activities (Fig. 8.4). Wild plants with a tendency to colonize disturbed habitats of all kinds are transformed through natural selection into weeds once they colonize areas disturbed by human activities. They may be recognized as usable resources, although they receive no special attention, or they may simply be tolerated, in which case they benefit from unconscious human selection. In either case, they still remain weeds. On the other hand, they may be encouraged and finally utilized to the extent that a conscious selection for certain preferred characteristics develops. Such plants may become "weed crops," cultivated but essentially nondomesticated plants, or they may become domesticates. Biologi-

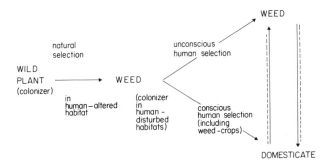

Fig. 8.4. Pathways to plant domestication. (After Bye 1981:119)

cally, however, some weeds may continue to influence the genetic composition of their domesticated relatives through frequent crossing and backcrossing in fields.

Ethnobotanical observations of the processes involved in incipient plant domestication tend to support such a scheme (Bye 1979). Although these processes are not clearly demonstrated in the archaeological record, they may have occurred again and again "in different places and at different rates" (Galinat 1985:275).

Flannery advances the following conclusion:

It would seem that agriculture arose in the context of *efforts to deal locally* with long-term environmental variation. If this was the case, we could not fully understand incipient cultivation by subsuming it under a covering law, or a biological concept such as coevolution. We need to understand the system of late pre-agricultural strategies out of which incipient cultivation arose. (Flannery 1986:514; emphasis added)

This position may be a valid assumption, although I would argue that the interrelationships between humans and the plants they exploited were still more complex than Flannery implies here. In order to determine the subsistence strategies that developed into agricultural practices, we also need to have a better understanding of the biological relationships between plants and human populations that characterize incipient cultivation.

Summary and Conclusions

The origin of agriculture in Mesoamerica and its gradual development in the remainder of Central America can be viewed from several different but related perspectives. One aspect is the biological transformation of a series of plants, resulting occasionally in domesticated forms, and the subsequent role of these plants in human subsistence. Another consideration is the impact that human exploitation and manipulation had, and continues to have, on the adaptation and distribution of plants. For example, the proliferation of cultivated edible weeds or the possible development of wild populations from previously cultivated forms in some areas are significant phenomena. Still another aspect is the complex relationship among such factors as sedentarism, population growth, intensification of plant cultivation, and cultural diffusion, whose specific interactions may have varied greatly in different subregions of an area as broad as that comprising Mesoamerica and Central America.

In view of the present state of knowledge about early plant cultivation and domestication in Mesoamerica and of the incomplete nature of the data currently available, the systems model that Flannery proposes provides a framework within which important relationships can be conceptualized once their components are recognized. Such an ecological orientation, while reflecting the way things seem to be related in the real world, reduces the dependence of models on environmental or demographic causes and emphasizes the interplay among other important factors.

Any current discussion of plant domestication in Mesoamerica must examine the evidence for the sociocultural transition from a mobile hunter-gatherer subsistence base to a sedentary village organization with agriculture as a permanent and basic productive activity. While recent research has greatly illuminated the apparent characteristics of this transition, particularly as it must have taken place in the Tehuacan and Oaxaca valleys, many gaps remain in the picture. The exact nature of this relationship is still unresolved and will probably be shown to demonstrate many variants as other regional manifestations are uncovered.

Although prehistoric Mesoamerican and Central American subsistence agriculture incorporated a large variety of plants, several are outstanding because of the tremendous impact their cultivation—and, in some cases, their domestication—had upon human society. Among these plants are maize, amaranth, beans, squash, gourds, chile peppers, and avocado. The species utilized and their importance with respect to one another vary in relation to such environmental factors as altitude, climate, and soil conditions and with changing economic factors, such as the appearance of "commercial" resource distribution patterns. An ample selection of vegetable subsistence elements results from the combination of less significant cultigens together with a broad selection of gathered wild plants.

In the case of Mesoamerica—and Flannery (1985, 1986) has made this point quite clear—the present archaeological evidence does not support the concept of plant cultivation as a way of increasing food resources in marginal areas, which were colonized by human groups as a last resort following excessive population growth and consequent resource depletion in more optimal zones, where natural stands of preferred plants occurred. This model and others (e.g., Binford 1968; Flannery 1969; Meyers 1971) that have been proposed primarily in the context of Old World developments are not applicable to the Mesoamerican situation. It is only in later cultural phases, when agriculture is well established as the basic subsistence activity of sedentary communities, that we can postulate a definite relationship between population growth and increased agricultural production as a plausible response to resource stress. Even then, the causal direction of the hypothesized relationship is not at all clear. It may be that agricultural intensification, in fact, represented a response to other societal demands, at least in some instances. Examples might include the intensification of agricultural production in order to reduce or eliminate dependence on unreliable (ecologically or politically) food resources, or the perception of surplus production as a prestige symbol ("institutionalized waste"), both of which are difficult to detect archaeologically.

Future Research in Mesoamerica and Central America

The investigations carried out in the central part of Mesoamerica under the direction of MacNeish and Flannery have provided the most useful data with which to formulate and test hypotheses concerning the cultivation and domestication of particular plants and their contribution to the development of a sedentary, agriculturally based lifestyle. Familiarity with the geographic distribution of wild plant populations and, if relevant, their domesticated forms together with archaeobotanical evidence suggests certain areas where initial modifications in these plants may have occurred or, at the very least, areas that were probably unimportant insofar as these processes were concerned. An interpretive mosaic may be constructed from the bits and pieces of such evidence, with many unknowns that prevent us from completing the picture.

While economic limitations have hindered archaeological investigations, the main explanation for the absence of detailed evidence about agricultural beginnings in Mesoamerica and Central America is the failure to utilize effectively the techniques available for the recovery of different types of plant remains. In general, research project directors must make more serious efforts to recover archaeobotanical material systematically during the course of excavation, and data must be disseminated more effectively once they have been analyzed.

In particular, archaeological survey in Mesoamerica, especially in the state of Jalisco, should be continued and supported by the sophisticated techniques that have been developed in recent years to locate preceramic sites where archaeobotanical remains might be preserved. Similarly, it is hoped that a more concentrated effort will be made to broaden our knowledge of the introduction and spread of plants and their cultivation techniques into Central America, in order to formulate a more complete picture of the development of agricultural subsistence in relation to human settlement and sociocultural complexity in that relatively little known area.

Future investigations of agricultural origins must concentrate on expanding the picture of resource utilization by populations in various ecological surroundings within an appropriate chronological context. There are so many geographical and chronological gaps in the region that excavation in almost any area is potentially a valuable source of data. Clearly, it would be appropriate to try to correlate the areas of archaeological research with the range of distribution of plants that were and continue to be important subsistence elements in specific regions. The limitations of this approach, however, must be recognized: the modern range of many plants is greatly modified from prevailing conditions during the prehistoric past, and the wild ancestors of many important cultivated plants are still unrecognized.

Notes

1. No attempt is made here to evaluate the concept of a culture area in general or as it applies specifically to Mesoamerica, or to enter into a discussion about when such a culture area can be said to have come into existence (see Willey 1962:84; Litvak King 1975:186–87). Mesoamerica is considered to be a geographical entity, clearly defined in the literature (e.g., Kirchoff 1943; Sanders and Price 1968; Palerm and Wolf 1972; Jiménez Moreno 1977). Finally, no effort is made to address the chronological variation in the border areas of Mesoamerica, except to point out the so-called ecological limitations in relation to agricultural subsistence that Palerm and Wolf (1972) discuss.

2. Since this manuscript was written, a number of maize cobs from the excavations at Coxcatlan and San Marcos caves in the Tehuacan Valley have been dated directly by accelerator mass spectrometry (AMS). Previous dates for the appearance and development of maize in this region were based on their association with radiocarbon 14 dated strata. The new dates indicate that the maize specimens, originally thought to belong to the Coxcatlan and Abejas phases, are about 2,000 years more recent than was previously believed (Long et al. 1989). Furthermore, a careful study of the morphological characteristics of the cobs themselves reveals suggestive evidence that maize was still undergoing important evolutionary changes at this time (Benz 1988; Benz and Iltis 1990). Thus, the new dates have important implications for the antiquity of domesticated maize, as well as for models of changing settlement patterns associated with maize agriculture. Undoubtedly, a reevaluation of the chronology of the Tehuacan Valley sites is necessary. New dates for other botanical remains utilizing AMS and a careful comparison of possible discrepancies between traditional radiocarbon 14 and AMS dates for materials from the same archaeological contexts would be useful.

3. Since this chapter was submitted for publication, archaeological research directed toward locating preceramic and other sites in the state of Jalisco has resumed. Published results, however, are not yet available.

4. Smith (1987:92) refers to the presence of coyol among archaeobotanical remains recovered in Costa Rica but mentions no specific site (see Table 8.4).

5. Pollen and phytolith analyses are important complements to flotation and dry screening as a way to recover evidence for the use of root crops, as well as for seed, fruit, and other crops, which are not always preserved as macroremains in the archaeological record. It is necessary to coordinate these three kinds of evidence in the selection of sample proveniences in order to support cultural interpretations.

References Cited

Asch, D. L., and N. B. Asch.

1977   Chenopod as Cultigen: A Re-evaluation of Some Prehistoric Collections from Eastern North America. Mid-Continental Journal of Archaeology 2(1):3–45.

1985   Prehistoric Plant Cultivation in West-Central Illinois. In Prehistoric Food Production in North America, edited by R. I. Ford, pp. 149–203. Anthropological Papers No. 75. Museum of Anthropology, University of Michigan, Ann Arbor.

Barghoorn, E. S., M. K. Wolfe, and K. H. Clisby

1954   Fossil Maize from the Valley of Mexico. Botanical Museum Leaflets, Harvard University 16:229–40. Cambridge.

Beadle, George W.

1972   The Mystery of Maize. Field Museum of Natural History Bulletin 43(10):2–11.

1977   The Origin of Zea mays. In Origins of Agriculture, edited by C. Reed, pp. 615–35. Mouton Publishers, The Hague.

1982   El orígen del maíz comprobado por el polen. Información Scientífica y Technológica 4(72):20–28. Mexico.

Bender, Barbara.

1975   Farming in Prehistory: from Hunter-Gatherer to Food Producer. John Baker, London.

1978   Gatherer-hunter to Farmer: A Social Perspective. World Archaeology 10(2):204–22.

Benz, Bruce F.

1988   The Wild Maize from Tehuacan Revisited. Paper presented at the 53rd Annual Meeting of the Society for American Archaeology, Phoenix.

Benz, Bruce F., and Hugh H. Iltis

1990   Studies in Archaeological Maize I: The "Wild" Maize from San Marcos Cave Re-examined. American Antiquity 55(3):500–511.

Binford, Lewis R.

1968   Post-Pleistocene Adaptations. In New Perspectives in Archaeology, edited by S. Binford and L. Binford, pp. 313–41. Aldine, Chicago.

Braidwood, Robert J.

1960   The Agricultural Revolution. In Old World Archaeology: Foundations of Civilizaiton. Readings from Scientific American, edited by C. C. Lamberg-Karlovsky, pp. 71–79. W. H. Freeman and Company, San Francisco.

Bronson, Bennett

1966   Root Crops and the Subsistence of the Ancient Maya. Southwestern Journal of Anthropology 122(3):251–79.

Bye, Robert A., Jr.

1979   Incipient Domestication of Mustards in Northwest Mexico. The Kiva 44(2–3):237–56.

1981   Quelites—Ethnoecology of Edible Greens—Past, Present and Future. Journal of Ethnobiology 1(1):109–123.

Byers, Douglas (editor)

1967   The Prehistory of the Tehuacan Valley, vol. 1: Environment and Subsistence. University of Texas Press, Austin.

Callen, Eric O.

1963   Diet as Revealed by Coprolites. *In* Science in Archaeology. Pp. 186–94. Thames and Hudson, London.

1967   Analysis of the Tehuacan Coprolites. *In* Prehistory of the Tehuacan Valley, vol. 1: Environment and Subsistence, edited by D. Byers, 261–89. University of Texas Press, Austin.

Coe, Michael D., and Kent V. Flannery

1967   Early Cultures and Human Ecology in South Coastal Guatemala. Smithsonian Contributions to Anthropology No. 3. Smithsonian Institution, Washington, D.C.

Cohen, Mark Nathan

1981   La crisis alimentaria de la prehistoria. Alianza Universidad 291. Madrid.

Cowgill, U. M., and G. E. Hutchinson

1962   The Chemical History of Laguna de Pentexil, Peten, Guatemala. Memoir 17. Connecticut Academy of Arts and Sciences, New Haven.

Cutler, Hugh C., and Thomas W. Whitaker.

1961   History and Distribution of the Cultivated Cucurbits in the Americas. American Antiquity 26:469–85.

1967   Cucurbits from the Tehuacan Caves. *In* The Prehistory of the Tehuacan Valley. vol. 1: Environment and Subsistence, edited by D. Byers, pp. 212–19. University of Texas Press, Austin.

Flannery, Kent V.

1968   Archaeological Systems Theory and Early Mesoamerica. *In* Anthropological Archaeology in the Americas, edited by B. Meggers, pp. 67–87. Anthropological Society of Washington, Washington, D.C.

1969   Origins and Ecological Effects of Early Domestication in Iran and the Near East. *In* The Domestication and Exploitation of Plants and Animals, edited by P. Ucko and G. Dimbleby, pp. 73–100. Duckworth, London.

1972   The Origins of the Village as a Settlement Type in Mesoamerica and the Near East. *In* Man, Settlement, and Urbanism, edited by P. Ucko, R. Tringham, and G. Dimbleby, pp. 23–53. Duckworth, London.

1973   The Origins of Agriculture. Annual Review of Anthropology 2:271–310.

1976   Empirical Determination of Site Catchments in Oaxaca and Tehuacan. *In* The Early Mesoamerican Village, edited by K. Flannery, pp. 103–117. Academic Press, New York.

1985   Los orígenes de la agricultura en México: las teorías y la evidencia. *In* Historia de la Agricultura: Época Prehispánica-Siglo XVI, edited by T. Rojas-Rabiela and W. T. Sanders, pp. 237–65. Colección Biblioteca del INAH. Instituto Nacional de Antropología e Historia, Mexico.

Flannery, Kent V. (editor)

1976a   The Early Mesoamerican Village. Academic Press, New York.

1982   Maya Subsistence: Studies in Memory of Dennis E. Puleston. Academic Press, New York.

1986   Guilá Naquitz: Archaic Foraging and Early Agriculture in Oaxaca, Mexico. Academic Press, New York.

Flannery, K. V., A. V. T. Kirkby, M. J. Kirkby, and A. W. Williams, Jr.

1967   Farming Systems and Political Growth in Ancient Oaxaca. Science 158(3800):445–53.

Ford, Richard I.

1976   Carbonized Plant Remains. In Fábrica San José and Middle Formative Society in the Valley of Oaxaca, edited by Robert D. Drennan, pp. 261–68. Prehistory and Human Ecology of the Valley of Oaxaca, Vol. 4, No. 8. Memoirs of the Museum of Anthropology, University of Michigan, Ann Arbor.

1981   Gardening and Farming before A.D. 1000: Patterns of Prehistoric Cultivation North of Mexico. Journal of Ethnobiology 1(1):6–27.

1985   Patterns of Prehistoric Food Production in North America. *In* Prehistoric Food Production in North America, edited by R. I. Ford, pp. 341–64. Anthropological Papers No. 75. Museum of Anthropology, University of Michigan, Ann Arbor.

Galinat, Walton C.

1971   The Origin of Maize. Annual Review of Genetics 5:447–78.

1985   Domestication and Diffusion of Maize. *In* Prehistoric Food Production in North America, edited by R. I. Ford, pp. 245–78. Anthropological Papers No. 75. Museum of Anthropology, University of Michigan, Ann Arbor.

Gonzalez Quintero, L.

1986   Orígen de la domesticación de los vegetales en México. Historia de México 1:77–92. Salvat, Mexico.

Guzmán, Rafael

1981   El Teosinte en Jalisco. Su Distribución y Ecología. Manuscript on file at Laboratorio de Paleoetnobotánica, Instituto de Investigaciones Antropológicas Universidad Nacional Autónoma de México, Mexico.

Hawkes, J. G.

1969   The Ecological Background of Plant Domestication. *In* The Domestication and Exploitation of Plants and Animals, edited by P. Ucko and G. W. Dimbleby, pp. 17–29. Aldine, Chicago.

Iltis, Hugh H.

1983   From Teosinte to Maize: The Catastrophic Sexual Transmutation. Science 222(4626):886–94.

Iltis, Hugh H. and John F. Doebley.

1984   *Zea*—A Biosystematic Odyssey. *In* Plant Biosystematics, edited by W. F. Grant, pp. 587–616. Academic Press, New York.

Jiménez Moreno, Wigberto.

1977   Mesoamérica. Enciclopedia de México 8:471–83. Mexico.

Kaplan, Lawrence.

1965   Archaeology and Domestication in American *Phaseolus* (Beans). Economic Botany 19(4):358–68.

1967   Archaeological *Phaseolus* from Tehuacan. *In* The Prehistory of the Tehuacan Valley, vol. 1: Environment and Subsistence, edited by D. Byers, pp. 201–11. University of Texas Press, Austin.

1981   What is the Origin of the Common Bean? Economic Botany 35(2):241–54.

1986   Preceramic *Phaseolus* from Guilá Naquitz. *In* Guilá Naquitz: Archaic Foraging and Early Agriculture in Oaxaco, Mexico, edited by K. Flannery, pp. 281–84. Academic Press, New York.

Kaplan, Lawrence, and Richard S. MacNeish.

1960   Prehistoric Bean Remains from Caves in the Ocampo Region of Tamaulipas, Mexico. Botanical Museum Leaflets, Harvard University 19(2):33–56. Cambridge.

Kirchoff, Paul.

1943   Mesoamérica. Acta Americana 1(1):92–107. Mexico.

Lentz, David L.

1983   Plant Remains from the Archaeological Sites of the Lower Sulaco River Drainage, Honduras. Paper presented at the 48th Annual Meeting of the Society for American Archaeology, Pittsburgh.

Lentz, David L.

1990   *Acrocomia mexicana:* Palm of the Ancient Mesoamericans. Journal of Ethnobiology 10(2):183–94.

Linares de Sapir, Olga, and Anthony J. Ranere

1971   Human Adaptation to the Tropical Forests of Western Panama. Archaeology 24(4):346–55.

Litvak King, Jaime

1975   En torno al problema de la definición de Mesoamérica. Anales de Antropología 12:171–95. Mexico.

Long, A., B. Benz, D. Donahue, A. Jull, and L. Toolin

1989   First Direct AMS Dates on Early Maize from Tehuacan, Mexico. Radiocarbon 31(3):1035–1040.

Lorenzo, José Luis

1956   Notas sobre arqueología y cambios climáticos en la Cuenca de México. *In* La Cuenca de México: Consideraciones Geológicas y Arqueológicas, edited by Fenerico Mooser, Sidney E. White, and José L. Lorenzo. Dirección de Prehistoria, Publicaciones 2. Instituto Nacional de Antropología e Historia. Mexico.

Lorenzo, José Luis, and Lauro González Q.

1970   El más antigüo teosinte. Boletín. Museo Nacional de Antropología e Historia 42:41–43. Mexico.

Ludlow-Weichers, Beatríz, Jose Luis Alvarado, and Mario Aliphat

1983   El polen de *Zea* (Maíz y teosinte): perspectivas para conocer el orígen del maíz. Biótica 8(3):235–58. Mexico.

McClung de Tapia, Emily

1975   The Origins of Agriculture: Hypotheses and Prehistoric Paleoethnobotanical Evidence from the Near East and Mesoamerica. Manuscript on file, Department of Anthropology, Brandeis University.

1985   Investigaciones arqueobotánicas en Mesoamérica y Centroamérica. Anales de Antropologia 22:133–57. Mexico.

1987   Patrones de subsistencia urbana en Teotihuacan. In Teotihuacan: Nuevos Datos, Nuevas Síntesis, Nuevos Problemas, edited by E. McClung de Tapia and E. C. Rattray, pp. 57–74. Instituto de Investigaciones Antropológicas. Universidad Nacional Autónoma de México. Mexico.

McClung de Tapia, Emily, Mari Carmen Serra Puche, and Amie Ellen Limon de Dyer

1986   Formative Lacustrine Adaptation: Botanical Remains from Terremote-Tlaltenco, D.F., Mexico. Journal of Field Archaeology 13:99–113.

MacNeish, Richard S.

1958   Preliminary Archaeological Investigations in the Sierra de Tamaulipas. Transactions of the American Philosophical Society No. 48, Part 6. Philadelphia.

1964a   The Food-gathering and Incipient Agriculture Stage of Prehistoric Middle America. In Handbook of Middle American Indians, vol. 1: Natural Environment and Early Cultures, edited by R. C. West, pp. 413–26. University of Texas Press, Austin.

1964b   Ancient Mesoamerican Civilizations. Science 143:531–37.

1967a   Introduction. In The Prehistory of the Tehuacan Valley, vol. 1: Environment and Subsistence, edited by D. Byers, pp. 3–13. University of Texas Press, Austin.

1967b   A Summary of the Subsistence. In The Prehistory of the Tehuacan Valley, vol. 1: Environment and Subsistence, edited by D. Byers, pp. 290–309. University of Texas Press, Austin.

1971   Speculation about How and Why Food Production and Village Life Developed in the Tehuacan Valley, Mexico. Archaeology 24(4):307–15.

1972   The Evolution of Community Patterns in the Tehuacan Valley of Mexico and Speculations about the Cultural Processes. In Man, Settlement and Urbanism, edited by P. Ucko, R. Tringham, and G. Dimbleby, pp. 67–93. Duckworth, London.

1981   Tehuacan's Accomplishments. In Supplement to the Handbook of Middle American Indians, vol. 1: Archaeology, edited by V. Bricker and J. Sabloff, pp. 31–47. University of Texas Press, Austin.

MacNeish, Richard S., and Frederick A. Peterson

1962   The Santa Marta Rockshelter, Ocozocuautla, Chiapas, Mexico. Papers of the New World Archaeological Foundation, No. 14. Publication No. 10. Provo, Utah.

Mangelsdorf, Paul C.

1947   The Origin and Evolution of Maize. Advances in Genetics 1:161–207.

1974   Corn: Its Origin, Evolution and Improvement. Harvard University Press, Cambridge.

1986   The Origin of Corn. Scientific American 22(2):72–79.

Mangelsdorf, Paul C., Richard S. MacNeish, and Walton C. Galinat

1967a   Prehistoric "Maize", Teosinte and *Tripsacum* from Tamaulipas, Mexico. Botanical Museum Leaflets, Harvard University 22(2):33–63. Cambridge.

1967b   Prehistoric Wild and Cultivated Maize. In The Prehistory of the Tehuacan Valley, vol. 1: Environment and Subsistence, edited by D. Byers, pp. 178–200. University of Texas Press, Austin.

Mangelsdorf, Paul C., Richard S. MacNeish, and Gordon R. Willey

1964   Origins of Agriculture in Middle America. In Handbook of Middle American Indians, vol. 1: Natural Environment and Early Cultures, edited by R. C. West, pp. 427–45. University of Texas Press, Austin.

Mangelsdorf, Paul C., and R. G. Reeves

1938   The Origin of Maize. Proceedings of the National Academy of Sciences 24:303–312.

Meyers, J. Thomas.

1971   The Origins of Agriculture: An Evaluation of Three Hypotheses. In Prehistoric Agriculture, edited by S. Struever, pp. 101–121. Natural History Press, Garden City, New Jersey.

Niederberger, Christine

1976   Zohapilco: Cinco Milenios de Ocupación Humana en un Sitio Lacustre de la Cuenca de México. Instituto Nacional de Antropología e Historia. Colección Científica 30. Mexico.

1979   Early Sedentary Economy in the Basin of Mexico. Science 203:131–42.

1986   Inicios de la vida aldeana en la América Media. Historia de México 1:93–120. Salvat, Mexico.

1987   Paleopaysages et Archéologie pre-Urbaine du Bassin de Mexico. Vols. I–II. Centre d'Etudes Mexicaines et Centramericaines, Mexico.

Palerm, Angel, and Eric Wolf

1972   Potencial Ecológico y desarrollo cultural de Mesoamérica. In Agricultura y Civilizacion en Mesoamérica, edited by A. Palerm and E. Wolf, pp. 149–205. SepSetentas 32, Secretaría de Educación Pública, Mexico.

Pearsall, Deborah M.

1982   Phytolith Analysis: Applications of a New Paleoethnobotanical Technique in Archaeology. American Anthropologist 84(4):862–71.

Pickersgill, Barbara, and Charles B. Heiser

1977   Origins and Distribution of Plants Domesticated in the New World Tropics. In Origins of Agriculture, edited by C. Reed, pp. 803–835. Mouton Publishers, The Hague.

Piperno, Dolores R.

1984   A Comparison and Differentiation of Phytoliths from Maize and Wild Grasses: Use of Morphological Criteria. American Antiquity 49(2):361–83.

1985a   Phytolithic Analysis of Geological Sediments from Panama. Antiquity 59(225):13–19.

1985b   Phytolith Taphonomy and Distributions in Archaeological Sediments from Panama. Journal of Archaeological Science 12(4):247–67.

Price, Barbara J.

1976   A Chronological Framework for Cultural Development in Mesoamerica. In The Valley of Mexico: Studies in Pre-Hispanic Ecology and History, edited by E. Wolf, pp. 13–27. University of New Mexico Press, Albuquerque.

Reed, Charles A. (editor)

1977   Origins of Agriculture. Mouton Publishers, The Hague.

Reyna Robles, Rosa María, and Lauro González Quintero

1978   Resultados de análisis botánico de formaciónes troncocónicas en "Loma Torremote", Cuauti-

tlan, Edo. de México. In Arqueobotánica (Métodos y Aplicaciones), coordinated by Fernando Sánchez, pp. 33–41. Colección Científica 63. Instituto Nacional de Antropología e Historia, Mexico.

Reynolds, Robert G.

1986   An Adaptive Computer Model for the Evolution of Plant Collecting and Early Agriculture in the Eastern Valley of Oaxaca. In Guilá Naquitz: Archaic Foraging and Early Agriculture in Oaxaca, Mexico, edited by K. Flannery, pp. 439–500. Academic Press, New York.

Rice, Don S.

1976   Middle Preclassic Maya Settlement in the Central Maya Lowlands. Journal of Field Archaeology 3:425–45.

Rindos, David

1984   The Origins of Agriculture: An Evolutionary Perspective. Academic Press, New York.

Rovner, Irwin

1984   The Cradle of Maize, Phase 1: Reconnaissance and Assessment of Preceramic Sites in the Rio Balsas Region of West Mexico. Research proposal submitted to the National Science Foundation.

Sanders, William T.

1965   The Cultural Ecology of the Teotihuacan Valley. Report on file, Department of Sociology and Anthropology, Pennsylvania State University.

Sanders, William T., and Barbara J. Price

1968   Mesoamerica: The Evolution of a Civilization. Random House, New York.

Sanders, William T., Jeffrey R. Parsons, and Robert S. Santley

1979   The Basin of Mexico: Ecological Processes in the Evolution of a Civilization. Academic Press, New York.

Sauer, Jonathan D.

1969   Identity of Archaeologic Grain Amaranths from the Valley of Tehuacan, Puebla, Mexico. American Antiquity 34(1):80–81.

Sears, Paul

1951   Pollen Profiles and Culture Horizons in the Basin of Mexico. Civilizations of Ancient America. XXIX International Congress of Americanists. University of Chicago Press, Chicago.

1982   Fossil Maize Pollen in Mexico. Science 216:932–34.

Sheets, Payson D
1982   Prehistoric Agricultural Systems in El Salvador. *In* Maya Subsistence: Studies in Memory of Dennis E. Puleston, edited by K. Flannery, pp. 99–118. Academic Press, New York.

Smith, Bruce D.
1984   *Chenopodium* as a Prehistoric Domesticate in Eastern North America: Evidence from Russell Cave, Alabama. Science 226:165–67.

Smith, C. Earle, Jr.
1966   Archaeological Evidence for Selection in Avocado. Economic Botany 20:169–75.

1967   Plant Remains. In The Prehistory of the Tehuacan Valley, vol. 1: Environment and Subsistence, edited by D. Byers, pp. 220–55. University of Texas Press, Austin.

1969   Additional Notes on Pre-Conquest Avocados in Mexico. Economic Botany 23:135–40.

1980   Plant Remains from the Chiríqui Sites and Ancient Vegetational Patterns. *In* Adaptive Radiations in Prehistoric Panama, edited by O. Linares and A. Ranere, pp. 151–74. Peabody Museum of Archaeology and Ethnology, Harvard University, Cambridge.

1986   Preceramic Plant Remains from Guilá Naquitz. *In* Guilá Naquitz: Archaic Foraging and Early Agriculture in Oaxaca, Mexico. Academic Press, New York.

1987   Current Archaeological Evidence for the Beginning of American Agriculture. *In* Studies in the Neolithic and Urban Revolutions, The V. Gordon Childe Colloquium, edited by L. Manzanilla, pp. 81–101. BAR International Series 349. British Archaeological Reports, Oxford.

Stark, Barbara L.
1981   The Rise of Sedentary Life. *In* Supplement to the Handbook of Middle American Indians, vol. 1: Archaeology, edited by V. Bricker and J. Sabloff, pp. 345–72. University of Texas Press, Austin.

Struever, Stuart (editor)
1971   Prehistoric Agriculture. Natural History Press, Garden City, New Jersey.

Tolstoy, Paul
1978   Western Mesoamerica before A.D. 900. *In* Chronologies in New World Archaeology, edited by R. E. Taylor and C. W. Meighan, pp. 241–84. Academic Press, New York.

Ucko, P., and G. W. Dimbleby (editors)
1969   The Domestication and Exploitation of Plants and Animals. Duckworth, London.

Ucko, P., R. Tringham, and G. W. Dimbleby (editors)
1972   Man, Settlement, and Urbanism. Duckworth, London.

Wagner, Philip L.
1964   Natural Vegetation of Middle America. *In* Handbook of Middle American Indians, vol. 1: Natural Environment and Early Cultures, edited by R. C. West, pp. 216–63. University of Texas Press, Austin.

West, Robert C.
1964   The Natural Regions of Middle America. *In* Handbook of Middle American Indians, vol. 1: Natural Environment and Early Cultures, edited by R. C. West, pp. 363–83. University of Texas Press, Austin.

Whitaker, Thomas W., and Hugh C. Cutler
1971   Pre-Historic Cucurbits from the Valley of Oaxaca. Economic Botany 25(2):123–27.

1986   Cucurbits from Preceramic Levels at Guilá Naquitz. *In* Guilá Naquitz: Archaic Foraging and Early Agriculture in Oaxaca, Mexico, edited by K. Flannery, pp. 275–79. Academic Press, New York.

Whitaker, Thomas H., Hugh C. Cutler, and Richard S. MacNeish
1957   Cucurbit Materials from 3 Caves near Ocampo, Tamaulipas. American Antiquity 22(4):351–58.

Willey, Gordon R.
1962   Mesoamerica. *In* Courses toward Urban Life: Archaeological Considerations of Some Cultural Alternates, edited by R. J. Braidwood and G. R. Willey, pp. 84–101. Viking Foundation Publications in Anthropology 32, New York.

Wilkes, H. Garrison.
1972   Maize and its Wild Relatives. Science 177:1071–1077.

Wilson, Hugh D
1980   Artificial Hybridization among Species of *Chenopodium* sect. *Chenopodium*. Systematic Botany 5(3):253–63.

Wiseman, Frederick M.

1978   Agricultural and Historical Ecology of the Maya Lowlands. *In* Pre-Hispanic Maya Agriculture, edited by P. D. Harrison and B. L. Turner, pp. 63–115. University of New Mexico Press, Albuquerque.

Zurita Noguera, Judith.

1987   Análisis de Fitolitos de Muestras de Suelos del Sitio Arqueológico de Guachimontón, Teuchitlán, Jalisco. Undergraduate thesis. Escuela Nacional de Antropología e Historia, Mexico.

9

The Origins of Plant Cultivation in South America

At the time of European contact, a number of well-developed agricultural systems existed in South America, adapted to the diverse ecological habitats of that continent. Among these were low-altitude systems based on the cultivation of root crops, such as manioc (*Manihot esculenta*), maize (*Zea mays*), and *Canavalia* beans; mid-altitude Andean systems dominated by maize and legumes, such as the peanut (*Arachis hypogaea*) and *Phaseolus* beans; and high-altitude systems dominated by the potato (*Solanum tuberosum*), a diversity of minor tubers, and pseudocereals, such as *quinoa* (*Chenopodium quinoa*). The development of these systems and the origins of the crops that constituted them are areas of current research interest in South American archaeology.

In an earlier article (Pearsall 1978a), I reviewed the state of paleoethnobotanical research in western South America. That paper summarizes identifications of archaeological plant materials published since Margaret Towle's 1961 work, *The Ethnobotany of Pre-Columbian Peru,* and emphasizes data from hunting-and-gathering and early agricultural sites. The present paper updates that article for the period covering 1978 to 1985 and focuses on cultivated plant remains. I review our knowledge of the origins of plant cultivation in South America by discussing the paleoethnobotanical data available for the major crops domesticated in South America, by tracing the history of the introduction of maize into the continent, and by presenting models for the evolution of the agricultural systems observed at contact.

N. I. Vavilov (1951) included the South American continent as one of the eight centers of plant domestication in the world. His South American center had three parts: the central Andes, the southern Andes, and southern Brazil. Harlan (1975) has suggested that South America be considered a "noncenter" of plant domestication, i.e., a large geographic area in which different plants were domesticated in various areas. Table 9.1 summarizes the major crops of the South American region. An already complex situation is further complicated by the fact that many crops were moved out of their probable areas of origin early in prehistory. Abundant plant remains are recovered only from arid areas, while archaeobotanical material from most of the continent is limited to charred remains. This not only limits the quantity of material available for study, but also biases the archaeological record toward certain geographic areas (desert west coast) and certain crops (seed crops).

DEBORAH M. PEARSALL

Table 9.1. Major Crops of the South American Region

Latin Name	Common Name
Pseudocereals	
Amaranthus caudatus	amaranth
Chenopodium quinoa	quinoa
Pulses	
Arachis hypogaea	peanut
Canavalia plagiosperma	jack bean
Lupinus mutabilis	tarwi
Phaseolus lunatus	lima bean
P. vulgaris	common bean
Vegetables and Spices	
Capsicum baccatus	aji
C. chinense	
C. frutescens	
Cucurbita ficifolia	squash
Cucurbita maxima	
Cucurbita moschata	
Fiber and Utility	
Bixa oxellana	achiote
Gossypium barbadense	cotton
Lagenaria siceraria	gourd
Fruits	
Ananas comosus	pineapple
Annona cherimolia	cherimoya
Bunchosia armeniaca	
Carica candicans	papaya
Inga Feuillei	pacae
Persea americana	avocado
Psidiumgua java	guava
Roots and Tubers	
Canna edulis	achira
Ipomoea batatas	sweet potato
Lepidium meyenii	maca
Manihot esculenta	yuca, manioc
Oxalis tuberosa	oca
Pachyrrhizus tuberosus	jicama
Solanum tuberosum	potato
Tropaeolum tuberosum	mashua
Ullucus tuberosus	ullucu
Drugs and Fatigue Plants	
Erythroxylon coca	coca
E. novogranatense	
Ilex paraguariensis	mate
Nicotiana rustica	tobacco
N. tabacum	

Geographic Area

The South American continent is an area of great diversity of climate, topography, and natural resources. For my purpose here, however, the continent can be divided into five major zones (Fig. 9.1): the western coastal plain, the Andean mountain chain, the Amazon and Orinoco lowlands, the old uplands fringing the lowland forest to the north and south (Guyana and Brazilian highlands), and the *pampas* and *llanos* of the south (Gran Chaco, La Plata Basin) (Willey 1971). Most of the archaeological sites to be discussed are located on the western coast and in the Andean mountains (Fig. 9.2). This focus on western South America is an artifact of the history of archaeological research in the continent, rather than a statement of the preeminence of this region as a source of cultivated plants.

The western coast of South America is an extremely arid desert for much of its length. Desert conditions prevail from northern Peru to central Chile. Along the Peruvian coast, where many sites yielding desiccated botanical remains are located, the narrow coastal plain is cut by forty or so seasonally flowing rivers. These river valleys were the focus of human occupation on the coast for much of prehistory (Moseley 1983). Some early sites were also located in fog oases, or *lomas* areas, where seasonal vegetation is maintained by dense clouds blanketing portions of the coast and abutting the dry Andean foothills or coastal hills (Benfer 1982).

In Ecuador, the coastal plain widens and becomes more complex in topography and climate (Feldman and Moseley 1983). Rainfall increases from south to north, and desert conditions grade gradually into forest until the dense tropical forests of coastal northern Ecuador and Colombia are reached. Rainfall also increases west to east, or from the coast inland. The western Andean slopes in this region are moist and forest covered, as are broad river valleys in the western coastal plain, such as the Daule and Babahoya in Ecuador.

The inter-Andean area is a mosaic of diverse climatic and vegetation zones (Willey 1971). The Andes are actually made up of several mountain chains, interwoven to form a complex topography of basins, deeply incised valleys, and high rolling grasslands (*paramo* or *puna*). The Colombian Andes are characterized by two deep parallel troughs, in which the Cauca and Magdalena rivers flow from south to north. The Ecuadorian Andes, by contrast, consist of ten or so basins, bounded on the east and west by high mountain peaks and separated by lower altitude spurs. Southward into Peru, Bolivia, and northern Chile and Argentina, topographic diversity and elevation increase. Deeply incised valleys, most of which eventually flow east into the Amazon, are separated by expanses of *puna* grassland and high peaks. High-altitude lakes, such as

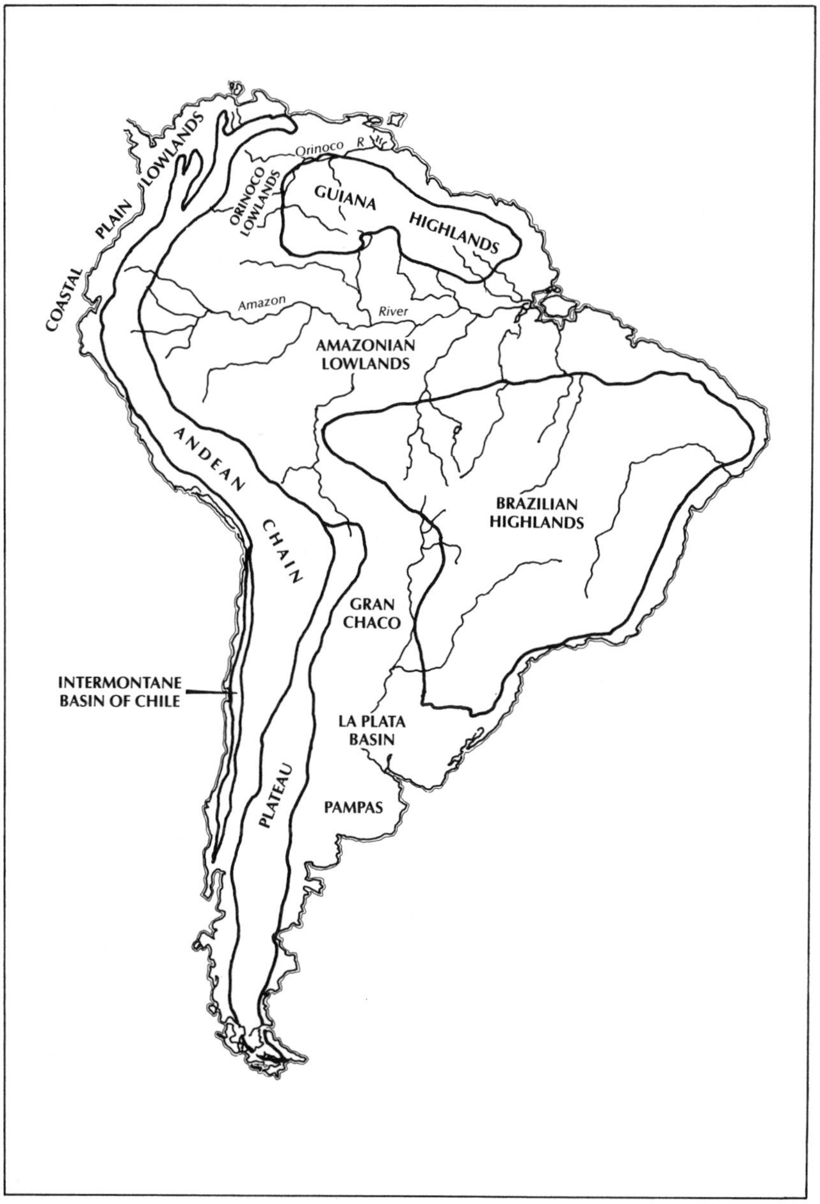

Fig. 9.1. Major physiographic regions of South America.

Lake Titicaca and Lake Junin, are common in the central Andes.

To the east of the Andes lies the Amazon Basin and its adjacent highlands, the Guiana and Brazilian uplands (Willey 1971). The eastern Andean slopes, or *montaña*, are characterized by high rainfall and dense vegetation cover. In general, these conditions prevail throughout the Amazon and Orinoco river basins, but all areas experience a dry season, and forest cover is thinner on the older soils of the Guiana and Brazilian uplands. The two eastward flowing rivers, the Amazon and its many tributaries and the Orinoco,

are broad and meandering, supporting a diversity of fish, reptile, and mammalian life. Archaeology in this region is very difficult due to the dense vegetation cover and the huge quantities of silt dropped by the meandering rivers (Lathrap 1970; Meggers and Evans 1983). Sites can be buried by tens of meters of alluvium. South of the Amazon, between the Brazilian highlands and the Andes, is another lowland area, the Gran Chaco and the La Plata Basin. The Gran Chaco is a seasonally inundated grassland or *llanos*. Massive earthworks attest to the occupation of this zone prehistorically, but little is known of the cultures occu-

Fig. 9.2. Location of South American sites with early plant remains.

pying this area (Willey 1971). South of the La Plata Basin lie the dry *pampas* and plateaus of Argentina.

History of Archaeological and Botanical Investigations

Archaeological research focused on the investigation of agricultural beginnings is a relatively recent development in South American archaeology. Although numerous plant remains were recovered during ex-

cavations on the coast of Peru carried out during the 1940s and 1950s (Towle 1961), those investigations were not focused on subsistence but on basic issues of chronology, settlement, and ceramic typology. These early excavations were valuable in demonstrating the remarkable preservation of botanical materials on the coast and for establishing that cultivated plants were present during the Late Preceramic period (ca. 2000–1500 B.C.). Abundant remains of cotton (*Gossypium barbadense*) gave this period its name: the Cotton Preceramic. Pickersgill (1969), in

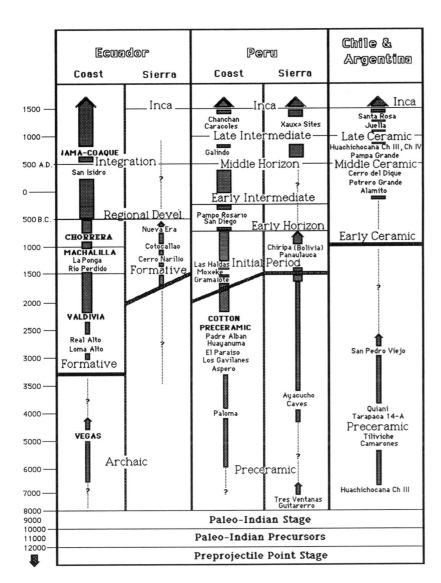

Fig. 9.3. Summary of cultural chronology for Ecuador, Peru, and the southern Andes (Chile and Argentina). Bold type, phrase names; small letters, site names. Width of vertical lines is roughly proportional to the amount of cultivated plants in diet. (Chronology after Lynch 1983; Moseley 1983. Composed by E. White.) [*See p. 180, infra.*]

reconsidering many of these early data, argues convincingly that most, if not all, of the cultivated plants recovered from coastal Peruvian sites were not domesticated there but were introduced from the east. The Andean *montaña* and eastern Bolivia were suggested as source areas for a number of crops, including common and lima beans (*Phaseolus vulgaris, P. lunatus*), chile peppers (*Capsicum* spp.), and *achira* (*Canna edulis*). Northwestern Argentina was considered a possible source area for peanut.

It became clear that excavations outside the western coastal area would be necessary to investigate the early history of crops recovered in a fully domesticated form on the coast. There has, however, been little work done in the areas suggested by Pickersgill in 1969 as likely source areas of native South Amer-

ican domesticated plants. Few of the sites excavated and reported from Argentina are early enough to be relevant to this question. The number of archaeological projects carried out in eastern Bolivia and the eastern Andean slopes of Peru, Ecuador, and Colombia are few. This situation may be changing, however, at least in Peru, where interest in economy and subsistence in the *montaña* zone is increasing (i.e., Hastings 1981).

Most South American research focused on subsistence, and incorporating techniques such as fine sieving or water flotation to recover botanical materials, has been carried out since the late 1960s. Areas studied include coastal Ecuador, coastal and highland Peru, and to a limited extent Colombia. Botanical data are also available from sites in Chile and Argen-

Table 9.2. Archaeological Plant Remains from the Peruvian Coast. (Adapted from Pearsall 1978)

Time Period	Region Site	Cultivated Plant	Comments
8000–6000 B.C.	**Central Coast** Chilca Caves	jicama, potato, ullucu, manioc, sweet potato	Engel 1973
5800 B.C.	**Central Coast**	*Phaseolus lunatus*	Heiser 1965
6000–4000 B.C.	**Far North Coast** Siches Complex	bottle gourd	Richardson 1972
4200–2500 B.C.	**Central Coast** Chilca Caves	gourd, *Phaseolus lunatus*, cotton (?)	Lanning 1967; Engel 1963, 1973; dated 3800–2650 B.C.
	Pampa	*Cucurbita moschata, C. ficifolia, C. andreana*	Lanning 1967
	Yacht Club	cotton, gourd, aji, guava	Lanning 1967; Moseley 1975
	Encanto	gourd, squash	Pickersgill 1969; Moseley 1975
	South Coast Cabeza Larga	cotton, *Phaseolus lunatus, P. vulgaris* (?)	Pickersgill 1969; Lanning 1967
2500–1800 B.C.	**North Coast** General pattern	*Cucurbita ficifolia, C. moschata, Phaseolus lunatus, Canavalia plagiosperma, Capsicum baccatum, Canna*, lucuma, ciruela de fraile, cotton, gourd	Pickersgill 1969; Lanning 1967; Sauer and Kaplan 1969
	North Central General pattern	*Cucurbita ficifolia, C. moschata, Phaseolus lunatus, Canavalia plagiosperma, Capsicum baccatum, Canna*, lucuma, ciruela de fraile, cotton, gourd, pacae, avocado, tobacco, guava, *Zea mays*	Lanning 1967; Kelley and Bonavia 1963; Moseley and Willey 1973
	Central Coast General pattern	cotton, *Capsicum baccatum*, gourd, guava, *Canna*, sweet potato, potato (?), *Canavalia, Cucurbita moschata, C. ficifolia, C. andreana*, beans, lucuma, *Phaseolus lunatus*, pacae, jicama	Pickersgill 1969; Lanning 1967; Patterson and Moseley 1968; Stephens and Moseley 1974; Willey 1971
	South Coast General pattern	cotton, squash, gourd, pacae	Lanning 1967; Willey 1971; Engel 1963
1800–1200 B.C.	**All Regions**	earlier occurring plants continue and spread; maize spreads throughout coast; peanut, manioc(?) added; quantities of plant remains increase	Lanning 1967
1200–800 B.C.	**All Regions**	manioc, *Capsicum chinense* appear; *Phaseolus vulgaris*, coca, llacon, pepino, potato, *Cucurbita maxima*, cherimoya, papaya, pineapple, guanabana, granadilla added	Towle 1961; Pickersgill 1969; Whitaker and Cutler 1965

Table 9.3. Archaeological Plant Remains from the Meridional Andean Sierra. (Adapted from Pearsall 1978)

Time Period	Region Site	Cultivated Plant	Comments
2500–500 B.C.	**Quebrada de Humahuaca** Huachichocana Cave	maize, *Capsicum*, *Phaseolus*, *Solanum tuberosum*	Schobinger 1974
	Quebrada de Inca Cave	gourd	Schobinger 1974
	Valliserrana Puente del Diablo	*Phaseolus*	Schobinger 1974
	Mendoza-Neuquén La Gruta I	squash	Nuñez A. 1974
500 B.C.–A.D. 0	**Valliserrana** Tafí	maize, *Phaseolus vulgaris*, quinoa, *Amaranthus caudatus*	(This complex also dated A.D. 0–300) González and Pérez 1966, 1968
300–100 B.C.	**Mendoza-Neuquén** Atuel II	guaraní maize, beans, squash, quinoa, *Zea mays* var. *minima*	This complex also dated A.D. 700–1000) Lagiglia 1968; Nuñez A. 1974
300 B.C.–A.D. 200	Los Morrillos III	maize, beans, squash	Nuñez A. 1974
A.D. 0–200	**Valliserrana** Cienega I	*Zea mays* var. *microsperma*, other maize	González and Pérez 1968
A.D. 200–400	Cienega II	*Zea mays* var. *microsperma* (perla), *Cucurbita maxima*, *Bixa orellana*, peanut	González and Pérez 1968
A.D. 350–500	Alamito II	*Zea mays* var. *microsperma*, peanut (?), bean, squash	Nuñez R. 1971
A.D. 700–900	Aguada Culture	potato, *Zea mays* var. *microsperma*, popcorn, *Capia*, *Cucurbita maxima*	González and Pérez 1968
A.D. 1000–1450	**Valliserrana** General pattern	quinoa, peanut gourd, potato, *Canna*(?), maize: Morocho, Pisingallo, Perla, Capia, Chulpi, Rosero; *Phaseolus vulgaris*, *Cucurbita pepo*, *C. maxima*	González and Pérez 1968; Cigliano 1968; Raffino 1973
	Mendoza-Neuquén Atuel I	*Zea mays* var. *amilacea*, (Capia, Guarani, Culli types), *Zea mays* var. *minima*, *Cucurbita*	Lagiglia 1968

tina but access to detailed reports of this work is often limited. If more information from locally published or unpublished reports could be added to this overview, more data would be available for these nations, as well as for Brazil, Colombia, and Venezuela. From Venezuela comes the only published report of systematically recovered botanical remains from the eastern lowlands: the Parmana project (Roosevelt 1980).

It should be clear that many gaps exist in the ar-chaeological record of plant domestication in South America. Each new archaeological project raises more questions than it answers. There is no crop for which we have a complete archaeological record of cultivation history. Many sites are still excavated without the use of flotation or other systematic botanical recovery methods. Recent work in Ecuador and Venezuela has demonstrated that information on plant utilization can be recovered from the moist lowlands, if techniques such as flotation and phytolith

Table 9.4. Archaeological Plant Remains from the Chilean Coast and Sierran Oases of the Atacama Region. (Adapted from Pearsall 1978)

Time Period	Region Site	Cultivated Plant	Comments
	San Pedro de Atacama Oasis		
1760 B.C.	Tulan Cave	maize	Popper 1977; from just above sterile
1760 B.C.–A.D. 770	Tulan Cave	maize	Popper 1977; from levels VIII and IX.
	Chilean Coast		
2500 B.C.–A.D. 0	Quiani II	cotton, quinoa, maize, gourd	Willey 1971; maize and gourd from top of midden
	Rio Loa Area		
500–200 B.C.	Vega Alta II	gourd	Pollard 1971
A.D. 100–400	Loa II	gourd, maize: Capio Chico Chileño, Polulo, Chutucuno	Pollard 1971; Mangelsdorf and Pollard 1975
A.D. 400–1450	Lasana Complex	beans, quinoa, aji, gourd, maize: Capio Chico Chileño, Polulo, Chutucuno Chico; potato	Pollard 1971; Mangelsdorf and Pollard 1975
	Chilean Coast		
360 B.C.	Conanoxa	maize	Nuñez A. 1974
A.D. 0	Pichalo III	gourd, maize, cotton	Willey 1971
500 B.C.–A.D. 800	Various Sites	maize, squash, gourd, quinoa, sweet potato, bean, algarrobo, pacae, potato, pallar, peanut, cotton	Nuñez A. 1974
A.D. 800–1000	Arica, Pica	intensive agriculture, irrigation	Nuñez A. 1974
	San Pedro de Atacama Oasis		
600–800 A.D.	Coyo Oriental	*Cucurbita maxima, Cucurbita* sp., maize, gourd	Popper 1977
1000–1300 A.D.	Catarpe 2	maize	Popper 1977

analysis are employed, yet data remain scarce from the wetter areas of the continent. Tables 9.2 through 9.9 present a summary of archaeological botanical remains recovered from sites in South America. For other overviews of agricultural origins in South America, refer to Bray 1976, Heiser 1979, Pickersgill and Heiser 1977, and Simmonds 1976.

Culture History and the History of Cultivation

The date of the first peopling of South America is an issue currently under debate. Most researchers feel there is good evidence for occupation by 12,000 B.C. (For a review of the evidence, see Lynch 1983.) Less universally accepted evidence exists for occupation much earlier, by 35,000 B.C. or before (Bryan 1983; Dillehay 1985; Guidon 1985; MacNeish 1976). By about 8000 B.C., this early Paleoindian lifeway, based on hunting now extinct Pleistocene fauna and smaller animals and on gathering wild plants, had given way to a diversity of Archaic (later Preceramic) adaptations (Fig. 9.3). In the Andean region, subsistence based on hunting camelids and deer and gathering wild plants is documented from a number of sites in the Junin *puna* of Peru (Lavallee et al. 1984; Moore 1985; Pearsall 1984, 1989a; Rick 1980, 1984; Wheeler Pires-Ferreira et al. 1976; Wing 1977). In the mid-altitude inter-Andean valleys, the first indications of plant cultivation date to 8000–7500 B.C. (Kaplan 1980; Kaplan et al. 1973; Lynch et al. 1985; Smith

Table 9.5. Archaeological Remains of Cultivated Plants from Highland Peru and Bolivia

Site	Cultivated Plant (first appearance)	Date	Comments
Guitarrero Cave	*Phaseolus lunatus* *P. vulgaris* *Oxalis tuberosa* *Capsicum chinensis* *Lucuma bifera*	8000–7500 B.C.	Revised short chronology; Complex II, *Zea mays* appears in Complex III, which may be disturbed. Kaplan 1980; Lynch 1980; Lynch et al. 1985; Smith 1980
Ayacucho Caves	*Lagenaria siceraria* *Chenopodium quinoa* *Cucurbita andina*	5800–4400 B.C.	MacNeish et al. 1980
	Phaseolus vulgaris *Lucuma bifera* *Erythroxylum* spp. *Solanum* spp. *Zea mays*	4400–3100 B.C.	Maize appears at end of phase
	Zea mays *Cucurbita* spp. *Gossypium* spp. *Capsicum* spp. *Canna edulis*	3100–1750 B.C.	
Panauluaca Cave	*Solanum* spp.	1600 B.C.	One tuber. Pearsall 1988a, 1989a; Pearsall and Moore 1985; Rick 1984
	Lepidium spp. *Chenopodium* spp. *Zea mays*	1600 B.C. to A.D. 1200 A.D. 1200 ca. A.D. 1200	Primitive cultivars One kernel
Chupacancha Cave	*Zea mays*	ca. 200–300 B.C.	One kernel. B. Bocek, pers. com. concerning date, 1985
	Zea mays *Phaseolus vulgaris*	A.D. 205–850 A.D. 205	Various maize remains One bean
Huachumachay Cave	*Lagenaria siceraria*	A.D. 390	One rind fragment
Xauxa Valley	*Zea mays* *Chenopodium* spp. *Lupinus mutabilis*	A.D. 650 to A.D. 1530	Recovered from a series of late sites. D'Altroy and Hastorf 1984; Hastorf 1983
	Phaseolus spp. *Solanum* spp. *Oxalis tuberosa* *Ullucus tuberosus* *Tropaeolum tuberosum*		
Tres Ventanas	*Solanum tuberosum* *Ullucus* spp. *Manihot* spp. *Ipomoea* spp. *Pachyrrhizus* spp.	8000 B.C.	Potato confirmed as domesticated species. Engel 1973; Ugent et al. 1982
Chiripa, Bolivia	*Amaranthus* spp. *Solanum* spp. *Chenopodium spp.*	1350 B.C.–A.D. 50	Browman 1989; Erickson 1976, 1977

Table 9.6. Archaeological Remains of Cultivated Plants from Coastal Peru: New Data

Site	Cultivated Plant	Date (first appearance)	Comments
1. Moche Valley			
Padre Alban	*Cucurbita* spp. *Lagenaria siceraria* *Gossypium barbadense*	1980–1729 B.C.	Early Cotton Preceramic. Pozorski 1983; Pozorski and Pozorski 1979
Alto Salaverry	*Phaseolus vulgaris* *P. lunatus* *Capsicum* spp. *Cucurbita* spp. *Persea americana* *Inga Feuillei* *Bunchosia aremeniaca* *Psidium guajava* *Lucuma bifera* *Lagenaria siceraria* *Gossypium barbadense*	No C-14 date	Late Cotton Preceramic
Gramalote	*Phaseolus vulgaris* *P. lunatus* *Capsicum* spp. *Cucurbita* spp. *Persea americana* *Inga Feuillei* *Bunchosia armeniaca* *Psidium guajava* *Lucuma bifera* *Lagenaria siceraria* *Gossypium barbadense* *Arachis hypogaea* *Zea mays*	1590–1100 B.C.	Initial Period
Galindo	Same as Gramalote	ca. 600–900 A.D.	*P. vulgaris* and *Zea mays* abundant. Pozorski 1976, 1982
Chanchan	Same as Gramalote & *Annona muricata*	A.D. 1000–1500	Tree fruits abundant
Caracoles	Same as Gramalote & *Annona muricata* *Cyclanthera pedata*	A.D. 1000–1500	
Cerro la Virgen	Same as Gramalote & *Annona muricata* *Cyclanthera pedata* *Ipomoea batatas*	A.D. 1000–1500,	
Choroval	Same as Cerro la Virgen	A.D. 1000–1500	
2. Casma Valley			
Huaynuma	*Canna edulis* *Solanum tuberosum* *Ipomoea batatas* *Gossypium barbadense* *Lagenaria siceraria* *Capsicum* spp. *Cucurbita ficifolia* *C. maxima* *Lucuma bifera*	2250–1775 B.C.	Preceramic. Ugent et al. 1981, 1982, 1984
Pampa de las Llamas Moxeke	*Canna edulis* *Solanum tuberosum* *Ipomoea batatas* *Gossypium barbadense* *Lagenaria siceraria* *Lucuma bifera* *Arachis hypogaea* *Persea americana.*	1785–1120 B.C.	Initial Period

Table 9.6.—*Continued next page*

Table 9.6.—*Continued*

Site	Cultivated Plant	Date (first appearance)	Comments
Pampas de las Llamas Moxeke (*continued*)	*Phaseolus lunatus* *P. vulgaris* *Cucurbita maxima* *Capsicum* spp. *Bunchosia armeniaca*		
Tortugas	*Solanum tuberosum* *Pachirrhyzus* spp. *Ipomoea batatas*	No C-14 date	Initial Period
Las Haldas	*Canna edulis* *Solanum tuberosum* *Gossypium barbadense* *Lagenaria siceraria* *Persea americana* *Phasolus vulgaris* *Arachis hypogaea* *Lucuma bifera* *Zea mays*	1040–895 B.C.	
Pampa Rosaria	*Canna edulis* *Zea mays* *Gossypium barbadense* *Lagenaria siceraria* *Arachis hypogaea* *Phaseolus vulgaris* *Persea americana* *Psidium guajava* *Bunchosia armeniaca* *Inga Feuillei* *Sapindus saponaria* *Capsicum* spp. *Cucurbita maxima* *Manihot esculenta*	810–450 B.C.	
San Diego	*Canna edulis* *Zea mays* *Phaseolus vulgaris* *Canavalia* spp. *Arachis hypogaea* *Persea americana* *Gossypium barbadense* *Lagenaria siceraria* *Capsicum* spp. *Cucurbita maxima* *Manihot esculenta* *Bunchosia armeniaca* *Inga Feuillei*	550–295 B.C.	

3. Chilca Valley

Site	Cultivated Plant	Date (first appearance)	Comments
La Paloma	*Begonia geraniifolia* *Lagenaria siceraria* *Psidium guajava* *Cucurbita ficifolia* *Phaseolus* spp.	5700–3000 B.C.	Begonia is a local domesticate. Benfer 1982, 1984; Dering and Weir 1979, 1981; Weir and Dering 1986
Chilca I	*Phaseolus* spp. *Canavalia* spp. *Lagenaria siceraria* *Cucurbita* spp. *Cyclanthera* cf. *pedata* *Canna edulis* *Begonia geraniifolia* *Pachyrrhizus* spp. *Gossypium barbadense*	3700–2400 B.C.	Cotton is from uppermost level, or intrusive. Jones 1988.

Table 9.6.—Continued next page

Table 9.6.—*Continued*

Site	Cultivated Plant	Date (first appearance)	Comments
Asia	*Cyclanthera* cf. *pedata* *Lagenaria siceraria* *Psidium guayaba* *Cucurbita* spp. *Canavalia* spp. *Phaseolus lunatus* *Gossypium barbadense* *Zea mays*		Cotton Preceramic; maize from an intrusive Chavin burial
4. Supe Valley			
AS8A	*Lagenaria siceraria* *Cucurbita* spp.	4135 B.C.	Feldman 1980
Aspero (AS1) midden	*Lagenaria siceraria* *Cucurbita* spp. *Gossypium barbadense* *Psidium guajava* *Zea mays*	2410–2000 B.C.	Maize is from area with ceramic mix
Aspero architectural units	Same as midden & *Inga* spp. *Canna edulis* *Capsicum* spp. *Phaseolus* spp.	2410–2000 B.C.	Maize is from later or mixed contexts
Aspero AS1J area	*Lagenaria siceraria* *Cucurbita* spp. *Gossypium barbadense*	2410–2000 B.C.	
Puerto de Supe AS2A	*Lagenaria siceraria* *Cucurbita* spp. *Gossypium barbadense* *Zea mays* *Canavalia* spp. *Phaseolus* spp. *Inga* spp. *Arachis hypogaea* *Lucuma bifera* *Persea americana* *Capsicum* spp. *Psidium guajava* *Bunchosia armeniaca* *Manihot esculenta*	No C-14 date	Early Horizon
5. Other Sites			
El Paraiso de Chuquitanta	*Canna edulis* *Cucurbita* spp. *C. moschata* *Gossypium barbadense* *Inga Feuillei* *Lagenaria siceraria* *Lucuma bifera* *Pachyrrhizus* spp. *Phaseolus vulgaris* *P. lunatus* *Psidium guajava* *Sapindus* spp. *Arachis hypogaea* *Zea mays*	850–1050 B.C.	Preceramic; peanuts and maize from unit with ceramic mix. Pearsall and Ojeda 1988; Quilter et al. 1991
Los Gavilanes	*Canna* spp. *Arachis hypogaea* *Inga Feuillei* *Lagenaria siceraria*	2700–2200 B.C.	Direct dates on maize (200–800 B.P.) are not accepted by Bonavia (1982: 73). Bonavia and Grobman 1979; Grobman and Bonavia 1978; Popper 1982

Table 9.6.—*Continued next page*

Table 9.6.—*Continued*

Site	Cultivated Plant	Date (first appearance)	Comments
Los Gavilanes (*cont.*)	*Persea americana* *Manihot esculenta* *Psidium guajava* *Capsicum* spp. *Phaseolus* spp. *P. lunatus* *Cucurbita moschata* *Cucurbita* spp. *Gossypium barbadense* *Annona cherimolia* *Pachyrrhizus* spp. *Ipomoea* spp. *Zea mays*		
Garagay	*Inga Feuillei* *Phaseolus vulgaris* *Vicia faba* *Lucuma bifera* *Cucurbita pepo* *C. maxima* *Zea mays* *Arachis hypogaea* *Canna edulis* *Gossypium* spp.	1400–600 B.C.	All plant material is from fill; late occupation is present (Middle Horizon, later). Ravines et al. 1984
Huaca Prieta	*Canna edulis* *Canavalia plagiosperma* *Phaseolus lunatus* *P. vulgaris* *Bunchosia armeniaca* *Gossypium barbadense* *Psidium guajava* *Lucuma bifera* *Capsicum baccatum* *C. chinense* *C. frutescens* *Cucurbita ficifolia* *C. moschata* *C. maxima* *Lagenaria siceraria*	2400–1200 B.C.	Bird et al. 1985
La Galgada	*Canna* spp. *Persea americana* *Canavalia* spp. *Inga* spp. *Phaseolus lunatus* *P. vulgaris* *Bunchosia* spp. *Sapindus saponaria* *Gossypium barbadense* *Passiflora* spp. *Psidium* spp. *Pouteria* spp. (*Lucuma*) *Capsicum* spp. *Lagenaria siceraria* *Cucurbita moschata* *C. maxima*	2662–2000 B.C.	Preceramic. Grieder 1988; Smith 1988
	Zea mays *Persea americana* *Arachis* spp. *Inga* spp. *Bunchosia* spp.	2085–1395 B.C.	Initial Period

Table 9.6.—*Continued next page*

Table 9.6.—*Continued*

Site	Cultivated Plant	Date (first appearance)	Comments
La Galgada (*cont.*)	*Gossypium barbadense* *Passiflora* spp. *Pouteria* spp. *Capsicum* spp. *Lagenaria siceraria* *Cucurbita moschata* *C. maxima*	2085–1395 B.C.	Initial Period
Cardal	*Arachis hypogaea* *Phaseolus* spp. *Inga Feuillei* *Psidium guajava* *Lucuma bifera* *Cucurbita maxima* *C. moschata* *Zea mays* *Capsicum* spp.	1100–850 B.C.	Initial Period. Umlauf 1988

1980). On the South American coastlines, evidence of occupation increases as populations begin to exploit the resources of the marine environment. Early coastal Archaic sites excavated to recover plant remains also often reveal the presence of cultigens, as at the Vegas site in Ecuador (Piperno 1981a, 1981b; Stothert 1977, 1983, 1985) and the Paloma site in Peru (Benfer 1982, 1984; Dering and Weir 1979; Jones 1988; Weir and Dering 1986). In the Late, or Cotton, Preceramic of coastal Peru, large sites with ceremonial architecture appear, and cultivated gourds, squashes, cotton, and beans become widespread (Moseley 1983).

Beginning about 3300 B.C., the ceramic-producing Valdivia culture appears on the coast of Ecuador. The ceramics are associated with village construction and ceremonial structures (Lathrap et al. 1975; Meggers et al. 1965). It has been argued that the Valdivia and later Formative cultures of Ecuador were basically agricultural in their adaptation (Lathrap et al. 1975; Lathrap et al. 1977; Pearsall 1979). The earliest sites producing ceramics in Colombia are Puerto Hormiga, dating to 3000–2200 B.C. (Feldman and Moseley 1983), and Monsu, dated to 3300 B.C. (Reichel-Dolmatoff 1985). Sites of the Saladoid ceramic tradition appear in Venezuela by 2100 B.C. (Roosevelt 1980). Pottery appears relatively late on the coast of Peru and in the highlands of Peru and Ecuador, dating to 1800–1200 B.C., depending on the area. The earliest ceramic phase in Peru is called the

Initial period. Formative stage cultures are later still in the southern Andean area, appearing around 1000 B.C. in northwestern Argentina and Chile (Kolata 1983).

The later prehistoric periods in Peru are characterized by alternating episodes of regional differentiation and consolidation. Three horizons are defined: the Early Horizon (Chavin, 900–200 B.C.), the Middle Horizon (Huari, 600–1000 A.D.), and the Late Horizon (Inca, 1476–1534 A.D.) (Moseley 1983; Willey 1971). Between these periods of more widespread political control or influence, regional chiefdoms and, later, states occur. In Ecuador and Colombia, regional development at the complex chiefdom or state level continues throughout most of prehistory, with large polities not appearing until 1000 A.D. or after (Feldman and Moseley 1983). Social and political development in the eastern tropical forest is not well understood. Basic understanding of chronology and the nature of settlement are still lacking for many areas of the Amazon.

Comparing Table 9.1, which lists the major crops of the South American region, to Tables 9.2–9.9, which list archaeological plant remains, quickly reveals the many gaps in our knowledge. Sufficient quantities of material have been recovered, however, to make it difficult to discuss in detail all the remains of cultivated plants presented in these tables. Instead, I point out several broad patterns in these data, specifically the sequence of appearance of

Table 9.7. Archaeological Remains of Cultivated Plants from the Valliserrana Region, Argentina: New Data

Site	Cultivated Plant	Date (first appearance)	Comments
Inca Cueva	*Lagenaria siceraria*	2130 B.C.	Tarrago 1980
Huachichocana ChIII	*Capsicum* spp. *Phaseolus vulgaris* *Zea mays*	7670–6720 B.C.	Level E3; *C. baccatum* or *C. chacoense*
	No information	1450–500 B.C.	Level E2; Preceramic
	Zea mays *Lagenaria siceraria* *Oxalis tuberosus* *Arachis hypogaea*	A.D. 730	Level E1
	Zea mays	A.D. 1100–1300	Level D
	Zea mays *Arachis hypogaea* *Capsicum* spp. *Lagenaria siceraria* *Canna* spp. *Curcubita* spp. *Phaseolus* spp.	A.D. 1535	Level C, Incaic
Huachichocana CHIV	*Erythroxylum* spp. *Zea mays* *Arachis hypogaea* *Canna edulis* Cucurbitaceae	ca. A.D. 730	
Puente de Diablo	Cucurbitaceae	Preceramic	Seeds
Cerro del Dique	*Zea mays* *Lagenaria siceraria*	A.D. 260	
Potrero Grande	*Zea mays* *Lagenaria siceraria*	A.D. 240	
Campo Colorado	*Zea mays*	A.D. 55	
Alamito	*Zea mays* *Arachis hypogaea* *Phaseolus* spp. *Cucurbita* spp.	A.D. 250–450	
Palo Blanco	*Zea mays*	A.D. 0–500	
Punta Colorada	*Cucurbita* spp. *Zea mays*	A.D. 720	
Pampa Grande	*Chenopodium quinoa* *Amaranthus caudatus* *Zea mays* *Phaseolus vulgaris* *P. lunatus* *Lagenaria siceraria* *Cucurbita* spp. *C. maxima*	A.D. 700	
Juella	*Cucurbita* spp. *Zea mays*	A.D. 1335	
Santa Rosa de Tastil	*Zea mays* *Canna* spp. *Phaseolus vulgaris* *Cucurbita pepo*	A.D. 1362–1439	
Morohuasi	*Zea mays* *Cucurbita pepo*	No C-14	Late Ceramic
La Gruta del Inca	*Zea mays* *Lagenaria siceraria*	No C-14	Late Ceramic

Table 9.8. Archaeological Remains of Maize from Chile: New Data

Site	Date (first appearance)	Comments
Tiliviche 1-b	5900–4110 B.C.	Rivera 1980; leaves, cobs
Camarones 14	5470–4665 B.C.	Association of maize and date not clear; dried remains
Quiani Shell Mound	4220–3680 B.C.	Dried remains
San Pedro Viejo Pichasca	2750 B.C. 425 B.C. A.D. 665	Maize associated with all dates; dried remains
Canamo-1	860 B.C.	Stratum I; dried remains
Caleta Huelen-43	450 B.C.–A.D. 820	Dried remains
Conanoxa, CXA E-6	320 B.C.	Dried remains
Conanoxa, CXA E-1	A.D. 800	Dried remains
ChiuChiu (RANL-100)	A.D. 105	Dried remains
Guatacondo	A.D. 775 A.D. 60–775	Dried remains in coprolites
Tarapaca 14-A	4880–2830 B.C.	In coprolites
Tarapaca 12	2740 B.C.	In coprolites
Tarapaca 15	ca. A.D. 1350	Pollen, coprolites
Tarapaca 13-A	ca. A.D. 1450–1550	Pollen, coprolites
Caserones, Unidad 1	ca. 300 B.C.	Pollen, coprolites
Azapa-83	A.D. 560–760	In coprolites

the native South American cultigens and the timing of the introduction of maize (*Zea mays*) to the continent.

The Earliest Cultivars

The macroremains from the Ayacucho caves, Guitarrero Cave, the Tres Ventanas site, the Paloma *lomas* site, the coastal Siches complex (all in Peru), the earliest level of the Huachichocana site (CHIII) (Argentina), the Tiliviche, Camarones, and Tarapaca sites (all in Chile), and the phytoliths recovered from the upper strata of the Vegas site (in western Ecuador) provide evidence for the cultivation of the following crops by the fifth millennium B.C.: maize (*Zea mays*), beans (*Phaseolus vulgaris, P. lunatus*), gourd (*Lagenaria siceraria*), squash (*Cucurbita ficifolia*), *aji* (*Capsicum chinense, Capsicum* sp.), quinoa (*Chenopodium quinoa*), potato (*Solanum tuberosum*), guava (*Psidium guajava*), and, perhaps, *oca* (*Oxalis tuberosa*) and *Begonia geraniifolia* (Fig. 9.4). The *oca* tubers from Guitarrero Cave could be wild, as could other tubers (*jicama, ullucu,* sweet potato, manioc) recovered from Tres Ventanas. The begonia tubers from Paloma are considered local domesticates (Weir and Dering 1986). On the basis of starch grain analysis, Ugent et al. (1982) have recently confirmed the domesticated status of the early potato remains from Tres Ventanas.

Leaving aside for the moment the issue of Preceramic maize in South America, one sees that evidence is accumulating for plant cultivation many millennia before the beginning of the Peruvian Cotton Preceramic (2000 B.C.). Most of the earlier finds (Guitarrero, Ayacucho, Tres Ventanas, Huachichocana) are sierran rather than coastal, strengthening the suggestion of an Andean or eastern *montaña* origin for a number of the crops in question (Heiser 1979; Pickersgill 1969; Pickersgill and Heiser 1977). It is important to emphasize, however, that most of these early finds of cultivated plants in South America are not without interpretive problems. Dry caves, such as Guitarrero, the caves excavated during the Ayacucho project, Tres Ventanas, and Huachichocana are very complex to excavate and interpret. Desiccated food remains attract burrowing animals and can be redeposited by them. The late prehistoric pattern of burial in caves can cause mixing of upper cave strata. Ves-

Table 9.9. Archaeological Remains of Cultivated Plants from Ecuador and Venezuela

Site	Cultivated Plant	Date (first appearance)	Comments
1. Ecuador, coast			
Vegas (OGSE 80)	*Zea mays*	6000–4500 B.C.	Phytoliths. Pearsall and Piperno 1990; Piperno 1988b; Stothert 1983, 1985, 1988
Real Alto	*Canavalia plagiosperma* *Gossypium* spp. *Zea mays*	3300–1500 B.C.	Charred beans, cotton; maize phytoliths. Damp et al. 1981; Pearsall 1978a, 1979, 1982; Pearsall and Piperno 1990
	Canna spp.	2300–1500 B.C.	Phytoliths
San Pablo	*Zea mays*	2000–1800 B.C.	Kernel fragment in vessel. Zevallos et al. 1977
Loma Alta	*Canavalia* spp. *Zea mays*	3000–2700 B.C.	Charred remains. Pearsall 1988b
San Isidro	*Zea mays* *Canavalia* spp. *Lagenaria siceraria*	1700–1500 B.C.	Late Valdivia; maize phytoliths. Pearsall n.d. Veintimilla et al. 1985
	Zea mays *Canavalia* spp.	500 B.C.–A.D. 1500	Jama-Coaque, charred remains
La Ponga	*Zea mays* *Canavalia* spp.	1200–800 B.C.	Charred remains. Lippi et al. 1984; Stemper 1980
Rio Perdido	*Zea mays*	1200–800 B.C.	Phytoliths. Pearsall 1979
Vessel looted near Chacras site	*Zea mays*	ca. 800 B.C.	Date based on vessel style; charred kernels. Pearsall 1980a
2. Ecuador, sierra			
Cerro Narrio	*Zea mays*	2000–1800 B.C.	Charred remains. Braun 1971; Collier, pers. com. confirming identification, 1978; Collier and Murra 1943
Nueva Era	*Zea mays*	670–500 B.C.	Charred remains. Isaccson, pers. com. concerning date, 1984; Pearsall 1986
Cotocollao	*Zea mays* *Phaseolus* spp.	1500–500 B.C.	Charred remains. Pearsall 1984; Peterson and Rodriquez 1977
Site 48, Quito	*Phaseolus vulgaris* *Zea mays*	150 B.C.	Charred remains. Pearsall 1977
3. Ecuador, Amazon			
Ayauch[i] Lake core	*Zea mays*	3300 B.C.	Pollen, phytoliths. Bush et al. 1989
4. Venezuela			
Parmana area, various sites	*Zea mays*	800 B.C.–A.D. 400	Charred remains. Roosevelt 1980
	Zea mays *Canavalia ensiformis* *Manihot esculenta*	A.D. 400–1500	Charred remains
El Tiestal	*Zea mays*	200 B.C.	Mangelsdorf and Sanoja 1965

cellius (1981a, 1981b) has discussed in detail some problems of this nature that he saw in the Guitarrero Cave excavations. In this instance, however, recent direct dating by accelerator mass spectrometry of wooden artifacts and cord associated with the cultivars provides additional support for the early dates (Lynch et al. 1985). Direct dating of actual cultivated material, after thorough study and reporting, would resolve a number of similar controversies, especially for early maize (the maize controversy is discussed below; see also Bird 1984). The early evidence for cultivars is further weakened by incomplete study of botanical specimens and by data reporting that often does not detail the specific associations of plant material and imperishable artifacts, features, and stratigraphy. Guitarrero is the notable exception. Having to rely on unpublished or incompletely published data or on secondary sources of information is a recurring problem.

It is interesting to note that the beans, chili peppers, and potato remains reported from Guitarrero or Tres Ventanas are all considered fully domesticated (Smith 1980; Kaplan 1980, 1981; Ugent et al. 1982). The early history of the manipulation of these plants by humans thus may predate the seventh or eighth millennium B.C.

Cultivated Plants of the Later Preceramic in Peru

During the period from about 4000 B.C. to the end of the Peruvian Preceramic (1800–1200 B.C.), there is increasingly abundant evidence for cultivated plants. Plants recovered from sites spanning the earlier Preceramic periods (8000–4000 B.C.) occur in the following millennia as well (Fig. 9.4). *Phaseolus* beans, gourd, squashes, chili peppers, and *guava* become common on the coast. Maize remains, although not widespread, are reported from the Ayacucho caves in the sierra and from a number of coastal sites. Feldman (1980), however, believes that all of the maize occurring in Preceramic levels at the Aspero site (3055–2533 B.C.) is intrusive, and the maize from El Paraiso de Chuquitanta occurs only in a small portion of the site reoccupied in later times (Quilter et al. 1991). *Quinoa* and, possibly, potato are reported from the Chihua phase (4400–3100 B.C.) at Ayacucho. Recent work by Donald Ugent and Sheila and Tom Pozorski (1981, 1982, 1984) in the Casma valley documents additional finds of confirmed domesticated potato in the Late Preceramic (2250–1775 B.C.) and after. A possible charred *Solanum* sp. tuber was recovered from the terminal Preceramic level at Panaulauca Cave.

A variety of new crops appear in the interval from about 4000 B.C. to the end of the Peruvian Preceramic (Fig. 9.5). These include: cotton (*Gossypium barbadense*), avocado (*Persea americana*), *Inga Feuillei*, *Bunchosia armeniaca*, *Lucuma bifera*, achira (*Canna edulis*), sweet potato (*Ipomoea batatas*), jicama (*Pachyrrhizus* sp.), *Sapindus* sp., *Canavalia plagiosperma*, peanut (*Arachis hypogoea*), manioc (*Manihot esculenta*), coca (*Erythroxylon* sp.), and *cherimoya* (*Annona cherimolia*).

Peanut is noted in the Preceramic only from Los Gavilanes. Peanuts occur at El Paraiso only in a limited area of Ceramic period reoccupation at that predominantly Preceramic site.

Most of the crops preserved in the archaeological record of the Peruvian coastal and Andean later Preceramic are tree fruits, legumes, tubers, or "industrial" plants. It is an interesting phenomenon of agricultural development in South America that no indigenous grain became a major cultivar. Sites on the Junin *puna* document use of large-grained grasses, such as *Festuca* sp., throughout the Preceramic (Pearsall 1980, 1988a, 1989a). Grass seeds of various types are reported for later sites (Hastorf 1983), but if these seeds were used as food, there is no evidence for their cultivation. Harlan (1975) lists *Bromus mango* as a high-elevation cultivated grass, but this minor Chilean crop has no archaeological record. Jones (1988) reports high frequencies of grass seeds in some coprolites from the Paloma and Chilca I sites on the coast of Peru. Pseudocereals, such as *Chenopodium quinoa*, filled this niche to some extent, especially at mid- and high altitudes. The main carbohydrate source remained root crops, however, until maize spread into the mid- and low-altitude zones.

Plants of the Ecuadorian Formative

The Early Formative, or first ceramic phase, on the coast of Ecuador begins around 3300 B.C. (Damp 1979, 1984). Given the diversity of crops known from Peru at this time period and after, the inventory of crops from coastal Ecuador appears meager indeed (refer to Table 9.9). This difference is due to the dramatically different contexts of preservation operating in these two areas. No dry caves have been excavated in Ecuador, nor is the coast a desert. Coastal Guayas has a very desolate aspect today, but this is largely due to almost total deforestation. Tropical tree taxa have been identified from the Real Alto site (Pearsall 1979, 1983), and seasonal rains are regular, if not yearly, phenomena in the area. Plant remains are preserved only by virtue of being charred, and once charred, must be sturdy enough to hold up over the millennia. Plants eaten raw (many tree fruits), used for industrial purposes, or fragile in the charred state (most tubers) are likely to be underrepresented or absent.

It is clear that jack bean (*Canavalia plagiosperma*) is an early cultivated plant in Ecuador. It occurs earlier there than in Peru, which suggests a possible north-

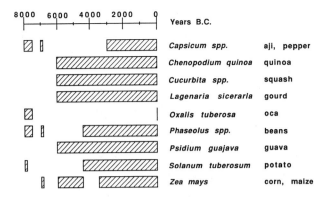

Years B.C.

Capsicum spp.	aji, pepper
Chenopodium quinoa	quinoa
Cucurbita spp.	squash
Lagenaria sicerarla	gourd
Oxalis tuberosa	oca
Phaseolus spp.	beans
Psidium guajava	guava
Solanum tuberosum	potato
Zea mays	corn, maize

Fig. 9.4. Early cultivars in South America.

ern origin for this crop (Damp et al. 1981). Charred remains of maize are few during the Valdivia period (3300–1500 B.C.), but are somewhat more common in the later Formative (1500–500 B.C.), both on the coast and in the sierra. Additional evidence for Formative period maize comes from the Real Alto site, where maize phytoliths were identified in strata spanning the entire Valdivia period. Common bean is present at least by 150 B.C. Phytoliths identified as Cannaceae, possibly *achira* (*Canna edulis*), were identified at Real Alto in strata dating to 2300 B.C. and after.

The Issue of Maize

What is our current understanding of the timing of maize introduction into South America? There are only a few maize remains that date prior to 4000 B.C. (e.g., Vegas phytolith data, Huachichocana dried remains, Chilean finds). Data for the subsequent 2,500

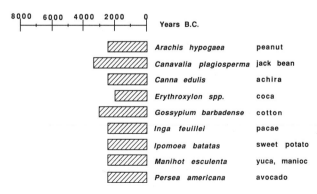

8000 6000 4000 2000   0

Years B.C.

Arachis hypogaea	peanut
Canavalia plagiosperma	jack bean
Canna edulis	achira
Erythroxylon spp.	coca
Gossypium barbadense	cotton
Inga feuillei	pacae
Ipomoea batatas	sweet potato
Manihot esculenta	yuca, manioc
Persea americana	avocado

Fig. 9.5. Later cultivars in South America.

years (4000–1500 B.C.), including the early Ecuadorian Formative and the middle to late Peruvian Preceramic, are also sparse: phytoliths from Real Alto, charred remains from Cerro Narrio, San Isidro, and San Pablo, and pollen and phytoliths from a core in Lake Ayuach[i], all in Ecuador; and dried remains from Ayacucho, Los Gavilanes, Culebras I, Rio Seco, and Los Cerrillos, all in Peru (Tables 9.5–9.9; also see Wilson 1981). During the Initial period in Peru (beginning between 1800 and 1200 B.C. and, depending on the area, ending around 900 B.C.), maize remains are still uncommon. Initial period sites with maize reported include Garagay, Gramalote, and Las Haldas, all located on the coast. Maize remains become fairly ubiquitous on the coast during the Early Horizon (ca. 900–200 B.C.). Charred maize remains are recovered from the Parmana site in Venezuela at 800 B.C. and from a number of sites in Ecuador between 1200 and 500 B.C.

I suggest that maize was introduced into South America before 5000 B.C. I have argued elsewhere (Pearsall 1988b) that the variability of charred maize remains recovered from Preceramic and early Formative contexts in Ecuador is difficult to explain if a late introduction of maize into South America is envisioned. The presence of maize at the Vegas and Real Alto sites has been confirmed by a reanalysis of phytolith data using Piperno's (1984) three-dimensional cross classification system and employing discriminant analysis to segregate "wild" and "maize" cross assemblages (Pearsall and Piperno 1990).

There is debate, however, about the validity of the earliest maize finds. For example, Bird, who has examined most maize macroremains from Peru and Ecuador, proposes that a multi-row small-grained maize was introduced to South America about 3000 B.C. and evolved into the races present by 1000 B.C. (Bird 1984). He believes that a number of the earlier maize finds, such as those discussed by Rivera (1980) from Chile, are morphologically similar to local modern races or types known from later prehistory. As was discussed earlier, a number of these early finds are also plagued by incomplete study or publication.

Additionally, phytolith evidence for maize is not accepted by all. It is beyond the scope of this paper to describe in detail the phytolith maize identification technique and the strong evidence for its validity (refer to Pearsall 1978, 1979, 1982, 1989b; Pearsall and Piperno 1990; Piperno 1983, 1984, 1988). It is no longer possible to ignore the findings of phytolith analysis nor to dismiss the technique as of no more strength than indirect data, such as kernel impressions, artistic representations, or the presence of metates (i.e., Lippi et al. 1984). Phytoliths, like pollen grains, are microbotanical remains. Critics of phytolith analysis (i.e., Dunn 1983; Roosevelt 1984) tend to overstate the problems and minimize the utility of the approach. Legitimate questions concerning the integrity of strata sampled for phytolith analysis can be addressed by examining the archaeological data for the sites in question. In the case of Real Alto, for example, there is no evidence whatsoever for contamination of archaeological strata by modern deposition. Real Alto was abandoned by 1500 B.C. and no later prehistoric or modern settlement reoccupied the site. All strata sampled for phytolith analysis are clearly Formative (refer to Lathrap et al. 1977; Marcos 1978; Pearsall 1979, for data on this site). At the Vegas site, OGSE 80, phytoliths identifiable as maize occur only in the upper part of the sequence (6000–4500 B.C.), rather than from the beginning (8800 B.C.). If deposits at this site were mixed, more random occurrences might be expected.

If we accept the phytolith evidence for an early (at least 5000 B.C.) introduction of maize into South America, it appears that the crop was not widespread, nor was it likely to have made an important contribution to the diet. In fact, it is likely that maize continued to be a relatively minor component of subsistence throughout the Preceramic and the Early Formative, as compared to root crop cultivation, use of other cultivated plants, and/or reliance on terrestrial hunting and gathering or marine/riverine fishing. If we are guided by the coastal Peruvian data, which are less subject to preservation bias than the other available data, maize does not become ubiquitous until 1500–1000 B.C. or slightly later. For example, paleoethnobotanical work by Sheila and Thomas Pozorski in the Casma and Moche valleys revealed that none of the three Late Preceramic sites had maize, and only two of four Initial period sites had it (refer to Table 9.6). Although ubiquity is only a rough estimator of the overall importance of a cultigen, these data suggest that maize assumed the preeminent role in subsistence observed at contact only late in the scheme of plant cultivation in South America. There is little support for Wilson's (1981) contention that maize played an important role in subsistence on the Peruvian coast during the Late Preceramic and Formative.

Patterns Later in Prehistory

In coastal Peru, the diversity and quantity of crop plants present in the archaeological record increase throughout the Initial period and Early Horizon (1500–200 B.C.). A number of new taxa, mostly fruits, are added to the assemblage. Manioc and peanuts also become more widespread, and new species of *aji* and squash appear. As discussed above, maize remains become more common.

In the high-altitude Lake Junin area of Peru, incipient cultivation of a local root crop, *maca* (*Lepidium meyenii*), is documented for the period between 1700 B.C. and A.D. 1200 (Pearsall 1989a). This crop never spread far from the central Peruvian Andes but formed a component of the *puna* system of llama herding and root crop cultivation.

Discussion

Although basic data are still lacking to do much more than suggest a rough chronology for the origins of plant cultivation in South America, some insight can be gained into the development of the agricultural systems observed at the time of European contact. To address this issue, I present a brief summary of the available data for each system, and then describe what I believe to be the more interesting and potentially useful models for their evolution.

Table 9.10 presents a summary of available data on the probable areas of origin of the major South American cultivated plants. The arrangement of taxa into complexes is based on Harlan (1975). The lowland complex is composed of tropical forest cultivars, which can be grown from sea level to about 1500 m elevation (somewhat higher near the equator). The mid-elevation complex characterizes the *kichwa* zone (1500–2500 m), a zone of mild temperatures and the primary grain-producing zone (Brush 1977). The transition between maize and root crop cultivation occurs between 2500 and 3000 m in the *templado* zone, a transitional zone between the warmer lower valley and the cooler upper valley. Most cultivation of crops of the high-elevation complex takes place from just under 3000 m to just over 3500 m in the *jalka* zone, an alpine grassland and the primary tuber-producing zone. Above 3500 m in the *jalka fuerte* zone (Brush 1977), cultivation gives way to pasture, although certain frost-resistant tubers can be grown even above 4000 m.

Table 9.10. Proposed Areas of Origin of Major South American Cultivated Plants. (Complexes after Harlan 1975)

Latin Name (common name)	Location; Reference
Andean High-Elevation Complex	
Chenopodium quinoa (quinoa)	Peruvian Andes; Simmonds 1976b
Lepidium meyenii (maca)	Peruvian Andes; Leon 1964
Oxalis tuberosa (oca)	Andean area; Smith 1976
Solanum tuberosum (potato)	Central-southern Andes; Pickersgill and Heiser 1977; Simmonds 1976c
Tropaeolum tuberosum (mashua)	Andean area; Smith 1976
Ullucus tuberosus (ullucu)	Andean area; Smith 1976
Andean Mid-Elevation Complex	
Amaranthus caudatus (amaranth)	Central-southern Andes; Sauer 1976
Arachis hypogaea (peanut)	Non-Andean Bolivia/Argentina; Gregory and Gregory 1976
Bunchosia armeniaca	Mid-elevation Andes; Harlan 1975
Erythroxylum coca (coca)	Peruvian montaña; Plowman 1984
Lupinus mutabilis (tarwi)	Andean area; Smith 1976
Pachyrrhizus ahipa (jicama)	Andean Bolivia/Argentina; Smith 1976
Phaseolus lunatus (lima bean)	Northern South America; montaña; Pickersgill and Heiser 1977
P. vulgaris (common bean)	Northern Argentina/Bolivia/Western Brazil; Central Andes; Evans 1976; Pickersgill and Heiser 1977
Psidium guajava (guava)	Mid-elevation Andes; Harlan 1975
Lowland Complex	
Annona cherimolia (cherimoya)	Ecuador/Peru Andes; Smith 1976
A. muricata (guanabana)	Lowlands; Smith 1976
Ananas comosus (pineapple)	Xerophytic lowlands; Pickersgill 1976
Bixa orellana (achiote)	Lowlands; Harlan 1975
Canavalia plagiosperma	Western Ecuador; Central Andean montaña; Pickersgill and Heiser 1977
Canna edulis (achira)	Montaña; Bolivian Andes; Pickersgill and Heiser 1977
Capsicum baccatum (chili pepper)	Bolivia/Western Brazil; Pickersgill 1984; Pickersgill and Heiser 1977
C. chinense	Amazonia; Pickersgill 1984; Pickersgill and Heiser 1977
C. frutescens	Amazonia; Pickersgill 1984; Pickersgill and Heiser 1977
C. pubescens	Andes; Pickersgill 1984; Pickersgill and Heiser 1977
Carica papaya (papaya)	Northern Andes/Venezuela; Prance 1984
Cucurbita maxima (squash)	Western Ecuador/Argentina/Bolivia; Pickersgill and Heiser 1977
Gossypium barbadense (cotton)	Southwestern Ecuador/Northern coast Peru, with secondary center in Amazonia; Phillips 1976
Inga Feuillei (pacae)	Lowlands; Harlan 1975
Ilex paraguariensis (mate)	Northern Argentina-Southern Brazil; Smith 1976
Ipomoea batatas (sweet potato)	Lowlands: Northwestern South America, Amazon, or Mexico/Central America; Pickersgill and Heiser 1977; Yen 1976
Manihot esculenta (yuca)	Northern South America; Northeastern Brazil-Paraguay; Jennings 1976; Pickersgill and Heiser 1977
Nicotiana rustica (tobacco)	Andes, Ecuador-Bolivia; Gerstel 1976
N. tabacum	Montaña, Northern Argentina; Gerstel 1976
Persea americana (avocado)	Northwestern Colombia-Southern Mexico; Bergh 1976
Other cultivated plants, not listed by complex	
Cucurbita ficifolia (mid/high-altitude)	Probable South American origin, with spread into Mesoamerica; Pickersgill and Heiser 1977
Cyclanthera pedata (low/mid-altitude)	Mexico-Bolivia; Towle 1961
Lucuma bifera (low/mid-altitude)	Peru; Towle 1961
Sapindus saponaria (low/mid-altitude)	No information
Non-indigenous cultivated plants observed archaeologically	
Canavalia ensiformis	Late prehistoric/early Colonial introduction, Mesoamerica; Sauer and Kaplan 1969
Cucurbita moschata	Southern Mexico/Guatemala origin, early spread to South America; Pickersgill and Heiser 1977
C. pepo	Mesoamerican cultivar, not present in South America in the pre-Conquest period; Pickersgill and Heiser 1977
Lagenaria siceraria	African origin, naturalized in Eastern South America; Pickersgill and Heiser 1977
Zea mays	Early introduction from Mesoamerica; Pickersgill and Heiser 1977

Lowland Complex

The system that in many ways presents the greatest challenge to the archaeologist is the low-elevation agricultural system. As Table 9.10 shows, most of the crops of the lowland complex were brought under domestication in the tropical forest. Yet most archaeological data on these crops come from the desert of coastal Peru, where floodplain and, later, irrigation agriculture incorporating most of these cultivated plants was practiced. As a result of more than 40 years of systematic archaeological research, abundant dried plant remains are available from this zone. By contrast, in the tropical forest, sites are difficult to find and excavate, and the preservation of botanical material is often limited. We are faced with the paradoxical situation of having virtually no data on the lowland agricultural complex, except that from outside the hearth area.

One of the major crops of this system is *yuca,* or manioc (*Manihot esculenta*). Archaeological data, however, indicate that cultivation of manioc is predated by sweet potato (*Ipomoea batatas*) and *achira* (*Canna edulis*), at least on the Peruvian coast. *Achira* and sweet potato are present on the coast in Late Preceramic or Initial period sites (2300 B.C. and after), while manioc appears generally in the Early Horizon (900 B.C. and after). Exceptions are the occurrences of manioc at Los Gavilanes (2700–2200 B.C.) and Tres Ventanas (8000 B.C.). It appears, therefore, that the tuber crops of the lowland system were not introduced into the coast as a complex. This may reflect a different antiquity of domestication for each crop but is perhaps more likely a reflection of differences in trade or movement of peoples between lowland areas and the west coast, and different areas of origin for these cultivated plants.

Initial domestication of these root crops, probably occurring in the eastern or northern lowlands, must date to the late third millennium B.C. or before. The only archaeological manioc remains recovered from the tropical forest are fairly late: post-400 A.D. at Parmana. Roosevelt (1980), however, suggests that manioc cultivation dates to the beginning of the Parmana cultural sequence, the La Gruta tradition (2100–1600 B.C.). Her work demonstrates that data useful for understanding agricultural evolution can be obtained from the tropical forest region. Roosevelt proposes that agricultural intensification and population growth in the Parmana region was based on the cultivation of maize, which begins at 800 B.C., rather than on the earlier cultivation of manioc or other root crops. Lathrap (1970) has proposed a more pivotal role for manioc in the prehistoric tropical forest system. This issue cannot be decided on the basis of the first systematically recovered botanical remains from the eastern lowlands. The relative importance of root and seed crops, the antiquity of maize cultivation, and the broader issue of population growth and sociopolitical development in the Amazon and Orinoco regions are wide-open fields of inquiry.

Canavalia plagiosperma is a traditional component of the lowland agricultural complex. Evidence for cultivation of this crop dates to 3300 B.C. in western Ecuador, where remains identified as probable cultivated *C. plagiosperma* occur at the Real Alto site. Another species, *C. ensiformis,* was recovered from post-400 A.D. levels at Parmana. *C. ensiformis* has been considered a late prehistoric to early Colonial period introduction from Mesoamerica (Sauer and Kaplan 1969), but the Parmana find suggests a greater antiquity for this introduction, at least in northern South America.

Few data are available on the domestication of the numerous tree fruits of the lowland agricultural complex. Avocado (*Persea americana*) and *pacae* (*Inga Feuillei*) are present in Late Preceramic or Initial period contexts in Peru; finds of *guanabana* (*Annona muricata*), *cherimoya* (*A. cherimolia*), pineapple (*Ananas comosus*), and papaya (*Carica papaya*) are all later, with the exception of *cherimoya* from Los Gavilanes (2700–2200 B.C.). Archaeological data are lacking for many other minor domesticated and semidomesticated tree taxa. Cotton (*Gossypium barbadense*) was probably domesticated in western South America, either in northern coastal Peru or southwest Ecuador, with a secondary center of development in Amazonia. Early occurrences include the Ayacucho caves (3100–1750 B.C. strata), Los Gavilanes (2700–2200 B.C.), and numerous finds dating to between 2400 and 2000 B.C. It appears that cotton had been brought under domestication and had spread at least as far as the sierra at about the time crops were being introduced from the east.

Cucurbits and chili peppers round out the inventory of lowland cultivars for which we have archaeological data. Finds of *Cucurbita* are fairly common on the Peruvian coast by the Late Preceramic. A number of sites have older *Cucurbita* remains: ASA8 (4135 B.C.), Paloma (5700–3000 B.C.), Los Gavilanes (2700–2000 B.C.), Chilca (3800–2650 B.C.), and Ayacucho (3100–1750 B.C.). In general, gourd (*Lagenaria siceraria*) also occurs in early contexts, including the coastal Siches complex (6000–4000 B.C.). Although squashes and gourds are not the only cultivated plants dating to the fifth millennium B.C. or earlier, they are the most commonly occurring early cultivated plants (Fig. 9.4). It is likely that all of them were introduced to Peru from outside the central Andes. Chile peppers (*Capsium*) also occur in early Preceramic contexts. The earliest find of *C. chinense* at Guitarrero Cave (8000–7500 B.C.) is another example of an early introduction from the east.

The demonstrated antiquity in coastal or Andean Peru of a number of components of the lowland agricultural complex points to an early development of this system. There is evidence for settled village life in the tropical forests of the Amazon and Orinoco by around 2000 B.C. (Lathrap 1970; Meggers and Evans 1983; Roosevelt 1980), with even greater antiquity (3300–3000 B.C.) in Colombia and Ecuador (Feldman and Moseley 1983). The early manipulation of plants, leading to eventual domestication, must predate the tropical forest Formative by many millennia.

Although only a few crops were brought into domestication on the western coast of South America, the excellent preservation of botanical materials in the coastal Peruvian desert has led to considerable interest in subsistence issues among archaeologists working in this region. For example, there is currently a lively debate over the role of agriculture in the rise of complex societies on the coast (e.g., Moseley 1975; Quilter and Stocker 1983; Raymond 1981; Richardson 1980; Weir and Benfer 1985; Wilson 1981). Although the record is not without bias, there are sufficient data to allow the Peruvian coast to be used as a laboratory of state origins. Stated in broad terms, the debate concerns whether sedentary, complex societies on the coast evolved on a subsistence base of agricultural resources, or whether marine resources played the dominant role. Coastal data have also been used as a test case for the role of population pressure in the origins of agriculture (Cohen 1977a, 1977b).

Andean Mid-Elevation Complex

Few botanical data are available from archaeological sites located in the mid-elevations (1500–2500/3000 m) of the Andes. Exceptions are Guitarrero Cave and a number of the sites investigated during the Ayacucho project in Peru, and a few sites in northwest Argentina and Chile. As was the case for the lowland complex, most of the early data on mid-elevation cultivars come from Peruvian coastal sites.

Only one tuber crop in this complex, *jicama* (*Pachyrrhizus ahipa*), is known archaeologically. Data are lacking on two others, *arracacha* (*Arracacia xanthorrhiza*) and *yacon* (*Polymnia sonchifolia*). Although *achira* (*Canna edulis*) is considered part of the lowland complex, it is cultivated today to a limited extent in the Peruvian Andes (Gade 1966) and may have been domesticated in the *montaña* or southern Andes (Pickersgill and Heiser 1977). Remains of *jicama* occur on the Peruvian coast beginning in the Late Preceramic (Los Gavilanes, El Paraiso de Chuquitanta) or Initial period. None are reported from Guitarrero or the Ayacucho caves. If this crop was initially domesticated in the Bolivian or Argentinian Andes, it may have been introduced to the coast by a route that bypassed the central sierra of Peru.

The mid-elevation complex contains a number of domesticated legumes: peanut (*Arachis hypogaea*), common bean (*Phaseolus vulgaris*), lima bean (*P. lunatus*), and lupine, or *tarwi* (*Lupinus mutabilis*). The altitude ranges of these crops vary somewhat, with peanuts also adapted to lower elevations and *tarwi* to higher. Common and lima beans have been recovered archaeologically from sites in the mid-elevation Andes. Remains date to 4400–3100 B.C. at Ayacucho and to 8000–7500 B.C. at Guitarrero. *P. vulgaris* has also been identified from the earliest levels at the Huachichocana site in Argentina (7670–6720 B.C.). The pattern of early occurrence of the Phaseolus beans is repeated on the Peruvian coast where remains date to the third millennium (Aspero and Chilca sites). By contrast, peanut is a relatively late arrival on the Peruvian coast, dating to the Initial period, with the exception of a Late Preceramic occurrence at Los Gavilanes. No early remains of lupine are reported. It is known from the Xauxa area beginning at A.D. 650. These data suggest that the legumes associated with the mid-elevation agricultural system did not evolve as a complex. *Phaseolus* beans were domesticated many millennia before any other legume, including *Canavalia*. They were introduced to the Peruvian coast together with early low-elevation crops, such as squash and gourd. Interestingly, neither lima nor common bean occur early in coastal Ecuador: only *Canavalia* beans have been identified there. This distribution lends some support to a central or southern Andean area of origin for the *Phaseolus* beans.

There are a number of tree fruits in the mid-elevation complex, including *Bunchosia armeniaca*, guava (*Psidium guajava*), and *lucuma* (*Lucuma bifera*). No remains of these fruits occur at mid-elevation archaeological sites. *Guava* is the first to arrive on the Peruvian coast, beginning in the Preceramic (5700–3000 B.C. at Paloma; 2410–2000 B.C. at Aspero). *Bunchosia* and *lucuma* occur in later Preceramic sites.

One of the Andean pseudocereals, *Amaranthus caudatus,* is considered a mid-elevation cultivar. Few archaeological data exist for this crop. Remains are reported from the Tafi site in the Valliserrana area of Argentina, dated to 0 B.C. or perhaps earlier (Table 9.3). Seeds identified only as *Amaranthus* sp. occur at Chiripa in the Lake Titicaca basin between 1500 and 500 B.C., and at sites on the Junin *puna* (e.g., Panaulauca, Pachamachay).

The final mid-elevation cultivar for which archaeological evidence exists is *coca.* Recently, cultivated *coca* has been reclassified into two species, each of which has two varieties: *Erythroxylum coca* var. *coca* (Huanuco *coca*), *E. coca* var. *ipadu* (Amazonian *coca*), *E. novogranatense* var. *truxillense* (Trujillo *coca*), and *E. novogranatense* var. *novogranatense* (Colombian *coca*) (Plowman 1984). The ancestral form, *E. coca,* occurs wild on the eastern slopes of the Andes. Plowman (1984) suggests that initial cultivation occurred in the Peruvian *montaña,* giving rise to domesticated Huanuco *coca.* The spread of this cultivar into the Amazon eventually resulted in the evolution of the Amazonian strain. When Huanuco *coca* was moved out of the moist *montaña* into drier areas of the mid-elevation Andes, it developed into a quite distinctive type, Trujillo *coca.* It was this type that was eventually established under cultivation in the arid western Andes. All early archaeological *coca* (remains dating from 2500 to 1800 B.C.) from coastal Peru reexamined by Plowman (1984) proved to be Trujillo *coca.* The northward spread of Trujillo *coca* into Ecuador and Colombia resulted in the evolution of the fourth variety, Colombian *coca.* Plowman (1984) considers the Peruvian coastal remains the earliest firmly identified *coca.* He believes the Ayacucho find (4400–3100 B.C.) to be a mistaken identification.

Although we have archaeological botanical data from several sites in the mid-elevation Andean zone, our picture of the antiquity and course of development of the mid-elevation agricultural complex is far from clear. At 8000–7500 B.C. at Guitarrero Cave, three mid-elevation cultivars—common bean, lima bean, and lucuma—are found in association with a high-elevation tuber (*oca*) and a low-elevation pepper, *Capsicum chinense.* Even at this early date, the mid-elevation Andes were in contact with other zones. It seems clear that the crops of this system were brought under domestication over a wide geographic area and over the course of several millennia. The most important missing elements are the mid-elevation tubers, *jicama, arracacha,* and *yacon,* unknown archaeologically from the mid-elevation zone. These crops were undoubtedly important carbohydrate sources prior to the introduction of maize.

Andean High-Elevation Complex

The high-elevation agricultural complex is in some ways the most self-contained of the South American agricultural systems. The crops making up this system commonly are grown today in the Andes between just under 3000 m to just over 3500 m. There has been little movement of the taxa to other zones. This pattern appears to have considerable antiquity. Only the potato (*Solanum tuberosum*) has been found to any extent outside the high-elevation zone. Archaeological remains of potato are known from Tres Ventanas (8000 B.C.) and from several sites in the coastal Casma Valley (2250 B.C. and after). Potatoes are possibly present in the Ayacucho area between 4400 and 3100 B.C. The other tubers, oca (*Oxalis tuberosa*), mashua (*Tropaeolum tuberosum*), and ullucu (*Ullucus tuberosus*), spread eventually throughout the Andes, but occur rarely in mid-elevation or lowland settings. *Ullucu* occurs at Tres Ventanas and *oca* is known from the mid-elevation Guitarrero Cave site (8000–7500 B.C.). *Maca* (*Lepidium meyenii*) has a very limited distribution today (central Peru) and is known archaeologically only from the same zone. Remains of quinoa (*Chenopodium quinoa*) have been identified from the coast of Chile (Table 9.4), but have been found in Peru only at mid-elevation or high-elevation sites.

The lack of archaeological botanical data from early sites in that part of the high-altitude zone suited for agriculture hampers our understanding of the evolution of this complex. Above 3500–4000 m, successful cultivation becomes difficult. Only frost-resistant strains of potatoes are commonly grown, although cultivation of *maca* can extend agriculture to 4400 m or higher. Sites in this upper part of the high-elevation zone (Panaulauca, Pachamachay) show utilization of wild plant resources and some indication of cultivation (*Lepidium, Chenopodium*), but are located above the main zone of cultivation. Most early Andean sites with abundant cultivated plant remains, such as Guitarrero Cave and many of the sites in the Ayacucho area, are in the mid-elevation zone.

One of the interesting aspects of the issue of the development of the high-elevation agricultural system is the nature of the relationship between the processes of plant and animal domestication. The camelids (*llama, alpaca*) were brought under domestication in the high Andean area. The earliest archaeological data documenting the process of domestication come from sites above 4000 m (Wing 1977). This is above the main zone of plant cultivation. Was there no relationship between the domestication of the camelids and the evolution of the tuber-based, high-elevation agricultural complex? At Panaulauca Cave on the Junin *puna,* a linkage seems to exist between the corralling of camelids and the increased use of disturbed habitat plants, notably *Chenopodium* spp. and *Lepidium* spp. (Pearsall 1989a). This pattern of use eventually developed into cultivation. The time period involved—the second millennium B.C.—follows the period when potatoes, *oca,* and *ullucu* are known from lower altitudes, however. It may be the case that the process of camelid domestication is linked to the use of certain cultivars hardy enough to thrive at the limits of plant cultivation, but that this process was isolated to some extent from the processes of tuber domestication occurring in the high Andean valleys. If there were a very early evolution of the pattern of integration of *puna* and valley zones, as has been proposed (MacNeish 1977), then the Junin *puna* seems to be an exception, suggesting that the issue is still open.

Conclusions

We have gained considerable insight in recent years into the origins and dispersals of cultivated plants in South America. This is due in large part to the incorporation into archaeological research projects of specialized techniques for the recovery of botanical remains, namely, water flotation, fine sieving, and sampling for pollen and phytolith analyses. Water flotation, which concentrates dispersed charred plant remains, and phytolith analysis, the study of inorganic plant residues, have been especially important in increasing the data base for the moist regions of the lowlands and sierra.

Cultivated plants first appear in the archaeological record in South America at 8000 B.C. Common bean, lima bean, chili pepper, potato, and *oca* are the earliest species recorded to date. By 6000 B.C., the list expands to include *quinoa,* squash, gourd, *guava,* and the introduced grain, maize. The remaining major crops for which we have archaeological data appear before 2000 B.C. This pattern, an appearance of cultivated plants in the early Holocene, with a subsequent increase in the diversity of cultivated taxa, is not unique to the South American area (see other papers, this volume).

Documenting that cultivated plants occur in certain locations at certain time periods is only one aspect of the issue of the origins of plant cultivation in South America, however. How did the well-developed agricultural systems observed at the time of European contact evolve? How did the role of plant cultivation in subsistence change as these systems developed? What regional differences can be documented? Understanding the process of domestication and the evolution of agricultural systems in regional settings are ambitious goals, but ones that should guide future research.

In Figure 9.3, I indicated by width of arrow the estimated importance of plant cultivation in subsistence throughout prehistory in Ecuador, Peru, and the southern Andes (Chile and Argentina). This is based in part on the data reviewed in this paper and in part on applying a coevolutionary perspective to those data. The coevolutionary model of agricultural origins (Rindos 1980, 1984) describes domestication as an evolutionary process, the result of predator/prey relationships characterized by mutualism, a relationship in which both prey (plant) and predator (human) populations benefit. As a result of selective pressures associated with harvesting and sowing, plants undergo morphological changes that increase their productivity while reducing their ability to disperse seeds. Over time, increased productivity of domesticated plants leads humans to reduce their use of nondomesticates. Agriculture, a set of activities that affects the environment inhabited by domesticated plants, is an outgrowth of this relationship. Less productive resources are dropped in favor of creating productive environments for domesticates. This marks the beginning of the agricultural stage.

How this view of domestication structures the interpretation of the archaeological data reviewed here can be illustrated by considering the coastal Ecuadorian sequence. Subsistence during the Vegas and Valdivia periods in coastal Ecuador can be characterized as mixed, including hunting, fishing, shellfish gathering, plant gathering, and plant cultivation (Pearsall

1988b, n.d.). During this long period of relative stability in subsistence, new cultivated plants are added and maize becomes more common, but cultivation does not dominate subsistence (narrow arrow). However, before the end of the Valdivia period, perhaps around 2000 B.C., a transformation occurs. Evidence of landscape modification and appearance of village sites in previously unoccupied areas suggests an expanding agricultural system. I propose that late Valdivia is the beginning of the agricultural stage of human-plant relationships in western Ecuador (medium arrow). Dating of a second major transformation, to fully agricultural subsistence (thick arrow), is very tentative; I place it at 500 B.C. to reflect the evidence for major earth-moving projects, such as raised fields, to create new cropping areas.

Whether the suggested timing of these transformations in subsistence in South America is correct remains to be tested. Domestication occurred at many times and in many places in the South American "non-center," ultimately resulting in the three major crop complexes discussed above. Multiple regional sequences in the areas of origin of these complexes are needed to trace the history of their development. It is clear from the data available that these complexes did not evolve in isolation; cultivated plants were exchanged almost from their first appearance in the archaeological record. Agricultural systems are dynamic; prehistoric systems formed and re-formed as new plants were introduced and agricultural technology was developed and exchanged. Of special interest in this regard is the introduction of maize into South America. Of minor importance for thousands of years, it came to occupy a primary role in both the mid-altitude and low-altitude systems. The dynamics of this process are only incompletely understood.

Finally, although this chapter has focused on plant cultivation, the role of animal resources in subsistence should not be overlooked. Rick (1980) has proposed, for example, that hunting populations occupying the high-altitude *puna* zone of Peru became sedentary long before the appearance of cultivated plants. Similarly, Moseley (1975) has argued that it was the rich marine resources of the Peruvian littoral rather than agriculture that underlay the development of early complex societies there. The emergence of plant food production is only part of the picture of the evolution of the subsistence systems that supported prehistoric cultural development in South America.

References Cited

Benfer, Robert A.
1982   Proyecto Paloma de la Universidad de Missouri y el Centro de Investigaciones de Zonas Áridas. Zonas Áridas 2:34–73.

1984   The Challenges and Rewards of Sedentism: the Preceramic Village of Paloma, Peru. *In* Paleopathology at the Origins of Agriculture, edited by M. N. Cohen and G. J. Armelagos, pp. 531–58. Academic Press, New York.

Bergh, B. O.
1976   Avocado. *Persea americana* (Lauraceae). *In* Evolution of Crop Plants, edited by N. W. Simmonds, pp. 148–51. Longman, London.

Bird, Junius B., John Hyslop, and Milica Dimitrijevic Skinner
1985   The Preceramic Excavations at the Huaca Prieta, Chicama Valley, Peru. Anthropological Papers, Vol. 62, Part 1. American Museum of Natural History, New York.

Bird, Robert Mck.
1984   South American Maize in Central America? *In* Pre-Columbian Plant Migration, edited by Doris Stone, pp. 39–65. Papers of the Peabody Museum of Archaeology and Ethnology, Vol. 76.

Bonavia, Duccio (editor)
1982   Precerámico peruano. Los Gavilanes. Mar, desierto, y oasis en la historia del hombre. Corporación Financiera de Desarrollo S.A. (COFIDE). Oficina de Asuntos Culturales; Instituto Arqueológico Aleman, Comisión de Arqueología General y Comparada. Editorial Ausonia-Talleres Gráficos, Lima.

Bonavia, Duccio, and Alexander Grobman
1979   Sistema de Depositos y Almacenamiento durante el Período Precerámico en la Costa del Perú. Journal de la Société des Américanistes 66:21–42.

Braun, Robert
1971   The Formative as Seen from the Southern Ecuadorian Highlands. *In* Primer Simposio de Correlaciones Antropológicas Andino-Mesoamericano, edited by Jorge G. Marcos y Presley Norton, pp. 41–99. Escuela Superior Politécnica del Litoral, Guayaquil, Ecuador.

Bray, Warwick
1976   From Predation to Production: the Nature of Agricultural Evolution in Mexico and Peru. *In* Problems in Economic and Social Archaeol-

ogy, edited by G. de G. Sieveking, I. H. Longworth, and K. E. Wilson, pp. 73–95. Duckworth, London.

Browman, David L.
1989   Chenopod Cultivation, Lacustrine Resources, and Fuel Use at Chiripa, Bolivia. *In* New World Paleoethnobotany: Collected Papers in Honor of Leonard W. Blake, edited by Eric E. Voigt and Deborah M. Pearsall. Missouri Archaeologist 47:137–72

Brush, Stephen B.
1977   Mountain, Field, and Family: the Economy and Human Ecology of an Andean Valley. University of Pennsylvania Press, Pennsylvania.

Bryan, Alan L.
1983   South America. *In* Early Man in the New World, edited by Richard Shutler, Jr., pp. 137–46. Sage Publications, Beverly Hills.

Bush, Mark B., Dolores R. Piperno, and Paul A. Colinvaux
1989   A 6000 Year History of Amazonian Maize Cultivation. Nature 340:303–305.

Cigliano, Eduardo Mario
1968   Sobre algunos vegetales hallados en el yacimiento arqueológico de Santa Rosa de Tastil, Depto. Rosario de Lerma. Revista Antropología 7(38):15–23. Museo de la Universidad Nacional, La Plata.

Cohen, Mark N.
1977a   Population Pressure and the Origins of Agriculture: An Archaeological Example from the Coast of Peru. *In* Origins of Agriculture, edited by Charles A. Reed, pp. 135–77. Mouton Publishers, The Hague.

1977b   The Food Crisis in Prehistory. Yale University Press, New Haven.

Collier, Donald, and John Murra
1943   Survey and Excavation in Southern Ecuador. Anthropological Series, Vol. 35. Field Museum of Natural History, Chicago.

D'Altroy, Terence N., and Christine A. Hastorf
1984   The Distribution and Contents of Inca State Storehouses in the Xauxa Region of Peru. American Antiquity 49(2):334–49.

Damp, Jonathan E.
1979   Better Homes and Gardens: The Life and Death of the Early Valdivia Community. Ph.D. dissertation, Department of Archaeology, University of Calgary.

1984   Architecture of the Early Valdivia Village. American Antiquity 49(3):573–95.

Damp, Jonathan E., Deborah M. Pearsall, and Lawrence T. Kaplan
1981   Beans for Valdivia. Science 212:811–12.

Dering, Phillip, and Glendon H. Weir
1979   Appendix II. Analysis of Plant Remains from the Preceramic Site of Paloma, Chilca Valley, Peru. Manuscript on file, Department of Anthropology, University of Missouri-Columbia.

1981   Preliminary Plant Macrofossil Aanlysis of La Paloma Deposits. Paper presented at the 46th Annual Meeting of the Society for American Archaeology, San Diego.

Dillehay, T. D.
1985   The Early Pebble Tool Culture at Monte Verde, Chile. Paper presented at the 50th Annual Meeting of the Society for American Archaeology, Denver.

Dunn, Mary E.
1983   Phytolith Analysis in Archaeology. Midcontinental Journal of Archaeology 8(2):287–97.

Engel, Frederic Andre
1963   Datations a l'aide du radio-carbone 14, et problemes de la prehistoire du Perou. Journal de la Société des Américanistes 52:101–31.

1973   New Facts about Pre-Columbian Life in the Andean Lomas. Current Anthropology 14(3): 271–80.

Erickson, Clark L.
1976   Chiripa Ethnobotanical Report: Flotation-Recovered Archeological Remains from an Early Settled Village on the Altiplano of Bolivia. Manuscript on file, Department of Anthropology, University of Missouri-Columbia.

1977   Subsistence Implications and Botanical Analysis at Chiripa. Paper presented at the 42nd Annual Meeting of the Society for American Archaeology, New Orleans.

Evans, Alice M.
1976   Beans. *Phaseolus* spp. (Leguminosae-Papilionatae). *In* Evolution of Crop Plants, edited by N. W. Simmonds, pp. 168–72. Longman, London.

Feldman, Robert A.
1980   Aspero, Peru: Architecture, Subsistence Economy, and Other Artifacts of a Preceramic Mar-

itime Chiefdom. Ph.D. dissertation, Department of Anthropology, Harvard University.

Feldman, Robert A., and Michael E. Moseley
1983   The Northern Andes. *In* Ancient South Americans, edited by Jesse D. Jennings, pp. 138–77. W. H. Freeman and Company, San Francisco.

Gade, Daniel W.
1966   *Achira,* the Edible *Canna,* Its Cultivation and Uses in the Peruvian Andes. Economic Botany 20(4):407–15

Gerstel, D. U.
1976   Tobacco. *Nicotiana tabacum* (Solanaceae). *In* Evolution of Crop Plants, edited by N. W. Simmonds, pp. 273–77. Longman, London.

Gonzalez, Alberto Rex, and José Antonio Perez
1966   El Área Andina Meridional. Actas y Memórias, 36th Congreso Internacional de Americanistas, Spain (1964). Vol. 1, pp. 241–65. Ecesa, Seville.

1968   Una Nota Sobre Etnobotánica del N.O. Argentina. Actas y Memorias, 37th Congreso Internacional de Americanistas, Argentina (1966), vol. 2, pp. 209–28. Buenas Aires.

Gregory, W. C., and M. P. Gregory
1976   Groundnut. *Arachis hypogaea* (Leguminosae-Papilionatae). *In* Evolution of Crop Plants, edited by N. W. Simmonds, pp. 151–54. Longman, London.

Grieder, Terence
1988   Radiocarbon Measurements. *In* La Galgada Peru. A Preceramic Culture in Transition. Terence Grieder, Alberto Bueno Mendoza, C. Earle Smith, Jr., and Robert M. Malina, pp. 68–72. University of Texas Press, Austin.

Grobman, A., and D. Bonavia
1978   Pre-ceramic Maize on the North Central Coast of Peru. Nature 276:386–87.

Guidon, Niede
1985   Early Man in Piaui, Brazil. Paper presented at the 50th Annual Meeting of the Society for American Archaeology, Denver.

Harlan, Jack R.
1975   Crops and Man. American Society of Agronomy, Madison, Wisconsin.

Hastings, Charles M.
1981   Territorial Verticality in the Eastern Andes of Central Peru. Paper presented at the 80th Annual Meeting of the American Anthropological Society, Los Angeles.

Hastorf, Christine A.
1983   Prehistoric Agricultural Intensification and Political Development in the Jauja Region of Central Peru. Ph.D. dissertation, Department of Anthropology, University of California, Los Angeles.

Heiser, Charles B.
1965   Cultivated Plants and Cultural Diffusion in Nuclear America. American Anthropologist 67(4):930–49.

1979   Origins of Some Cultivated New World Plants. *In* Annual Review of Ecology and Systematics 10:309–26.

Jennings, D. L.
1976   Cassava. *Manihot esculenta* (Euphorbiaceae). *In* Evolution of Crop Plants, edited by N. W. Simmonds, pp. 81–84. Longman, London.

Jones, John G.
1988   Middle to Late Preceramic (6000–3000 B.P.) Subsistence Patterns on the Central Coast of Peru: The Coprolite Evidence. M.A. thesis, Department of Anthropology, Texas A & M University.

Kaplan, Lawrence
1980   Variation in the Cultivated Beans. *In* Guitarrero Cave. Early Man in the Andes, edited by Thomas F. Lynch, pp. 145–48. Academic Press, New York.

1981   What is the Origin of the Common Bean? Economic Botany 35(2):240–54.

Kaplan, Lawrence, Thomas Lynch, and C. Earle Smith
1973   Early Cultivated Beans (*Phaseolus vulgaris*) from an Intermontane Peruvian Valley. Science 179:76–77.

Kelley, David H., and Duccio Bonavia
1963   Evidence for Pre-ceramic Maize on the Coast of Peru. Nawpa Pacha 1:39–42.

Kolata, Alan L.
1983   The South Andes. *In* Ancient South Americans, edited by Jesse D. Jennings, pp. 240–85. W. H. Freeman and Company, San Francisco.

Lagiglia, Humberto
1968   Plantas Cultivadas en el Área Centro-Andino y su Vinculación Cultural Contextual. Actas y Memorias, 37th Congreso Internacional de Americanistas, Argentina (1966). Vol. 2, pp. 229–33. Buenas Aires.

Lanning, Edward P.
1967   Peru before the Incas. Prentice-Hall, Englewood Cliffs, New Jersey.

Lathrap, Donald W.
1970   The Upper Amazon. Praeger, New York.

Lathrap, Donald W., Donald Collier, and Helen Chandra
1975   Ancient Ecuador: Culture, Clay, and Creativity, 3000–300 B.C. Field Museum of Natural History, Chicago.

Lathrap, Donald W., Jorge G. Marcos, and James A. Zeidler
1977   Real Alto: An Ancient Ceremonial Center. Archaeology 30(1):2–13.

Lavallee, Daniele, Michele Julien, and Jane Wheeler
1984   Telermachay: Niveles Precerámicos de Ocupación. Revista del Museo Nacional 46(1982):55–133. Lima.

Leon, Jorge
1964   Plantas Alimentícias Andinas. Instituto Interamericano de Ciéncias Agrícolas Zona Andina. Boletín Técnico, No. 6.

Lippi, Ronald D., Robert Mck. Bird, and David M. Stemper
1984   Maize Recovered at La Ponga, an Early Ecuadorian Site. American Antiquity 49(1):118–24.

Lynch, Thomas F.
1983   The Paleo-Indians. In Ancient South Americans, edited by Jesse D. Jennings, pp. 86–137. W. H. Freeman and Company, San Francisco.

Lynch, Thomas F. (editor)
1980   Guitarrero Cave. Early Man in the Andes. Academic Press, New York.

Lynch, Thomas F., R. Gillespie, John A. J. Gowlett, and R. E. M. Hedges
1985   Chronology of Guitarrero Cave, Peru. Science 229:864–67.

MacNeish, Richard S.
1976   Early Man in the New World. American Scientist 63(3):316–27.

1977   The Beginnings of Agriculture in Central Peru. In The Origins of Agriculture, edited by Charles A. Reed, pp. 753–801. Mouton Publishers, The Hague.

MacNeish, Richard S., Antoinette Nelken-Terner, and Robert K. Vierra
1980   Introduction. In Prehistory of the Ayacucho Basin, Peru. Vol. 3. Nonceramic Artifacts. R. S. MacNeish, R. K. Vierra, A. Nelken-Terner, and C. J. Phagen, pp. 1–34. University of Michigan Press, Ann Arbor.

Mangelsdorf, Paul C., and G. C. Pollard
1975   Archaeological Maize from Northern Chile. Botanical Museum Leaflets, Harvard University 24(3):49–64.

Mangelsdorf, Paul C., and M. Sanoja O.
1965   Early Archaeological Maize from Venezuela. Botanical Museum Leaflets, Harvard University 21:105–11.

Marcos, Jorge G.
1978   The Ceremonial Precinct at Real Alto: Organization of Time and Space in Valdivia Society. Ph.D. dissertation, Department of Anthropology, University of Illinois, Urbana.

Meggers, Betty J., and Clifford Evans
1983   Lowland South America and the Antilles. In Ancient South Americans, edited by Jesse D. Jennings, pp. 286–335. W. H. Freeman and Company, San Francisco.

Meggers, Betty J., Clifford Evans, and Emilio Estrada
1965   The Early Formative Period on Coastal Ecuador: The Valdivia and Machalilla Phases. Smithsonian Contributions to Anthropology, Vol. 1. Smithsonian Institution, Washington, D.C.

Moore, Katherine M.
1985   Hunting and Herding Economies on the Junin Puna: Recent Paleoethnozoological Research. Paper presented at the 50th Annual Meeting of the Society for American Archaeology, Denver.

Moseley, Michael E.
1975   The Maritime Foundations of Andean Civilization. Cummings Publishing Company, Menlo Park, California.

1983   Central Andean Civilization. In Ancient South Americans, edited by Jesse D. Jennings, pp. 178–239. W. H. Freeman and Company, San Francisco.

Moseley, Michael E., and Gordon R. Willey
1973   Aspero, Peru: A Reexamination of the Site and Its Implications. American Antiquity 38(4):452–68.

Nuñez A., Lautaro
1974   La Agricultura Prehistórica en los Ándes Meridionales. Editorial Obre, Santiago, Chile.

Nuñez R., Victor A.
1971   La Cultura Alamito de la Súbarea Valliserrana del Noroeste Argentina. Journal de la Société des Américanistes 60:7–64. Paris.

Patterson, Thomas C., and Michael E. Moseley

1968   Late Pre-Ceramic and Early Ceramic Cultures of the Central Coast of Peru. Nawpa Pacha 6:115–33.

Pearsall, Deborah M.

1977   Maize and Beans in the Formative Period of Ecuador: Preliminary Report of New Evidence. Paper presented at the 42nd Annual Meeting of the Society for American Archaeology, New Orleans.

1978a   Paleoethnobotany in Western South America: Progress and Problems. *In* The Nature and Status of Ethnobotany, edited by R. I. Ford, pp. 389–416. Anthropological Papers No. 67. Museum of Anthropology, University of Michigan, Ann Arbor.

1978b   Phytolith Analysis of Archeological Soils: Evidence for Maize Cultivation in Formative Ecuador. Science 199:177–78.

1979   The Application of Ethnobotanical Techniques to the Problem of Subsistence in the Ecuadorian Formative. University Microfilms, Ann Arbor.

1980a   Analysis of an Archaeological Maize Kernel Cache from Manabi Province, Ecuador. Economic Botany 34(4):344–51.

1980b   Pachamachay Ethnobotanical Report: Plant Utilization at a Hunting Base Camp. *In* Prehistoric Hunters of the High Andes, John W. Rick, pp. 191–231. Academic Press, New York.

1982   Phytolith Analysis: Application of a New Paleoethnobotanical Technique in Archeology. American Anthropologist 84(4):862–71.

1983   Evaluating the Stability of Subsistence Strategies by Use of Paleoethnobotanical Data. Journal of Ethnobiology 3(2):121–37.

1984   Informe del Análisis de Fitolitos y Semillas Carbonizadas del Sítio Cotocollao, Província de Quito, Ecuador. Manuscript on file, Department of Anthropology, University of Missouri-Columbia.

1986   Final Report on Analysis of Plant Macroremains and Phytoliths from the Nueva Era site, Pichicha Province, Ecuador. Manuscript on file, Department of Anthropology, University of Missouri-Columbia.

1988a   Interpreting the Meaning of Macroremain Abundance: The Impact of Source and Context. *In* Current Paleoethnobotany: Analytical Methods and Cultural Interpretations of Archaeological Plant Remains, edited by Christine Hastorf and Virginia Popper, pp. 97–118. University of Chicago Press, Chicago.

1988b   An Overview of Formative Period Subsistence in Ecuador: Palaeoethnobotanical Data and Perspectives. *In* Diet and Subsistence: Current Archaeological Perspectives, edited by Brenda V. Kennedy and Genevieve M. LeMoine, pp. 149–64. Proceedings of the Nineteenth Annual Chacmool Conference. University of Calgary.

1989a   Adaptation of Prehistoric Hunter-gatherers to the High Andes: the Changing Role of Plant Resources. *In* Foraging and Farming. The Evolution of Plant Exploitation, edited by D. R. Harris and G. C. Hillman, pp. 318–32. Unwin Hyman, London.

1989b   Paleoethnobotany. A Handbook of Procedures. Academic Press, San Diego.

n.d.   Agricultural Evolution and the Emergence of Formative Societies in Ecuador. *In* Development of Agriculture and Emergence of Formative Civilizations in Pacific Central and South America, edited by Michael Blake.

Pearsall, Deborah M., and Katherine Moore

1985   Prehistoric Economy and Subsistence. *In* Hunting and Herding in the High Altitude Tropics: Early Prehistory of the Central Peruvian Andes, edited by John W. Rick. Manuscript on file at the Department of Anthropology, University of Missouri-Columbia.

Pearsall, Deborah M., and Bernardino Ojeda

1988   Final Report of Analysis of Botanical Materials from the El Paraiso Site, Peru. With a Study of Cucurbitaceae, by Andrea Hunter. Manuscript on file, Department of Anthropology, University of Missouri-Columbia.

Pearsall, Deborah M., and Dolores R. Piperno

1990   Antiquity of Maize Cultivation in Ecuador: Summary and Reevaluation of the Evidence. American Antiquity 55(2):324–37.

Peterson, Emil, and Eugenia Rodriquez

1977   Cotocollao, A Formative Period Village in the Northern Highlands of Equador. Paper presented at the 42nd Annual Meeting of the Society for American Archaeology, New Orleans.

Phillips, L. L.
1976   Cotton. *Gossypium* (Malvaceae). *In* Evolution of Crop Plants, edited by N. W. Simmonds, pp. 196–200. Longman, London.

Pickersgill, Barbara
1969   The Archeological Record of Chili Peppers (*Capsicum* spp.) and the Sequence of Plant Domestication in Peru. American Antiquity 34:54–61.

1976   Pineapple. *Ananas comosus* (Bromeliaceae). *In* Evolution of Crop Plants, edited by N. W. Simmonds, pp. 14–18. Longman, London.

1984   Migrations of Chili Peppers, *Capsicum* spp., in the Americas. *In* Pre-Columbian Plant Migrations, edited by Doris Stone, pp. 105–123. Papers of the Peabody Museum of Archaeology and Ethnology, Vol. 76.

Pickersgill, Barbara, and Charles B. Heiser, Jr.
1977   Origins and Distribution of Plants Domesticated in the New World Tropics. *In* Origins of Agriculture, edited by Charles A. Reed, pp. 803–35. Mouton Publishers, The Hague.

Piperno, Dolores
1983   The Application of Phytolith Analysis to the Reconstruction of Plant Subsistence and Environments in Prehistoric Panama. Ph.D. dissertation, Department of Anthropology, Temple University, Philadelphia.

1984   A Comparison and Differentiation of Phytoliths from Maize and Wild Grasses: Use of Morphological Criteria. American Antiquity 49(2):361–83.

1988a   Phytolith Analysis: An Archaeological and Geological Perspective. Academic Press, New York.

1988b   Primer Informe sobre los Filolitos de las Plantas del OGSE-80 y la Evidencia del Cultivo de Maíz en el Ecuador. *In* La Prehistória Temprana de la Península de Santa Elena, Ecuador: Cultura Las Vegas, by Karen E. Stothert, pp. 204–13. Miscelánea Antropológica Ecuatoriana. Serie Monográfica 10. Museos del Banco Central del Ecuador, Guayaquil.

Plowman, Timothy
1984   The Origin, Evolution, and Diffusion of Coca, *Erythroxylum* spp. in South and Central America. *In* Pre-Columbian Plant Migration, edited by Doris Stone, pp. 124–63. Papers of the Peabody Museum of Archaeology and Ethnology, Vol. 76.

Pollard, Gordon C.
1971   Cultural Change and Adaptation in the Central Atacama Desert of Northern Chile. Nawpa Pacha 9:41–64.

Popper, Virginia S.
1977   Prehistoric Cultivation in the Puna de Atacama, Chile. Undergraduate Honors Thesis, Department of Anthropology, Harvard University.

1982   Análisis General de las Muestras. *In* Precerámico peruano. Los Gavilanes. Mar, desierto, y oasis en la historia del hombre, edited by Duccio Bonavia, pp. 148–56. Corporación Financiera de Desarrollo S.A. (COFIDE). Oficina de Asuntos Culturales; Instituto Arqueológico Aleman, Comisión de Arqueología General y Comparada. Editorial Ausonia-Talleres Gráficos, Lima.

Pozorski, Sheila G.
1976   Prehistoric Subsistence Patterns and Site Economics in the Moche Valley, Peru. Ph.D. dissertation, Department of Anthropology, University of Texas, Austin.

1982   Subsistence Systems in the Chimu State. *In* ChanChan: Andean Desert City, edited by Michael E. Moseley and Kent C. Day, pp. 177–96. University of New Mexico Press, Albuquerque.

1983   Changing Subsistence Priorities and Early Settlement Patterns on the North Coast of Peru. Journal of Ethnobiology 3(1):15–38.

Pozorski, Sheila, and Thomas Pozorski
1979   An Early Subsistence Exchange System in the Moche Valley, Peru. Journal of Field Archaeology 6:413–32.

Prance, Ghillean T.
1984   The Pejibaye, *Guilielma gasipaes* (HBK) Bailey, and the Papaya, *Carica papaya* L. *In* Pre-Columbian Plant Migrations, edited by Doris Stone, pp. 85–104. Papers of the Peabody Museum of Archaeology and Ethnology, Vol. 76.

Quilter, Jeffrey, and Terry Stocker
1983   Subsistence Economies and the Origins of Andean Complex Societies. American Anthropologist 85(3):545–62.

Quilter, Jeffrey, Bernardino Ojeda E., Deborah M. Pearsall, Daniel H. Sandweiss, John G. Jones, and Elizabeth S. Wing
1991   Subsistence Economy of El Paraíso, an Early Peruvian Site. Science 251:277–83.

Raffino, Rodolfo A.
1973 Agricultura Hidráulica y Simbiosis Económica Demográfica en la Quebrada de Toro, Salta, Argentina. Revista del Museo de la Plata (Nueva Serie), Sección Antropología 7(49):297–332.

Ravines, Rogger, Helen Engelstad, Victoria Palomino, and Daniel H. Sandweiss
1984 Materiales Arqueológicos de Garagay. Revista del Museo Nacional 46(1982):135–234. Lima.

Raymond, J. Scott
1981 The Maritime Foundations of Andean Civilization: A Reconsideration of the Evidence. American Antiquity 46(4):806–21.

Reichel-Dolmatoff, Gerardo
1985 Arqueología de Colombia. Fundación Segunda Expedición Botánica, Bogota, Colombia.

Richardson, James B. III
1972 The Pre-Columbian Distribution of the Bottle Gourd (Lagenaria siceraria): A Re-evaluation. Economic Botany 26(3):265–73.

1980 Modeling the Development of Sedentary Maritime Economies on the Coast of Peru: A Preliminary Statement. Annals of the Carnegie Museum 50:139–50.

Rick, John W.
1980 Prehistoric Hunters of the High Andes. Academic Press, New York.

1984 Structure and Style at an Early Base Camp in Junin, Peru. Paper presented at the 49th Annual Meeting of the Society for American Archaeology, Portland.

Rindos, David
1980 Symbiosis, Instability, and the Origins and Spread of Agriculture: A New Model. Current Anthropology 21(6):751–72.

1984 The Origins of Agriculture. An Evolutionary Perspective. Academic Press, Orlando.

Rivera D., Mario. A.
1980 Temas Antropológicos del Norte de Chile. Universidad de Chile, Antofagasta.

Roosevelt, Anna C.
1980 Parmana. Prehistoric Maize and Manioc Subsistence along the Amazon and Orinoco. Academic Press, New York.

1984 Problems Interpreting the Diffusion of Cultivated Plants. In Pre-Columbian Plant Migration, edited by Doris Stone, pp. 1–18. Papers of the Peabody Museum of Archaeology and Ethnology, Vol. 76.

Sauer, Jonathan D.
1976 Grain Amaranths. Amaranthus spp. (Amaranthaceae). In Evolution of Crop Plants, edited by N. W. Simmonds, pp. 4–7. Longman, London.

Sauer, Jonathan D. and Lawrence Kaplan
1969 Canavalia Beans in American Prehistory. American Antiquity 34(4), Part 1:417–24.

Schobinger, Juan
1974 Current Research, South America. American Antiquity 39(3):508–12.

Simmonds, N. W.
1976a Evolution of Crop Plants (editor). Longman, London.

1976b Quinoa and Relatives. Chenopodium spp. (Chenopodiaceae). In Evolution of Crop Plants, edited by N. W. Simmonds, pp. 29–30. Longman, London.

1976c Potatoes. Solanum tuberosum (Solanaceae). In Evolution of Crop Plants, edited by N. W. Simmonds, pp. 279–83. Longman, London.

Smith, C. Earle, Jr.
1980 Plant Remains from Guitarrero Cave. In Guitarrero Cave. Early Man in the Andes, edited by Thomas F. Lynch, pp. 87–119. Academic Press, New York.

1988 Floral Remains. In La Galgada, Peru. A Preceramic Culture in Transition. Terence Grieder, Alberto Bueno Mendoza, C. Earle Smith, Jr., and Robert M. Malina. Pp. 125–51. University of Texas Press, Austin.

Smith, P. M.
1976 Minor Crops. In Evolution of Crop Plants, edited by N. W. Simmonds, pp. 301–24. Longman, London.

Stemper, David
1980 Paleoethnobotanical Studies at La Ponga (OGSE-186), Guayas, Ecuador. Manuscript on file, Department of Anthropology, University of Missouri-Columbia.

Stephens, S. G., and Michael E. Moseley
1974 Early Domesticated Cottons from Archaeological Sites in Central Coastal Peru. American Antiquity 39(1):109–122.

Stothert, Karen E.
1983 Review of the Early Preceramic Complexes of the Santa Elena Peninsula, Ecuador. American Antiquity 48(1):122–27.

1985   The Preceramic Las Vegas Culture of Coastal Ecuador. American Antiquity 50(3):613-37.

1988   La Prehistoria Temprana de la Península de Santa Elena, Ecuador: Cultura Las Vegas. Miscelánea Antropológica Ecuatoriana. Serie Monográfica 10. Museos del Banco Central del Ecuador, Guayaquil.

Tarrago, Myriam N.
1980   El Proceso de Agriculturización en el Noroeste Argentina, Zona Valliserrana. In Actas de V Congreso Nacional de Arqueología Argentina, Vol. 1, pp. 181-217. Universidad Nacional de San Juan, Instituto de Investigaciones Arqueológicas y Museo.

Towle, Margaret
1961   The Ethnobotany of Pre-Columbian Peru. Aldine, Chicago.

Ugent, Donald, Sheila Pozorski, and Thomas Pozorski
1981   Prehistoric Remains of the Sweet Potato from the Casma Valley of Peru. Phytologia 49(5):401-15.

1982   Archaeological Potato Tuber Remains from the Casma Valley of Peru. Economic Botany 36(2):182-92.

1984   New Evidence for Ancient Cultivation of Canna edulis in Peru. Economic Botany 38(4):417-32.

Umlauf, Marcelle La Vieve
1988   Paleoethnobotanical Investigations at the Initial Period Site of Cardal, Peru. M.A. thesis, Department of Anthropology, University of Missouri-Columbia.

Vavilov, N. I.
1951   The Origin, Variation, Immunity, and Breeding of Cultivated Plants. K. Starr Chester, translator. Chronica Botanica 13(1-6).

Veintimilla, Cesar, Marcelle La Vieve Umlauf, and Deborah M. Pearsall
1985   Resultados Preliminares de Flotación y Análisis de Fitolitos por el Proyecto Arqueológico-Etnobotánico San Isidro. Informe presentado en el Congreso Internacional de Americanistas, Bogota. Manuscript on file, Department of Anthropology, University of Missouri-Columbia.

Vescelius, Gary S.
1981a   Early and/or Not-So-Early Man in Peru. The Case of Guitarrero Cave. Part 1. Quarterly Review of Archaeology 2(1):11-15

1981b   Early and/or Not-So-Early Man in Peru. Guitarrero Cave Revisited. Quarterly Review of Archaeology 2(2):8-13, 19-20.

Weir, Glendon H., Robert A. Benfer, and John G. Jones
1988   Preceramic to Early Formative Subsistence on the Central Coast. In Economic Prehistory of the Central Andes, edited by Elizabeth Wing and Jane Wheeler, pp. 56-94. British Archaeological Reports, International Series 427, Oxford.

Weir, Glendon H., and J. Philip Dering
1986   The Lomas of Paloma: Human-Environment Relationships in a Central Peruvian Fog Oasis: Archaeobotany and Palynology. In Andean Archaeology, edited by Ramiro Matos M., pp. 18-44. UCLA Monographs in Anthropology 27, Los Angeles.

Wheeler Pires-Ferreira, J., E. Pires-Ferreira, and P. Kaulicke
1976   Preceramic Animal Utilization in the Central Peruvian Andes. Science 194:483-90.

Whitaker, Thomas W., and Hugh C. Cutler
1965   Cucurbits and Cultures in the Americas. Economic Botany 19(4):344-49.

Willey, Gordon R.
1971   An Introduction to American Archaeology, vol. 2: South America. Prentice-Hall, Englewood Cliffs, New Jersey.

Wilson, David
1981   Of Maize and Men: A Critique of the Maritime Hypothesis of State Origins on the Coast of Peru. American Anthropologist 83(1):93-120.

Wing, Elizabeth S.
1977   Animal Domestication in the Andes. In Origins of Agriculture, edited by Charles A. Reed, pp. 837-59. Mouton Publishers, The Hague.

Yen, Donald E.
1976   Sweet Potato. Ipomoea batatas (Convolvulaceae). In Evolution of Crop Plants, edited by N. W. Simmonds, pp. 42-45. Longman, London.

Zevallos M., C., W. C. Galinat, D. W. Lathrap, E. R. Leng, J. G. Marcos, and K. M. Klumpp
1977   The San Pablo Corn Kernel and Its Friends. Science 196:385-89.

10

Some Concluding Remarks

C. WESLEY COWAN AND
PATTY JO WATSON

Introduction

In the discussion that follows, we attempt to draw together important points from the papers in this volume. We do not offer critiques of the individual papers, nor do we attempt to provide firm answers to questions about the origins of agriculture by formulating a general explanatory model. Indeed, we believe it unlikely that a single model will be satisfactory for all of the times and places represented in this book. It is possible, however, to define several common themes suggesting common processual pathways.

One of these recurrent themes is a seemingly trivial one. Humans, or human-like creatures, have been on the planet for nearly 3.5 million years. The advent of agriculturally based societies represents a very small proportion of that time continuum. Much of the evolution of the human species seems to have taken place during the Ice Age, or Pleistocene, and the evolution of agriculturally based societies took place in the Holocene, the post-Ice Age world. The chronology of agricultural origins, then, is one of the themes that is present in all the papers, and it is the last 10,000 to 15,000 years that are of most interest.

Another theme, which we expand upon below, is the *comparability* of the histories of agricultural origins from region to region and continent to continent. Even the most casual reader will have noticed that the evidence for agricultural beginnings varies significantly in quality and quantity from one part of the world to another. Nevertheless, it is possible to infer that similar processes were functioning in different places at different times.

We see these processes in two distinct settings. One may be referred to as agricultural origins in "pristine" contexts. In such areas, crop plants were gradually added to a foraging-based economy. This in situ process was one of coevolution of cultural and agricultural systems without the outside introduction of primary crop plants. We believe that the Near East, Mexico, and Eastern North America are good examples of this phenomenon.

In contrast are those situations where agriculturally based economies developed as "secondary" phenomena. Two sorts of secondary patterns may be identified. In one of these a suite of domesticated plants is introduced into an area where economies are previously dominated by foraging. Whether crop by crop, or by wholesale introduction, agriculture slowly replaces a foraging existence in part or all of the recipient region. Europe and the American Southwest are good examples of this sort of secondary introduction.

Secondary developments may also comprise the introduction of a new crop plant into cultural systems that are already, at least in part, dependent upon cultivated plants. Once this introduction occurs, the new crop plant may replace the old suite of plant foods to become the dominant focus of the economy. Such secondary occurrences took place in Eastern North America where the starchy seed plants of the Eastern Agricultural Complex were replaced by maize, and in Japan where the Jōmon pattern of multicrop usage eventually was replaced by rice cultivation.

Another observation concerns a problem that permeates each paper and casts a pall over the ultimate reliability of the patterns we can detect. Scholars interested in agricultural origins are dependent, in large part, upon plant remains recovered from archaeological contexts. Herein lies a paradox: precisely because plant tissues do not preserve well in the archaeological record, because the data that are recounted here were not collected in similar fashions, and because many of these data were collected more than two decades ago, the general histories from region to region are incomplete and, in many cases, in need of substantial modification.

The final theme is one that is not present in all the papers, but is nonetheless very important: for many areas of the world we now understand the chronology of the earliest cultigens, yet nowhere do we completely understand *why* people turned away from foraging as a way of life. These "why" questions are elusive, and, as discussed below, are likely to continue to be so.

Pleistocene Adaptations and the Origins of Agriculture

The impact of the Pleistocene can be measured in many ways: continental and mountain glaciers spread over the land mass of many areas of the world, forming vast ice sheets; the ice sheets and resultant cooler atmospheric conditions made some areas of the world moister and others drier; sea level on a global scale dropped by several hundred meters exposing what is now the continental shelf of many of the world's great land masses and allowing for the movement of

biological species from one region of the world to the next; and boundaries of present-day plant distributions shifted considerably in response to the pulsing of periods of intense cold during glacial periods and climatic amelioration during interglacials.

It was during this period of environmental flux that the first humans, or human-like creatures, emerged from an African homeland and radiated outward to fill niches worldwide. The relationships between the early hominids and humans and the plant world are virtually unknown. Historically, many archaeologists have assumed that the diets of our early ancestors were dominated by meat collected through hunting and/or scavenging. This belief is reinforced by the countless broken and splintered animal bones found during careful excavations of prehistoric transient camps. But the earliest hominids did not use fire, and, by and large, paleoethnobotany focuses on the study of charred remains. The plant foods our earliest ancestors ate—and surely, like ethnographically documented foragers everywhere, they depended on plants as a dietary mainstay—have not survived the ravages of time. In effect, then, the probability that direct evidence of Australopithecine plant use could even be found is negligible. Not until our ancestors mastered the use of fire do the probabilities increase substantially.

Comparatively few *Homo erectus* sites have been excavated, and most of these were investigated before the widespread use of flotation. Hackberry (*Celtis* sp.) stones were found in the ash beds of Choukoutien, China, but these were recovered during the general excavations, not from flotation. At the site of Terra Amata on the coast of southern France, all sediments were passed through fine-mesh screens, but no flotation was applied to any of the deposits (DeLumley 1969). Nevertheless, in saturated deposits at Kalambo Falls in northern Zambia remarkable finds of wooden artifacts and macrobotanical remains (including seeds of various tree fruits) were made. The excavator suggests that plant foods were probably the major source of food for the Acheulean occupants of this site (Clark 1969).

Our vision of the lifeways of Middle and Late Paleolithic peoples has been largely formed by archaeological excavation in Europe. It is possible that in many places outside of Europe (and even within it) where relevant sediments are still present, flotation would yield profitable results. At any rate, plants must have been important to all Paleolithic peoples, and, once fire came to be a routine part of their technological repertoire, the probability that plant remains could be preserved in the archaeological record increased dramatically.

With the melting of the glacial ice at the end of the Pleistocene and the concomitant warming of atmospheric temperatures worldwide, the stage was set for major Holocene transformations of human societies around the globe. It is at this juncture that archaeologists and paleoethnobotanists may profitably scrutinize the record of agricultural origins.

Origins of Agriculture in World Perspective: Pristine and Secondary Settings

As noted previously, each paper in this volume emphasizes that the record of agricultural origins is limited to the last 10,000 to 15,000 years. It is tempting to conclude that the climatic changes at the end of the Pleistocene were somehow the initial "kick" that eventually led to the development of agriculturally based economies. This may be true in a very general way, but it is insufficient to explain the variability manifest from continent to continent or even region to region in the record of agricultural developments.

Agricultural Origins in Pristine Settings

As the papers in this volume attest, agriculture evolved independently in many areas of the world. Where it is possible to document the development of farming systems based upon major crop plants in isolation from other such systems, we may refer to *pristine* origins. Areas identified within the context of this volume that meet this criterion include the Near East, East Asia and Japan, Eastern North America, and Mexico. Neither the quality of research nor the paleoethnobotanical data are comparable among these areas, yet all seem to share a common evolutionary trajectory that culminates in the development of agriculturally based societies.

Archaic economies dependent upon wild plant foods developed during the Early Holocene in each of these areas, and in each area the Archaic adaptation was dominated by a few staple species. In the Eastern Woodlands of North America the mast of a variety of nut-bearing trees was the dietary staple; in the Near East, wild grasses; in Japan, a variety of nuts and

weed seeds. In Mexico data from the Tehuacan and Oaxaca valleys suggest that a variety of plants was collected. Once dependence upon the various constellations of wild plant foods was achieved, the stage was set for the process of domestication.

When human populations were committed to wild plants as dietary staples, then some degree of residential stability was apparently a further, necessary prerequisite to the development of agriculture. One question that is not addressed in this collection of papers, however, is that of causation. Is residential stability sufficient to cause agricultural origins? Probably not. Ethnographically documented foragers in northern California provide good examples of societies that were basically sedentary and yet never became agriculturalists.

In many areas the plants that were eventually domesticated were not initially staple commodities. In fact, in every case, the threshold of economic dependency was not passed until many centuries after the initial stages of plant domestication. This affects our view of the process; in most cases, it is not possible to identify plants that are only "partially" domesticated. If this is true, then we must search for subtle patterns—changes in the frequency of plant usage, for example—to point to changes in the interrelationships between human and plant populations. It is important to remember that domestication is a product of cumulative changes in a plant species. For this simple reason, it is probably not possible to discover "when" plants were first domesticated.

Agricultural Origins in Secondary Settings

Besides those areas where agriculture developed in situ, or nearly so, there are a number of situations where crop plants were introduced and grew to play a dominant role in local and regional economies. These areas of secondary origins provide a contrast to pristine sequences. And, as pointed out in the papers in this volume, new interpretations of the primary data from these secondary areas force us to reevaluate some long-cherished beliefs.

Nowhere is this process better illustrated than in western Europe. As Dennell notes in his chapter, traditional wisdom has it that once cereals and legumes were introduced into the western Mediterranean, a "Neolithic revolution" resulted, spreading quickly from the Balkans, up the Danube, and then across the loess plains of western Europe. In the process, Mesolithic foragers quickly adopted the new cultigens (as well as domesticated livestock) and virtually overnight became Neolithic farmers. This interpretation was supported by the host of radiocarbon dates that seem to reveal the rapid spread of the Bandkeramik cultures. The picture is now known to be somewhat more complicated than this.

There is, in fact, no Neolithic "revolution." As more primary data from Mesolithic sites become available, it is increasingly clear that agriculture did not spring up suddenly throughout western Europe. It seems entirely likely that, at least initially, the introduction of crop plants had little impact at all. Over the course of many centuries, agriculture gradually replaced foraging as the dominant subsistence mode; in the current model, the Mesolithic pattern of gathering and hunting persisted long after domesticated plants were introduced. The history of agriculture in this secondary setting is one of gradual change, not revolution.

Paul Minnis's chapter provides another slant on the secondary development of agricultural economies. It is clear that Mesoamerica, and particularly northern Mexico, must have been the region from which maize, squash, beans, cotton, and other lesser domesticates were introduced to the Desert Borderlands of North America. Much of this history of introduction and selective adoption is virtually unknown, however. For more than a century, archaeologists have concentrated their efforts on later sites and have consistently ignored pre-Basketmaker data. These Archaic settlements hold the key to understanding later developments, yet a mere handful have been excavated.

Minnis's observations on the role of agricultural commodities in the Apache world suggest that horticulture need not initially disrupt older, non-horticultural economic pursuits. If more data were available from the Southwest, we could probably detect similarities to the European pattern wherein domesticated plants are selectively and creatively adopted.

The Japanese islands provide still another example of secondary agricultural development. As Crawford amply demonstrates, by the early Jōmon period there is good evidence that at least some Japanese peoples were cultivating and harvesting several plant species. Crawford suspects that the Jōmon period is probably characterized by increasing emphasis on cultivation of a number of small-seeded and fruit species. When

rice was introduced into southwestern Japan from mainland Asia, it gradually transformed the traditional multicrop late Jōmon system: within a few hundred years it was the major crop in southwestern Japan and later in the northern islands.

The North American Midwest provides a third example of how an existing horticultural system can be transformed by the introduction of a new crop. In this case, Smith outlines a 3,500-year-long record of plant manipulation and cultivation that culminated in an indigenous farming system known as the Eastern Agricultural Complex. The crops in this system were all small-seeded annuals of both starchy and oily varieties; nuts provided an important seasonal staple.

Maize was introduced into areas east of the Mississippi River by at least the first few centuries A.D. at a time when the Eastern Agricultural Complex was already well developed. Initially, this exotic cultigen was apparently hardly more than a garden curiosity. For the next six to seven centuries it was grown by Eastern Woodlands groups but at a level that is barely detectable in the paleoethnobotanical record. Around A.D. 900–1000, corn production exploded and rapidly replaced the old crop system in the Ohio River valley.

This nearly 1,000-year gap between the time when maize was first introduced and the time when it became the focal point of subsistence economies in parts of the East can probably be explained in genetic terms. Initially, maize was a subtropical domesticate, adapted to growing conditions quite different from those encountered in Eastern North America. Early maize remains from the Eastern Woodlands are not numerous, but they probably represent a very diverse gene pool. It was this diverse pool that various human populations in the East manipulated through selection for desired characteristics. By A.D. 1000, two major varieties of corn were being grown east of the Mississippi: one adapted to the cooler mean annual temperatures and shorter day-length of the Northeast and others adapted to the warmer Southeast. It was not until these various regional varieties were developed and the early gene pool stabilized that maize could be an effective competitor to the Eastern Complex annuals. Once this occurred (apparently ca. A.D. 900–1000), maize rapidly replaced the indigenous crop complex in the northerly portions of the East and then spread south and west.

Agricultural Origins in Other Areas

Virtually every continent witnessed the development of domesticated plants, yet for many of them the record of agricultural origins is either poorly known or completely unknown. In large part, this is due to a lack of archaeological research and, even more importantly, to a lack of concerted efforts to collect paleoethnobotanical data.

These problems are particularly acute in some of the areas where important crop plants eventually emerged. As Harlan notes, for example, the origins of sorghum and millet, two grain crops that feed millions of Africans and many other peoples as well, are a complete mystery. These two crop plants are hardly alone. Cucumber, watermelon, and a host of other important plant species were domesticated in Africa, but the histories of their emergence as crops are unknown.

In her review of South American developments, Pearsall is plagued by similar problems. The preservation of plant remains in the desert of the western coast is nothing short of spectacular, and yet most of the domesticates were initially brought into cultivation elsewhere. When and where did the potato, the peanut, or the manioc emerge as a major crop of the Andean area? Clearly, there was much movement of domesticated plants between various South American zones, but their history is largely unknown. These and other crops are shadowed in ignorance and are likely to remain so until more primary data are collected.

In part, these problems are related to a subtle (and sometimes not so subtle) form of bias in archaeological research. In many areas of the world where complex states eventually emerged, there has been a tendency for archaeologists to concentrate on the most spectacular sites. Cities, palaces, tombs, and similar monumental public works have often been the foci of archaeological attention to the exclusion of smaller, less obvious sites. Because agricultural development invariably preceded the emergence of state-level societies, the excavation of a thousand cities would do little to shed light on the beginnings of plant cultivation in the much smaller, early food-producing communities that preceded them.

We would be remiss if we did not mention another form of bias that has pervaded the study of the origins of plant cultivation in some parts of the world.

Ethnocentric and even blatantly racist attitudes, significantly fostered by Western colonialist governments, is surely one factor to blame for lack of interest in the archaeology and paleoethnobotany of such areas as sub-saharan Africa, the Indian subcontinent, parts of South America, and the Pacific. Fortunately, these attitudes are much attenuated and are slowly dying. As more research is conducted in Oceania, southern Asia, sub-saharan Africa, and South America, we anticipate vast new sets of information that will markedly alter current understandings both of particular cultural histories and of general processes.

As we rush into the twenty-first century, we must also be mindful that the story of agricultural origins may never be completely known for some parts of the world. In South America the vast Amazon Basin, home of one of the world's most diverse floras—and probably the area that gave us yams, manioc, and other important domesticates—is rapidly being radically transformed. Similar stories can be told for tropical rainforests in Malaysia, the Philippines, Africa, and other areas. As each hectare of these habitats is cleared, clues to agricultural origins disappear.

Two things are certain, however: (1) the origins of agriculture will remain a problem of anthropological interest for many generations to come, and (2) in spite of the important contributions that plant geographers and geneticists can make, the primary data pertaining to the problems of domestication must be recovered from archaeological contexts and interpreted by paleoethnobotanists. We do not necessarily advocate programs designed exclusively to search for agricultural beginnings. Indeed, because it now appears that agricultural economies did not have single-point origins in time and space, such expeditions are probably obsolete. What *is* necessary, however, is a change in attitude on the part of archaeologists worldwide.

It is fair to say that at least since the 1950s, many archaeologists have recognized the unique contributions they can make to questions centering on agricultural origins. Nowhere has this been more readily noticeable than in Eastern North America. In the last decade flotation has become widely accepted as a standard field procedure. Literally tons of archaeological sediments are "floated" annually in order to recover their plant components. As demonstrated by the work Smith summarizes in his chapter for this volume, an unexpectedly complex and wholly fascinating picture of horticultural and agricultural development has emerged.

Certainly, the diligence with which archaeologists in many regions (notably Eastern North America, Britain, and Switzerland, to name only a few) now collect information related to plant use in general and agricultural origins in particular could be matched in other parts of the world. As increasingly stronger commitments to paleoethnobotanical research are made, we will obtain the requisite data to enable a thoroughly comprehensive account of agricultural origins in an international perspective.

References Cited

Clark, J. Desmond
1969   Kalambo Falls Prehistoric Site. 2 vols. Cambridge University Press, Cambridge.

de Lumley, Henri
1969   A Paleolithic Camp at Nice. Scientific American 220:42–59.

Index

archaeology, 2, 3, 8; bias, 211; New, 74
Archaic period (New World), 8, 122,
  123, 125, 128, 129–130, 132, 135, 146,
  147, 177(fig.), 180, 186, 209, 210
Argemone sp., 150–151(table)
Argentina, 174, 176; archaeobotanical
  evidence, 179(table), 188, 195, 196;
  archaeological sites, 176(fig.), 179,
  187(table), 188; chronology, 177(fig.);
  crops, 177
Argissa (Greece), 77(table), 78(fig.)
Arica (Chile), 180(table)
arid climates, 3, 11, 41, 48, 53, 60, 122,
  125
Arizona, 125, 126–127(table), 128(fig.),
  129, 130
Arkansas, 104(fig.), 107, 112, 113
Armelagos, G. J., 50
Armyo Rockshelter (N.M.), 126–
  127(table), 128(fig.)
arracacha (*Arracacia xanthorriza*), 195, 196
Artemisia steppe, 7, 11
arundinaceum, 61, 62(fig.)
Asahikawa City Museum (Japan), 29
Asch, D. L., 155
Asch, N. B., 155
ash (nutritional), 109(table)
ash (tree), 49
Ash Cave (Ohio), 104(fig.), 107
ash lenses, 47
Asia (Peru), 176(fig.), 182–186(table)
Aspero (Peru), 190, 195
Aswad (Syria), 40(fig.), 42, 45(table),
  48, 52
Atlatl Cave (N.M.), 126–127(table),
  128(fig.)
aurochs, 84
Australia, 91
Austalopithecine, 208
Austria, 73
avocado (*Persea americana*), 132, 134,
  149–152(tables), 154, 157, 159, 163,
  174(table), 178(table), 182–186(table),
  190, 191(fig.), 193(table), 194
Ayacucho Caves (Peru), 181(table), 188,
  190, 191, 194, 195, 196
Ayauchi Lake (Ecuador), 176(fig.),
  189(table)
Azapa-83 (Chile), 188(table)
Azmak (Bulgaria), 77, 78(fig.), 79(ta-
  ble), 88
Aztec bean. *See* scarlet runner bean
Aztecs, 146

Babahoya River valley, 174
Bacon Bend (Tenn.), 103, 105(table)
Bahn, P., 87
Bailey, Liberty Hyde, 27, 28
Bakels, C. C., 80
Bambara groundnut (*Voandzeia subterra-
  nea*), 65(table), 66, 69(fig.)
bamboo rat, 12
Bandama River, 67
Banpo (China), 14, 15, 24, 29
Banpo Museum, 24
baobob (*Adansonia digitata*), 65
barbacoas, 114

Barker, G. W., 76, 79, 80, 84, 85
barley (*Hordeum vulgare*), 8, 9(table), 12,
  13, 18, 19, 20, 25, 30, 42, 43, 45(table), 49, 60, 68, 77, 79, 80, 82, 84,
  85, 86; center of diversity, 27; distribution, 21–22, 42, 46(fig.), 73, 74, 85,
  86; grain size, 22; hulled, 21, 48; little
  (*H. pusillum*), 108, 109, 113, 123,
  124(table); naked, 21, 49; and oats,
  68; subspecies (*humile*), 21; two-row
  (*H. distichum*), 44, 45(table), 77(table),
  81(table), 83(fig.); types, 21, 22, 44,
  81(table); wild (*H. spontaneum*), 21,
  39, 44, 46(fig.), 82, 83(fig.), 91
barnyard grass (*Echinochloa crusgalli*),
  9(table), 19, 23, 29
barnyard millet (*Echinochloa utiliis*), 8,
  9(table), 16, 22–23
barranca agriculture, 156
barranca horticulture, 156
Barriles (Panama), 144(fig.), 159
Bat Cave (N.M.), 125, 126–127(table),
  128, 132
Baudais-Lundstrom, K., 80
Beadle, George W., 147
beans, 8, 9(table), 12, 17, 18, 19, 20,
  26–27, 30, 132(table), 134(table), 150–
  152(tables), 189(table), 210; *Canaval-
  ius*, 173; common (*Phaseolus vulgaris*),
  111, 112, 113, 123, 124(table), 128,
  129, 131, 133, 134, 135, 147, 150–
  151(table), 153, 156, 160, 161, 163,
  173, 174(table), 177, 178–187(tables),
  188, 189(table), 190, 191(fig.), 193(table), 195, 196; seed size, 153; wild,
  153
Bedaux, R. M. A., 61
beech, 11
beefsteak plant (*Perilla frutescens* var.
  crispa), 9(table), 18, 19, 28, 30; uses,
  28
beer, 68
beeweed. *See* Rocky Mountain beeweed
Begonia geraniifolia, 182–186(table), 188
Beidha (Jordan), 40(fig.), 42(table),
  45(table)
Beizhuangcun (China), 11
Belgium, 73, 84
Belize, 144, 155, 159
Bemis, W. P., 27
Bender, B., 51
Benin, 59
Berglund, B. E., 86
Berry, Michael S., 122, 128
Bertsch, F., 74
Bertsch, K., 74
Beutler, John A., 27
bicolor race (sorghum), 62, 63, 65(ta-
  ble), 69(fig.)
Binford, Lewis R., 49, 162
birch, 12
Bird, Junius B., 191
birds, 157
bitter vetch (*Vicia ervilia*), 82, 83(fig.)
Bixa orellana, 179(table)
black beniseed (*Polygala butyracea*),
  65(table)

black fonio (*Digitaria iburua*), 65(table),
  69(fig.)
black gram (urd) bean (*Vigna mungo*),
  9(table), 18, 26, 29, 30
Black Mesa (Ariz.), 126–127(table),
  128(fig.)
boar, 82
Boat Axe culture, 87
Bocas (Panama), 160
Bodensee (lake), 80
Bogucki, P., 79
Bohrer, Vorsila L., 50
boiling pits, 106
Bolivia, 174; archaeological sites,
  176(fig.), 177, 181(table), 186; crops,
  177, 181(table), 196
Book of Odes, 23
Bottema, S., 77
bottle gourd (*Lagenaria siceraria*), 9(table), 18, 21, 28, 29, 30, 65(table),
  106, 123, 124(table), 133–134, 150–
  151(tables), 154, 174(table), 177(table),
  181–187(tables), 188, 189(table), 190,
  191(fig.), 193(table), 194, 195, 197
Bouqras (Syria), 40(fig.), 42(table),
  45(table)
Bowles (Ky.), 103, 104(fig.), 105(table)
Braidwood, Robert J., 2, 41, 49, 161
Bray, Warwick, 180
Brazil, 173, 176(fig.), 179
Brazilian highlands, 174, 175
British Academy Major Research Project
  in the Early History of Agriculture,
  41
British Isles, 73, 78(fig.), 84, 86, 89,
  212
"broad spectrum revolution," 42, 49,
  50, 52
Bromus mango, 190
bronze, 14(fig.), 16
Bronze Age: Europe, 88(fig.), 93; Korea, 13, 16, 24, 30; Spain, 84, 87
broomcorn (common) millet (*Panicum
  miliaceum*), 8, 9(table), 12, 14, 15, 16,
  24, 30; and barnyard millet, 23; center
  of diversity, 23; genome, 23, hulled,
  23
buckwheat (*Fagopyrum esculentum*), 8,
  9(table), 12, 16, 17, 19, 21, 28, 29;
  pollen, 19, 28; wild (*F. cymosum*),
  9(table), 28
Bulgaria, 72, 74, 76, 77, 78(fig.), 79(table), 88
Bunchosia armeniaca, 174(table), 182–
  186(table), 190, 193(table), 195, 196
burdock, 18. *See also* great burdock
Burgaschisee (lake), 80
Burial, 188; mounds, 14(fig.), 88(fig.),
  108
burrowing animals, 188
Buskirk, Winfred, 131
Bus Mardeh phase (Iran), 47, 48, 52
butternut squash (*Cucurbita moschata*),
  149(table), 174(table), 178(table),
  182–186(table), 193(table)
Bye, Robert A., Jr., 162
Byrosonima sp., 150–152(tables)